METHODS IN MOLECULAR BIOLOGY

For further volumes:
http://www.springer.com/series/7651

Cytokine Bioassays

Methods and Protocols

Edited by

Ivana Vancurova

Department of Biological Sciences, St. John's University, Queens, NY, USA

Humana Press

Editor
Ivana Vancurova
Department of Biological Sciences
St. John's University
Queens, NY, USA

ISSN 1064-3745 ISSN 1940-6029 (electronic)
ISBN 978-1-4939-0927-8 ISBN 978-1-4939-0928-5 (eBook)
DOI 10.1007/978-1-4939-0928-5
Springer New York Heidelberg Dordrecht London

Library of Congress Control Number: 2014939314

Printed on acid-free paper

Humana Press is a brand of Springer
Springer is part of Springer Science+Business Media (www.springer.com)

Preface

Cytokines are pleiotropic regulatory proteins involved in essentially all biological processes and associated with a wide range of diseases, including immune and inflammatory disorders, as well as many types of cancer and leukemia. Examples of cytokines include interleukins, chemokines, growth factors, interferons, and the tumor necrosis factor family. Information about the qualitative and quantitative nature of cytokine expression and release is essential for the understanding of physiological and pathological processes. However, the cytokine detection in biological and clinical samples faces many challenges that include the low abundance, the need to distinguish between active and latent cytokine forms, and the need to measure multiple cytokines in a single assay. This volume provides a comprehensive collection of classic and cutting-edge methodologies that are used to analyze and quantify cytokines and their biological activities in complex biological and clinical samples.

The chapters are divided into three main categories. The *first category* focuses on the *immunodetection of released cytokines* in tissue culture supernatants, plasma, serum, and whole blood samples by immunoassays. These immunoassays measure the total concentrations of released cytokines regardless of their biological activities, and include ELISPOT, immunoblotting, and ELISA. In addition, the complexity of cytokine responses has led to the development of multiplex technologies that can simultaneously detect large numbers of cytokines in small sample volumes. Thus, the first part also includes the recently developed multiplex arrays that allow the simultaneous measurement of multiple cytokines in small sample volumes with time-saving advantages and less cost compared to other immunoassays.

Since immunoassays cannot distinguish between biologically active and inactive cytokines, *bioassays* are used to measure different biological activities of cytokines, such as cytokine-induced cytokine release, chemotaxis, phagocytosis, proteasome activity, and immunoglobulin class switching. The *second part* focuses on the analysis of biologically active cytokines by bioassays using neutralizing antibodies, chemotaxis assay, cytokine-induced phagocytosis assay, proteasome activity assay, and analysis of cytokine-induced immunoglobulin class switching. In addition, since cytokines exert their functions through binding to cytokine receptors, this category also includes methods focused on the analysis of expression and function of cytokine receptors.

Often, there is a need to identify the cytokine-producing cells, or to analyze the intracellular cytokine protein levels or the mechanisms regulating cytokine expression. The *third part* focuses on the *analysis of intracellular cytokines* by flow cytometry, immunohistochemistry, immunofluorescence confocal microscopy, and western blotting. In addition, this category includes protocols for quantitative *analysis of cytokine gene expression* by real-time RT-PCR, luciferase assay, analysis of the cytokine promoter occupancy by chromatin immunoprecipitation, and analysis of alternative splicing of cytokine genes.

There is no one best method, since each has its own merits and limitations. Choosing the right method depends on the sample, purpose of the assay, and the available instrumentation. Often, using combined information from several different approaches can yield the most accurate information about the quantitative and qualitative nature of cytokine expression. The reliability of all the protocols has been tested in laboratories around the world.

Each chapter is appended by notes that navigate through the protocol and serve as a troubleshooting guide. By covering a broad variety of methods used in cytokine research and analysis, we hope that this book will be useful not only to biochemists, molecular biologists, and immunologists, but also to physician-scientists working in the field of cytokine research.

I would like to thank all the authors for their enthusiastic help and support in assembling this volume; I fully realize that in the highly competitive environment of academic research, many scientists are reluctant to commit their time to writing book chapters and method articles. I also would like to express my gratitude to the series editor, Dr. John Walker, and the outstanding staff of Humana Press for their support, help, and encouragement.

Queens, NY *Ivana Vancurova*

Contents

Contributors

MUMTAZ AKHTAR • *Department of Pharmaceutical Sciences, College of Pharmacy and Health Sciences, St. John's University, Queens, NY, USA*
KATRINA BAMBA • *Department of Pharmaceutical Sciences, College of Pharmacy and Health Sciences, St. John's University, Queens, NY, USA*
ABDOULLATIF BARAKET • *Institut de Sciences Analytiques, Université de Lyon, Villeurbanne, France*
FRANK A. BARILE • *Department of Pharmaceutical Sciences, College of Pharmacy and Health Sciences, St. John's University, Queens, NY, USA*
BRIAN T. BEATY • *Department of Anatomy and Structural Biology, Albert Einstein College of Medicine, Bronx, NY, USA*
THOMAS BLAUENFELDT • *Clinical Research Centre, Copenhagen University hospital of Hvidovre, Copenhagen, Denmark*
STEPHEN BYRNE • *Department of Biological Sciences, St. John's University, Queens, NY, USA*
MICHAEL CAMMER • *Office of Collaborative Science & Skirball Institute of Biomolecular Medicine, New York University Langone Medical Center, New York, NY, USA*
FRANCESCA CAMPABADAL • *Centro Nacional de Microelectrónica, Instituto de Microelectrónica de Barcelona, Barcelona, Spain*
JOSEPH J. CAMPO • *Barcelona Centre for International Health Research (CRESIB), Hospital Clínic - Universitat de Barcelona, Barcelona, Catalonia, Spain*
RAFFAELE CARUSO • *Consiglio Nazionale Ricerche, Clinical Physiology Institute, Milan, Italy*
TZU-PEI CHANG • *Department of Biology, St. John's University, Queens, NY, USA*
ZHE-SHENG CHEN • *Department of Pharmaceutical Sciences, College of Pharmacy and Health Sciences, St. John's University, Queens, NY, USA*
YUAN CHOU • *Division of Arthritis and Rheumatic Diseases, Oregon Health & Science University, Portland, OR, USA*
CONG-QIU CHU • *Division of Arthritis and Rheumatic Diseases, Oregon Health & Science University, Portland, OR, USA*
JOHN S. CONDEELIS • *Department of Anatomy and Structural Biology, Albert Einstein College of Medicine, Bronx, NY, USA*
DIANNE COX • *Department of Anatomy and Structural Biology, Gruss-Lipper Biophotonics Center, Albert Einstein College of Medicine of Yeshiva University, Bronx, NY, USA*
GREGORY DE CRESCENZO • *Department of Chemical Engineering, École Polytechnique de Montréal, Montréal, QC, Canada*
SASCHA A. DAVID • *Department of Nephrology and Hypertension, Medical School Hannover, Hannover, Germany*
DENNIS DAVIDSON • *Division of Neonatal-Perinatal Medicine, Stony Brook Long Island Children's Hospital, Stony Brook University, Stony Brook, NY, USA*
SANJAY L. DHOLAKIYA • *Department of Pharmaceutical Sciences, College of Pharmacy and Health Sciences, St. John's University, Queens, NY, USA*

CARLOTA DOBAÑO • *Barcelona Centre for International Health Research (CRESIB), Hospital Clínic - Universitat de Barcelona, Barcelona, Catalonia, Spain*
CAMILLA HELDBJERG DRABE • *Department of Pulmonary and Infectious Diseases, Copenhagen University hospital of Hillerød, Copenhagen, Denmark*
ROBERT J. EDDY • *Department of Anatomy and Structural Biology, Albert Einstein College of Medicine, Bronx, NY, USA*
ABDELHAMID ERRACHID • *Institut de Sciences Analytiques, Université de Lyon, Villeurbanne, France*
MARIA FARESJÖ • *School of Health Sciences, Department of Natural Science and Biomedicine, The Biomedical Platform, Jönköping University, Jönköping, Sweden*
CHARLES FORTIER • *Department of Chemical Engineering, École Polytechnique de Montréal, Montréal, QC, Canada*
GIULIA FREER • *Department of Translational Medicine, Retrovirus Center, University of Pisa, Pisa, Italy*
HIMAVANTH R. GATLA • *Department of Biology, St. John's University, Queens, NY, USA*
CHANDRA C. GHOSH • *Division of Nephrology, Center for Vascular Biology Research, Beth Israel Deaconess Medical Center and Harvard Medical School, Boston, MA, USA*
SAMIR GORASIYA • *Department of Pharmaceutical Sciences, College of Pharmacy and Health Sciences, St. John's University, Queens, NY, USA*
PRAJNA P. GUHA • *Roger Williams Medical Center, Cancer Immunotherapy, Providence, RI, USA*
LOUIS HODGSON • *Department of Anatomy and Structural Biology, Gruss-Lipper Biophotonics Center, Albert Einstein College of Medicine of Yeshiva University, Bronx, NY, USA*
JAMES HOY • *Sylvester Comprehensive Cancer Center, University of Miami Miller School of Medicine, Miami, FL, USA*
NICOLE JAFFREZIC-RENAULT • *Institut de Sciences Analytiques, Université de Lyon, Villeurbanne, France*
KAINAAT JAVED • *Department of Biological Sciences, St. John's University, Queens, NY, USA*
YANFANG JIANG • *Department of Central Laboratory, The First Hospital, Jinlin University Normal Bethune Medical College, Changchun, China*
ZHENYU JIANG • *Department of Rheumatology, The First Hospital, Jinlin University Normal Bethune Medical College, Changchun, China*
TAKAAKI KAMATANI • *Department of Oral and Maxillofacial Surgery, Showa University School of Dentistry, Tokyo, Japan*
PARUL KASHYAP • *Department of Biological Sciences, St. John's University, Queens, NY, USA*
MYRA KIM • *Department of Biology, St. John's University, Queens, NY, USA*
CHRISTOPHER J. KIRK • *Onyx Pharmaceuticals, South San Francisco, CA, USA*
CLARISSA LAM • *Department of Biological Sciences, St. John's University, Queens, NY, USA*
MICHAEL LEE • *Institut de Sciences Analytiques, Université de Lyon, Villeurbanne, France*
JIANHUA LI • *Laboratory of Biomedical Sciences, The Feinstein Institute for Medical Research, North Shore-LIJ Health System, Manhasset, NY, USA*
BENOÎT LIBERELLE • *Department of Chemical Engineering, École Polytechnique de Montréal, Montréal, QC, Canada*
BAL L. LOKESHWAR • *Department of Urology, University of Miami Miller School of Medicine, Miami, FL, USA*

LIN L. MANTELL • *Department of Pharmaceutical Sciences, College of Pharmacy and Health Sciences, St. John's University, Queens, NY, USA*
XIAO-LONG MEI • *Department of Cell Biology and Institute of Biomedicine, College of Life Science and Technology, Jinan University, Guangzhou, China*
EDMUND J. MILLER • *Center for Heart and Lung Research, The Feinstein Institute for Medical Research, Manhasset, NY, USA; Medicine and Molecular Medicine, Hofstra North Shore-LIJ Medical School, Hempstead, NY, USA*
VERONIKA MISKOLCI • *Department of Anatomy and Structural Biology, Albert Einstein College of Medicine of Yeshiva University, Bronx, NY, USA*
GEMMA MONCUNILL • *Barcelona Centre for International Health Research (CRESIB), Hospital Clínic - Universitat de Barcelona, Barcelona, Catalonia, Spain*
DANIEL MUÑOZ • *Research Service, Miami VA Medical Center, University of Miami Miller School of Medicine, Miami, FL, USA*
HARDIK PATEL • *The Feinstein Institute for Medical Research, Manhasset, NY, USA*
VIVEK PATEL • *Department of Pharmaceutical Sciences, College of Pharmacy and Health Sciences, St. John's University, Queens, NY, USA*
MAHMUD ARIF PAVEL • *Department of Biological Sciences, St. John's University, Queens, NY, USA*
VALERIE PERSAUD • *Department of Biology, St. John's University, Queens, NY, USA*
SAI A. PHYO • *Department of Biology, St. John's University, Queens, NY, USA*
VLADIMIR POLTORATSKY • *Department of Pharmaceutical Sciences, College of Pharmacy and Health Sciences, St. John's University, Queens, NY, USA*
SAUL R. POWELL • *Cardiac Metabolism Laboratory, The Feinstein Institute for Medical Research, Manhasset, NY, USA; Molecular Medicine and Science Education Hofstra North Shore-LIJ Medical School, Hempstead, NY, USA*
JIAN-GE QIU • *Department of Cell Biology and Institute of Biomedicine, College of Life Science and Technology, Jinan University, Guangzhou, China*
SHLOKA REDDY • *Department of Pharmaceutical Sciences, College of Pharmacy and Health Sciences, St. John's University, Queens, NY, USA*
JANET ROLLINS • *Division of Natural Sciences, College of Mount Saint Vincent, Riverdale, NY, USA*
MATTEO RUGGIU • *Department of Biological Sciences, St. John's University, Queens, NY, USA*
MORTEN RUHWALD • *Clinical Research Centre, Copenhagen University hospital of Hvidovre, Copenhagen, Denmark*
NICOLE SALAZAR • *Sylvester Comprehensive Cancer Center, University of Miami Miller School of Medicine, Miami, FL, USA*
ZAHRA SALEHI-NAJAFABADI • *Department of Biological Sciences, St. John's University, Queens, NY, USA*
SHANNON SANACORA • *Department of Biology, St. John's University, Queens, NY, USA*
FELITSIYA SHAKOLA • *Department of Biological Sciences, St. John's University, Queens, NY, USA*
LOKESH SHARMA • *Department of Pharmaceutical Sciences, College of Pharmacy and Health Sciences, St. John's University, Queens, NY, USA*
VED P. SHARMA • *Department of Anatomy and Structural Biology, Albert Einstein College of Medicine, Bronx, NY, USA*
ZHI SHI • *Department of Cell Biology and Institute of Biomedicine, College of Life Science and Technology, Jinan University, Guangzhou, China*

NATHAN SIEGELAUB • *Department of Pharmaceutical Sciences, College of Pharmacy and Health Sciences, St. John's University, Queens, NY, USA*
GURPREET SINGH • *Department of Biological Sciences, St. John's University, Queens, NY, USA*
BIPRADEB SINGHA • *Department of Biology, St. John's University, Queens, NY, USA*
RAVIKUMAR SITAPARA • *Department of Pharmaceutical Sciences, College of Pharmacy and Health Sciences, St. John's University, Queens, NY, USA*
DÉSIRÉE SPIERING • *Department of Anatomy and Structural Biology, Gruss-Lipper Biophotonics Center Albert Einstein College of Medicine of Yeshiva University, Bronx, NY, USA*
MARIA GIOVANNA TRIVELLA • *Consiglio Nazionale Ricerche, Clinical Physiology Institute and University of Pisa, Pisa, Italy*
VICTORIA VALENTINO • *Department of Biological Sciences, St. John's University, Queens, NY, USA*
IVANA VANCUROVA • *Department of Biological Sciences, St. John's University, Queens, NY, USA*
HAICHAO WANG • *Laboratory of Emergency Medicine, The Feinstein Institute for Medical Research, North Shore-LIJ Health System, Manhasset, NY, USA*
MAO WANG • *Department of Pharmaceutical Sciences, College of Pharmacy and Health Sciences, St. John's University, Queens, NY, USA*
JIAO WU • *Department of Pharmaceutical Sciences, College of Pharmacy and Health Sciences, St. John's University, Queens, NY, USA*
WENJUN WU • *Department of Pharmaceutical Sciences, College of Pharmacy and Health Sciences, St. John's University, Queens, NY, USA*
MAGGIE YEUNG • *Laboratory of Emergency Medicine, The Feinstein Institute for Medical Research, North Shore-LIJ Health System, Manhasset, NY, USA*
YONG YU • *Department of Biological Sciences, St. John's University, Queens, NY, USA*
TAMARA YUNUSOVA • *Department of Pharmaceutical Sciences, College of Pharmacy and Health Sciences, St. John's University, Queens, NY, USA*
MIGUEL ZABALA • *Centro Nacional de Microelectrónica, Instituto de Microelectrónica de Barcelona, Barcelona, Spain*
LIN ZHAO • *Laboratory of Emergency Medicine, The Feinstein Institute for Medical Research, North Shore-LIJ Health System, Manhasset, NY, USA*
LING ZHAO • *Department of Rheumatology, The First Hospital, Jinlin University Normal Bethune Medical College, Changchun, China*
SHU ZHU • *Laboratory of Emergency Medicine, The Feinstein Institute for Medical Research, North Shore-LIJ Health System, Manhasset, NY, USA*
NADIA ZINE • *Institut de Sciences Analytiques, Université de Lyon, Villeurbanne, France*
MICHELLE ZUR • *Department of Pharmaceutical Sciences, College of Pharmacy and Health Sciences, St. John's University, Queens, NY, USA*

Part I

Detection of Released Cytokines by Immunoassays

Chapter 1

The Challenge of Measuring Elusive Immune Markers by Enzyme-Linked Immuno-Spot (ELISPOT) Technique

Maria Faresjö

Abstract

The enzyme-linked immuno-spot (ELISPOT) technique is a sensitive method used for measurement of elusive immune markers in limited-volume samples. By virtue of the exquisite sensitivity of the ELISPOT assay, frequency analysis of rare cell populations (e.g., antigen-specific responses), which was not possible before, is now relatively easy. However, development of a method sensitive enough to pinpoint elusive immune markers at the single-cell level is a challenge since there are a number of demands that have to be fulfilled and traps to avoid, achieving a valuable outcome.

To optimize the environment for in vitro culture and analysis of immune spots by ELISPOT, a number of criteria have to be fulfilled: processing of sample and perhaps also cryopreservation of cells before analysis and, for the ELISPOT assay, optimal cell culture, positive and negative controls, antigen concentration, and, finally, development and readout of spots.

If these criteria are fulfilled for your ELISPOT assay, you will likely have the opportunity to pinpoint elusive immune markers at the single-cell level. This chapter describes the ELISPOT assay for detection of cytokines (e.g., IFN-γ and IL-4), with focus on the main criteria that affect the assay. However, this method could be easily adapted to measure other immune markers in small volumes of biological samples.

Key words ELISPOT, Cytokines, Cell separation, Cryopreservation, Control antigens, In vitro stimulation conditions, Development and readout of spots

1 Introduction

Enzyme-linked immuno-spot (ELISPOT) is a technique by which immune markers, e.g., cytokine and chemokine secretion, can be detected at the single-cell level since secreted cytokines are captured and accumulated in the ELISPOT plate. ELISPOT was initially performed for detection of antigen-secreting cells [1] and later adapted for enumeration of cytokine-producing cells at the single-cell level [2]. The ELISPOT assay provides both qualitative (type of immune protein) and quantitative (number of responding cells) information, as each spot that develops in the assay represents a single reactive cell.

Ivana Vancurova (ed.), *Cytokine Bioassays: Methods and Protocols*, Methods in Molecular Biology, vol. 1172,
DOI 10.1007/978-1-4939-0928-5_1, © Springer Science+Business Media New York 2014

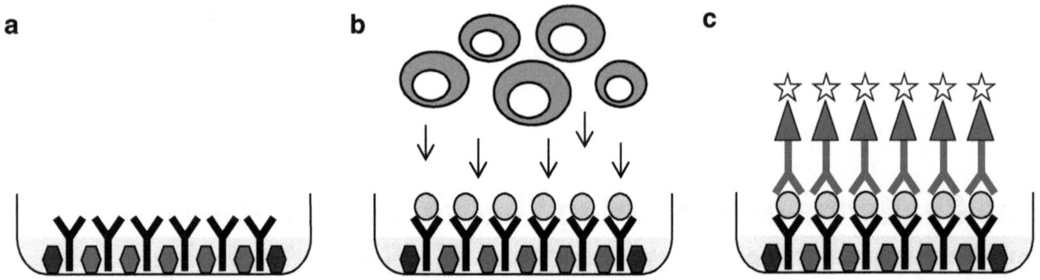

Fig. 1 Methodological principle of the ELISPOT assay. The wells are coated with a primary antibody, and thereafter nonspecific binding sites are blocked (**a**). Cells are added in the presence or the absence of specific stimulus. During incubation, cells become activated by the stimulus and start to produce and secrete a cytokine that binds to the capture antibody (**b**). Cells are removed, and a detection antibody, which may be directly conjugated with enzyme or biotinylated, is added (**c**)

In principle, the immune markers of interest, e.g., cytokines, are captured directly on the surface as a cytokine–antibody complex forming a spot around each cell secreting the cytokine of interest. The spot size and morphology reflect the kinetics and the quality of the cytokine production by individual cells over the entire test period.

The wells of the ELISPOT plate are coated with a primary antibody, and thereafter nonspecific binding sites are blocked (Fig. 1a). Cells are added in the presence or the absence of specific stimulus. During incubation, cells become activated by the stimulus, starting to produce and secrete a cytokine that binds to the capture antibody (Fig. 1b). Subsequently, cells are removed, and a detection antibody, which may be directly conjugated with enzyme or biotinylated, is added (Fig. 1c). Finally, a substrate will form a colored spot at the location of the secreting cell. By counting the number of spots in stimulated cultures and controls without stimulus, the frequency of responding cells is determined [3].

2 Materials

2.1 Isolation of Peripheral Blood Mononuclear Cells

1. Venous blood samples from healthy human volunteers or patients, drawn in tubes supplemented with sodium heparin.
2. RPMI-1640 medium without glutamine.
3. Washing buffer: RPMI-1640 without glutamine supplemented with 2 % heat-inactivated fetal calf serum (FCS).
4. Separation reagent: Ficoll-Paque density gradient.

2.2 Cellular Number and Viability

1. Türk's solution.
2. Trypan blue solution.

2.3 Cryopreservation of PBMC

1. Freezing medium: 50 % RPMI-1640 without glutamine, 40 % FCS, and 10 % dimethyl sulfoxide (Me_2SO).
2. Cryo 1 °C Freezing Container (Nalge Nunc International) containing isopropanol.
3. Vials appropriate for cryopreservation (Nalge Nunc International).

2.4 Thawing of PBMC

1. Washing buffer: RPMI-1640 without glutamine supplemented with 10 % heat-inactivated FCS.
2. Iscove's modification of Dulbecco's medium (Life Technologies) supplemented with 50 U/ml penicillin, 50 μg/ml streptomycin, 10 ml/l 100× nonessential amino acids, and 5 % FCS.

2.5 ELISPOT

1. Sterile PVDF membrane plate.
2. Ethanol (70 %).
3. MultiScreen™ Vacuum Manifold 96-well (Merck Millipore).
4. Sterile phosphate-buffered saline (PBS).
5. Sterile PBS-Tween (Life Technologies).
6. Iscove's modification of Dulbecco's medium (Life Technologies) supplemented with 50 U/ml penicillin, 50 μg/ml streptomycin, 10 ml/l 100× nonessential amino acids, and 5 % FCS.
7. Human IFN-γ or IL-4 ELISPOT kit including mouse anti-human (IFN-γ or IL-4) monoclonal antibody, biotinylated anti-human (IFN-γ or IL-4) monoclonal antibody, and streptavidin conjugated with alkaline phosphatase.
8. Alkaline phosphatase (AP) conjugate substrate kit.
9. AID ELISPOT Reader System.

3 Methods

3.1 Isolation of Peripheral Blood Mononuclear Cells

Separation of peripheral blood mononuclear cells (PBMC) should be handled at room temperature in a ventilated hood (*see* **Notes 1** and **2**).

1. Dilute venous heparinized blood in RPMI, 1:2 (i.e., half-volume blood, half-volume RPMI).
2. Invert the Ficoll-Paque bottle several times to ensure through mixing. Then, withdraw the required volume of Ficoll-Paque using aseptic technique. Ficoll-to-blood ratio should be 3:5 (i.e., 12 ml of Ficoll and 20 ml of diluted blood sample).
3. Transfer the required volume of Ficoll-Paque to a sterile centrifuge tube.
4. Carefully lay the diluted blood sample on the Ficoll layer to avoid mix between Ficoll and blood sample.

5. Centrifuge at 400 × *g* for 30 min, without brake, at 18–20 °C.
6. Carefully draw up the layer of PBMC using a clean Pasteur pipette, and transfer PBMC to a sterile centrifuge tube.
7. Resuspend PBMC in 20 ml of washing buffer.
8. Centrifuge at 400 × *g* for 10 min, with low brake, at 18–20 °C.
9. Remove the supernatant, and wash PBMC by repeating the **steps** 7 and **8** once more.
10. Remove the supernatant.

3.2 Cellular Number and Viability

1. Dissolve the PBMC pellet in 1 ml of washing medium.
2. For determination of the number of PBMC, dilute 10 μl of PBMC suspension with 90 μl of Türk's solution (1:10).
3. Place 10 μl of PBMC suspension in a Bürker's chamber. Count the number of PBMC using a light microscope (*see* **Note 3**).
4. For determination of membrane integrity, dilute 10 μl of PBMC suspension with 10 μl of trypan blue solution (1:1).
5. Place 10 μl of PBMC suspension in a Bürker's chamber. Count 100 PBMC in a light microscope (*see* **Note 4**).

3.3 Cryopreservation of PBMC (See Note 5)

1. Centrifuge the solution of PBMC at 400 × *g* for 10 min at 18–20 °C.
2. Remove the supernatant.
3. Resuspend PBMC in freezing medium (4 °C) by adding freezing medium dropwise, while the tube is continuously agitated, to a concentration of $5–10 \times 10^6$ PBMC/ml.
4. Transfer 1 ml of PBMC solution to vials suitable for cryopreservation.
5. Place the vials with PBMC in a precooled (4 °C) Cryo 1 °C Freezing Container containing isopropanol.
6. Place the container at −70 °C (the freezing rate is −1 °C/min).
7. Transfer the vials to liquid nitrogen (−196 °C) the following day.

3.4 Thawing of PBMC

1. Thaw PBMC, directly from −196 to +37 °C, in a water bath under continuous agitation.
2. Immediately thereafter, add washing buffer dropwise to the cells until a total volume of 10 ml is reached.
3. Centrifuge the solution of PBMC at 400 × *g* for 10 min at 18–20 °C.
4. Remove the supernatant.
5. Resuspend PBMC in 1 ml of Iscove's modification of Dulbecco's medium with supplements.

6. Count the cell number and determine the cell viability (*see* Subheading 3.2), and adjust the concentration of PBMC to 1×10^6/ml.

3.5 ELISPOT

All steps during days 1 and 2 should be handled in a ventilated hood.

3.5.1 Day 1: Coating of Plate

1. Pre-coat the sterile PVDF membrane plate with 100 μl/well of ethanol (70 %) for 30 min at room temperature.
2. Ethanol is removed from the wells by MultiScreen™ Vacuum Manifold 96-well.
3. Wash the plate by adding 100 μl/well of sterile water, and then empty out the wells by vacuum manifold. Repeat the washing step 4–6 times (*see* **Note 6**).
4. Dilute the mouse anti-human cytokine (e.g., IFN-γ or IL-4) monoclonal antibody to 15 μg/ml (1:67), in sterile PBS.
5. Coat the plate with 100 μl/well mouse anti-human cytokine (e.g., IFN-γ or IL-4) monoclonal antibody.
6. Place the plate in a humid chamber and incubate overnight at 4 °C.

3.5.2 Day 2: Washing and Blocking of Plate

1. Empty the coating antibody out of the wells by vacuum manifold.
2. Wash the plate by adding 100 μl/well of sterile PBS, and then empty out the wells by vacuum manifold. Repeat the washing step eight times.
3. Add 100 μl/well of Iscove's modification of Dulbecco's medium with supplements.
4. Incubate for 30 min in a humidified atmosphere with 5 % CO_2 to block nonspecific binding sites on the PVDF membrane.

3.5.3 Day 2: Stimulation of PBMC in the Plate

1. Empty Iscove's modification of Dulbecco's medium out of the wells.
2. Apply aliquots of 100,000 PBMC/well together with 100 μl of Iscove's modification of Dulbecco's medium, with supplement, and eventually also antigen as well as positive and negative controls (*see* **Notes 7–9**).
3. Culture PBMC undistributed for 48 h at 37 °C in a humidified atmosphere with 5 % CO_2.

3.5.4 Day 4: Development of Spots

1. Empty PBMC out of the wells.
2. Wash the plate by adding 100 μl/well of sterile PBS, shake the plate repeatedly, and remove PBS. Repeat the washing step once more.

3. Wash the plate by adding 100 μl/well of sterile PBS-Tween, shake the plate repeatedly, and remove PBS-Tween. Repeat the washing step once more.
4. Dilute the biotinylated anti-human cytokine (e.g., IFN-γ or IL-4) monoclonal antibody, 1:1,000 (to 1 μg/ml), in sterile PBS-Tween.
5. Add 100 μl/well of biotinylated anti-human cytokine (e.g., IFN-γ or IL-4) monoclonal antibody, and incubate the plate for 2–4 h, in a humidified chamber, in darkness at room temperature.
6. Empty the biotinylated antibody out of the wells.
7. Wash the plate by adding 100 μl/well of sterile PBS-Tween, shake the plate repeatedly, and remove PBS. Repeat the washing step three times.
8. Dilute streptavidin conjugated with alkaline phosphatase, 1:1,000, in PBS-Tween.
9. Add 100 μl/well of streptavidin conjugated with alkaline phosphatase, and incubate the plate for 60 min in a humidified chamber at room temperature.
10. Empty streptavidin conjugated with alkaline phosphatase out of the wells.
11. Wash the plate by adding 100 μl/well of sterile PBS-Tween, shake the plate repeatedly, and remove PBS-Tween. Repeat the washing step once more.
12. Wash the plate by adding 100 μl/well of sterile PBS, shake the plate repeatedly, and remove PBS. Repeat the washing step three times, but leave PBS from the last washing in the plate during **step 13**.
13. Dilute 0.1 ml of reagent A with 10 ml of AP color development buffer and 0.1 ml of reagent B.
14. Empty PBS out of the plate.
15. Add 100 μl/well of the color development mix. Spots will develop 5–10 min after adding the color development mix. Allow the reaction to proceed for another 5 min.
16. Stop the spot development by rinsing the wells with excessive amount of tap water. Finally, empty and dry the plate upside down, overnight, at room temperature.

*3.5.5 Day 5: Counting of Spots (See **Note 10**)*

1. Spots are counted automatically, under manual supervision, using the AID ELISPOT Reader System.
2. The median value of the quadruplicates is calculated for each stimulation. The value of each specific secretion from in vitro stimulation, after subtraction of spontaneous secreted spots, is calculated.

4 Notes

1. Time interval between blood sampling and sample processing: Of great importance is the time period between blood sampling and processing of the sample. Our own research group has found that blood samples left at room temperature for 24 h before separation of PBMC will increase the expression of CD3/CD4 (unpublished data). It has also been shown that PBMC handled and cryopreserved within 8 h from venipuncture has better viability, higher cell recovery, and higher concentration of IFN-γ, detected by ELISPOT, compared to PBMC handled after 8 h from blood sampling [4].
2. Enrichment of PBMC: There is a number of different ways to enrich PBMC. The most frequently used techniques for isolation of PBMC are density gradient separation by Ficoll or Lymphoprep or separation by vacutainer cell preparation tube (CPT). Ficoll is routinely used to isolate mononuclear cells from bone marrow, peripheral blood, and umbilical cord blood. In principle, PBMC are enriched from whole blood that is layered onto a density gradient. Gentle centrifugation at room temperature results in a buffy coat of monocytes and lymphocytes under a layer of plasma, with the remaining white blood cells together with red blood cells passing through the interface and collecting at the bottom of the tube. The PBMC interface is collected and washed for several times in either PBS or cell culture medium (washing buffer) to remove any contaminating separation medium.

   The vacutainer CPT is a single-tube system, supplemented with sodium citrate, for collection of whole blood and the separation of PBMC. In principle, blood is collected in the CPT tube, the tube is centrifuged, and the cell pellet is resuspended in PBS or cell culture medium before subsequent assay or procedure.
3. Cell number: In the most common design, the volume of each large square of a Bürker's chamber is 0.1 mm³. The cells in four large squares are counted, and cells over or touching the lines on top and on the left are counted, but cells over or touching the right or the bottom lines are ignored. The concentration in cells per ml will be calculated as cells in four large squares/4 × 10,000.
4. Cell recovery and viability: Viability thresholds should be used in clinical trials in order to obtain reliable results of functional assays. The most convenient way to measure PBMC viability is to manually count PBMC by light microscopy after staining the cells with trypan blue. In a viable cell, trypan blue is not

absorbed; however, it traverses the membrane of dead cells. Hence, dead cells appear as distinctive blue color dots using light microscopy. In principle, the numbers of live (white) versus dead (blue) cells are counted. The number of white cells/100 PBMC gives the viability in percentage. A more reliable way to measure the cell viability is detection by flow cytometry. In principle, PBMC should be resuspended in a flow cytometry staining buffer; thereafter, the cell count and viability analyses are performed. However, this requires more cells and thus may not be possible in limited sample volumes.

5. Cryopreservation of PBMC: The functional assays on cryopreserved PBMC are associated with viability of the cells. For practical reasons, it can be an advantage to collect cells during a limited period of time and perform all analyses collectively, post-cryopreservation, in order to overcome inter-assay variation in the methodology analyzing cytokine secretion by ELISPOT. Cryopreservation is a convenient way to handle PBMC. Actually, frozen cells are used in the great majority of all studies of cell-mediated immunity. We and others have shown that cryopreserved PBMC can be used, trustworthy, after cryopreservation for detection of cytokines and chemokines [5] and Treg-associated markers [6] and also that cryopreserved PBMC maintain a stable expression of the T-regulatory marker FOXP3 (% and MFI) in the CD4 + CD25hi cell population after cryopreservation [7]. Collectively, these studies show that cryopreserved PBMC can be used and still give trustworthy and useful information.

6. Washing step after incubation with ethanol: The repeated washing step (4–6 times), after incubation with ethanol, is recommended to ensure that all ethanol is removed.

7. Positive and negative controls: Often used positive controls in immunoassays are phytohemagglutinin (PHA) and tetanus toxoid. PHA is a mitogen, acting as a highly potent activator of both CD4+ and CD8+ cells with a resulting cytokine production [8–10]. It has been shown that PHA with advantage can be used as a positive control in the ELISPOT assay [11, 12]. Tetanus toxoid is another frequently used positive control in immuno-assays. Tetanus toxoid is an active immunizing agent prepared from detoxified tetanus toxin that produces an antigenic response in the body, conferring permanent immunity to tetanus infection. Thus, tetanus toxoid is another good example of an antigen that can be used as a positive control in the ELISPOT assay with the remark that people not vaccinated to this antigen will not respond.

   As a negative control, usually some wells on each ELISPOT plate can be incubated exclusively with cell culture medium, without cells, but otherwise treated as the other wells. In some cases, it can be relevant to include an antigen that does not cause any immune response. Finding an antigen, not inducing an immune response, can however be tricky. The keyhole limpet hemocyanin (KLH) has been shown as one example of an antigen that can be used as negative control since this large metalloprotein, found in the hemolymph of the giant keyhole limpet, is a novel antigen to most individuals. We have previously shown that this antigen does not induce any immune response in children, healthy or diagnosed with type 1 diabetes [13], and it has also been shown that KLH induces systemic T-cell tolerance [14].

8. Optimal concentration of antigens: It is important to evaluate the optimal concentration of antigens that will be used in the ELISPOT assay. In an initial experiment, antigens (proteins and/or peptides) should be diluted in tissue culture medium in a wide range of different concentrations. The antigen in each dilution should be tested by stimulating PBMC, followed by ELISPOT detection. It should be noted that it is not always the highest concentration of antigen that will induce the highest number of spots (or highest concentrations of an immune marker) [11].
9. Settings for antigen stimulation: When possible, apply all variants of differently stimulated/non-stimulated cells in quadruplicates. The external line of wells within the plate is preferable left with exclusively 200 μl of Iscove's modification of Dulbecco's medium, with supplements, to avoid dehydration.
10. Development and readout of spots: Plates are preferably blinded for identity to avoid any influence of the outcome of the observation. A major factor in the ELISPOT assay is to distinguish low-frequency antigen-specific T-cell responses from the nature of the signal-to-noise ratio. Thus, it is of great importance to take into account the technique for separation of cells, number of cells incubated, the window of cytokine capture, and the reagents used for capture and detection to reach a valuable outcome [12].

References

1. Czerkinsky CC, Tarkowski A, Nilsson LÅ et al (1984) Reverse enzyme-linked immunospot assay (RELISPOT) for the detection of cells secreting immunoreactive substances. J Immunol Methods 72: 489–496
2. Czerkinsky C, Andersson G, Ekre HP et al (1988) Reverse ELISPOT assay for clonal analysis of cytokine production. J Immunol Methods 110:29–36
3. Faresjö M (2012) Enzyme linked Immunospot; a useful tool in the search for elusive immune markers in common pediatric immunological diseases. Cells 1:141–152
4. Bull M, Lee D, Stucky J et al (2007) Defining blood processing parameters for optimal detection of cryopreserved antigen-specific responses for HIV vaccine trials. J Immunol Methods 322:57–69
5. Axelsson S, Faresjö M, Hedman M et al (2008) Cryopreserved peripheral blood mononuclear cells are suitable for the assessment of immunological markers in type 1 diabetic children. Cryobiol 57:201–208

6. Kivling A, Nilsson L, Faresjö M (2009) How and when to pick up the best signal from markers associated with T-regulatory cells? J Immunol Methods 345:29–39
7. Rydén A, Faresjö M (2011) Efficient expansionofcryopreservedCD4+CD25+CD127lo/− cells in type 1 diabetes. Res Immunol 1: 36–44
8. Nowell PC (1960) Phytohemagglutinin: an inhibitor of mitosis in cultures of normal human leukocytes. Cancer Res 20:462–466
9. Gowans JD (1961) The management of rheumatoid arthritis. GP 24:106–116
10. Kruisbeek AM, Shevach EM (1999) Proliferative assay for T cell function. In: Coligan JE, Kruisbeek AM, Margulis DH et al (ed) Curr Protoc Immunol (IN), Wiley, pp 1–14
11. Karlsson MGE, Sederholm Lawesson S, Ludvigsson J (2000) Th1-like dominance in high-risk first-degree relatives of type 1 diabetic patients. Diabetologia 43:742–749
12. Schloot NC, Meierhoff G, Karlsson Faresjö M et al (2003) Comparison of cytokine ELISpot assay formats for the detection of islet antigen autoreactive T cells. Report of the third immunology of diabetes society T-cell workshop. J Autoimmun 21:365–376
13. Karlsson Faresjö M, Ernerudh E, Ludvigsson J (2004) Cytokine profile in children during the first 3 months after the diagnosis of type 1 diabetes. Scand J Immunol 59:517–526
14. Husby S, Mestecky J, Moldoveanu Z et al (1994) Oral tolerance in humans. T cell but not B cell tolerance after antigen feeding. J Immunol 152:4663–4670

Chapter 2

Analysis of the Released Nuclear Cytokine HMGB1 in Human Serum

Haichao Wang, Lin Zhao, Jianhua Li, Shu Zhu, and Maggie Yeung

Abstract

A ubiquitous nuclear protein, the high-mobility group box 1 (HMGB1), is secreted by activated macrophages/monocytes and leaked passively from injured cells. HMGB1 functions as a mediator of infection- and injury-elicited inflammatory diseases. Here, we describe a semiquantitative immuno-blotting method to measure the released HMGB1 in human serum, in comparison with a commercially available HMGB1 ELISA technique.

Key words HMGB1, Western blot, ELISA, Serum, Antibody

1 Introduction

High-mobility group box 1 (HMGB1) is expressed constitutively in most cells, and a large "pool" of preformed HMGB1 is stored in the nucleus due to the presence of two lysine-rich nuclear localization sequences [1, 2]. It contains two internal repeats of positively charged domains (known as "A box" and "B box") in the N terminus and a continuous stretch of negatively charged (aspartic and glutamic acid) acidic tail in the C-terminus. These HMG boxes enable HMGB1 to bind chromosomal DNA and fulfill its nuclear functions such as maintaining the nucleosomal structure and stability and regulating gene expression [3]. In response to exogenous bacterial products (such as endotoxin or CpG-DNA) [1, 4], or endogenous inflammatory stimuli (e.g., TNF, IFN-γ, or hydrogen peroxide) [1, 5, 6], innate immune cells actively release HMGB1 in a dose- and time-dependent manner. In addition, HMGB1 can be released passively from damaged cells [7] and similarly trigger an inflammatory response [8]. The accumulated evidence has supported a pathogenic role for extracellular HMGB1 in infection- or injury-elicited inflammatory diseases [9–13]. Thus, it is important to measure plasma or serum HMGB1 levels in patients

Ivana Vancurova (ed.), *Cytokine Bioassays: Methods and Protocols*, Methods in Molecular Biology, vol. 1172,
DOI 10.1007/978-1-4939-0928-5_2, © Springer Science+Business Media New York 2014

with various inflammatory diseases. In this chapter, we describe two immunoassays for measuring HMGB1 in human serum samples: ELISA and Western blotting.

2 Materials

2.1 Sandwich ELISA Kit

Several companies have recently developed sandwich ELISA kits for measuring HMGB1 levels in plasma and serum samples. For instance, the Shino-Test Corporation HMGB1 ELISA kit contains the following components:

1. ELISA plate: 8-well strips coated with the capture polyclonal antibody generated against the peptide KPDAAKKGVVKAEK adjacent to the C-terminal acidic tail of HMGB1 [14, 15] (*see* **Notes 1** and **2**).
2. Peroxidase (POD)-conjugated detection antibody (lyophilized for making 12 ml solution): Peroxidase-linked anti-HMGB1 monoclonal antibody (generated against human HMGB1 protein).
3. HMGB1 standard: Pig HMGB1 protein.
4. Sample diluent (20 ml).
5. Conjugate solvent (12 ml).
6. Color reagent A (3, 3′, 5, 5′-tetramethyl-benzidine).
7. Color reagent B containing 0.005 M hydrogen peroxide.
8. Substrate solution: Equal volumes of color reagents A and B, which have been brought to room temperature, and mixed just before use.
9. Stop solution: 0.35 M Sulfuric acid.
10. 5× Washing buffer (200 ml).
11. ELISA plate seal (two sheets).

2.2 Western Blotting

1. 10× Phosphate-buffered saline (10× PBS): 1.37 M NaCl, 27 mM KCl, 100 mM Na_2HPO_4, 18 mM KH_2PO_4, pH 7.4.
2. 10× Tris/glycine/SDS buffer: 0.025 M Tris, 0.192 M glycine, 0.1 % SDS, pH 8.3.
3. 10× Tris/glycine buffer: 0.025 M Tris, 0.192 M glycine, pH 8.3.
4. SDS-PAGE gel: 4–20 % Mini-PROTEAN® TGX™ Precast Gel (Bio-Rad).
5. PageRuler Plus Prestained Protein Ladder 10–250 kDa molecular weight marker (Thermo Scientific).
6. Immuno-blot PVDF membrane.
7. TWEEN® 20.
8. Nonfat dry milk.

9. Methanol.
10. CL-XPosure TM Film.
11. Amersham™ ECL™ Western Blotting Detection Reagents.
12. Anti-HMGB1 antibody (OncoImmune Inc., MI, USA, Cat# OI0001A-05).
13. ECL Rabbit IgG, HRP-linked whole antibody from donkey.
14. Donkey anti-mouse IgG-HRP antibody.
15. Laemmli 2× buffer/loading buffer: 4 % SDS, 10 % 2-mercaptoethanol, 20 % glycerol, 0.004 % bromophenol blue, 0.125 M Tris–HCl, pH 6.8.
16. Running buffer (Tris-glycine/SDS): 25 mM Tris, 190 mM glycine, 0.1 % SDS, pH 8.3. Add 100 ml of 10× Tris/glycine/SDS buffer to 900 ml of ultrapure water, and mix thoroughly.
17. Transfer buffer: 25 mM Tris, 190 mM glycine, 20 % methanol, pH 8.0. Add 100 ml of 10× Tris/glycine buffer to 200 ml of methanol and 700 ml of ultrapure water, and mix well. Chill buffer at 4 °C.
18. Washing buffer: Add 500 μl of Tween 20–1,000 ml of 1× PBS, and mix thoroughly.
19. Blocking buffer: Add 5 g of nonfat dry milk to 100 ml of washing buffer. Mix thoroughly, filter, and store at 4 °C. Failure to filter can lead to "spotting" tiny dark grains that will contaminate the blot during color development.

3 Methods

3.1 HMGB1 ELISA

Below is a brief protocol adapted from the Shino-Test Corporation HMGB1 ELISA kit for measuring HMGB1 in plasma or serum samples; this method has been used by many investigators to measure HMGB1 levels in serum and plasma [16–26].

1. Prepare serum or plasma samples immediately after blood collection to prevent progressive HMGB1 leakage from blood cells (*see* **Note 3**).
2. Set each reagent in the ELISA kit at room temperature for at least 30 min before use.
3. Add 100 μl of sample diluent to each well and 10 μl of sample diluent to "zero" well.
4. Add 10 μl of HMGB1 standard and serum or plasma samples to the wells, and shake the plate with a plate mixer (*see* **Note 4**).
5. Incubate for 20–24 h at 37 °C.
6. Wash the plate five times with wash solution (400 μl/well) using microplate washer.

7. Add 100 μl of peroxidase-linked anti-HMGB1 MAb to each well, and incubate for 2 h at room temperature (25 °C).
8. Wash the plate five times with wash solution (400 μl/well) using microplate washer.
9. Add 100 μl of substrate solution to each well, and incubate for 30 min at room temperature.
10. Add 100 μl of stop solution to each well in the same sequence and time intervals as the addition of substrate solution.
11. Read the absorbance at 450 nm using a microplate reader within 60 min after adding stop solution.
12. Subtract the absorbance of "zero" well from the absorbance of each well.
13. Read the HMGB1 concentrations of unknown samples from the standard curves prepared from purified human HMGB1 protein (*see* **Notes 5–7**).

3.2 HMGB1 Western Blotting

Western blotting enables indirect detection of protein samples immobilized on a nitrocellulose or a PVDF membrane and serves as a useful tool to quantify HMGB1 in serum or plasma samples. Briefly, plasma or serum protein samples are first resolved by SDS-PAGE and then electrophoretically transferred to the membrane. Following a blocking step, the membrane is probed with a primary antibody raised against HMGB1. After subsequent washings, the membrane is incubated with an enzyme-conjugated secondary antibody that is reactive toward the primary antibody. The activity of the enzyme, such as alkaline phosphatase (AP) and horseradish peroxidase (HRP), is necessary for signal generation. Finally, the membrane is washed again and incubated with an appropriate enzyme substrate (e.g., chemiluminescent substrates for HRP), producing a recordable signal. All procedures are carried out at room temperature unless otherwise specified.

3.2.1 Sample Preparation

1. Use a small volume (50 μl) of serum or plasma to determine the protein concentration.
2. Mix serum or plasma samples with an equal volume of 2× Laemmli sample buffer (*see* **Notes 8** and **9**). We recommend denaturing the sample using the following method unless non-reducing and non-denaturing conditions were intended to investigate HMGB1/protein interactions (*see* **Note 10**).
3. To reduce and denature the proteins, boil the sample mixtures at 100 °C for 5 min and spin briefly in a microcentrifuge to collect the condensed water from the tube cap immediately before loading to SDS-PAGE gel.

3.2.2 Separation of Protein Samples by Gel Electrophoresis

1. Mount the precast SDS-PAGE gels onto the electrophoresis apparatus, and add running buffer to the top and bottom reservoirs. Remove any air bubbles.

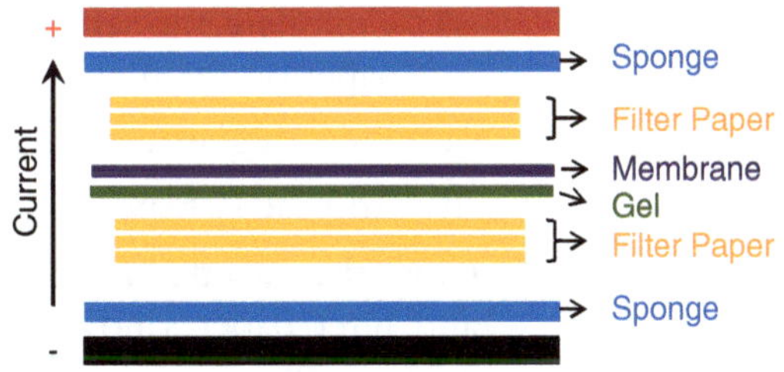

Fig. 1 Assembling the sandwich transfer stack

2. Load equal amounts of serum or plasma proteins (20–30 μg) into the wells of the SDS-PAGE gel, along with molecular weight markers (*see* **Note 11**), as well as purified HMGB1 protein at several concentrations (e.g., 1, 5, 20 ng/well).
3. Attach the electrophoresis apparatus to an electric power supply. The positive electrode should be connected to the bottom buffer reservoir.
4. Run the gel for 1–2 h at 110 V. The time and voltage may require some optimization. We recommend following the manufacturer's instructions. A reducing gel should be used unless non-reducing conditions were intended to investigate HMGB1/protein interactions (*see* **Note 10**).
5. Remove the plates from the electrophoresis apparatus, and place them on a paper towel. Using a spatula, pry the plates apart.
6. Remove the gel from the electrophoresis apparatus, and incubate it in transfer buffer for approximately 10 min to equilibrate the gel (*see* **Note 12**).

3.2.3 Transferring Proteins from the Gel to the Membrane

1. Cut the PVDF membrane to the dimensions of the gel, immerse it in 100 % methanol for 5 min, and rinse with transfer buffer before preparing the transfer stack (*see* **Note 13**). Mark and/or clip one corner for orientation. Handle only with flat forceps.
2. Soak the member, filter paper, and fiber pads in transfer buffer for 10 min.
3. Prepare the transfer stack as shown below (Fig. 1):
   (a) Place the cassette, with the gray side down, on a lean surface.
   (b) Place one pre-wetted sponge/fiber pad on the gray side of cassette.
   (c) Place a sheet of filter paper on the sponge pad.
   (d) Place the equilibrated gel on the filter paper.
   (e) Place the pre-wetted PVDF membrane on the gel.

(f) Complete the sandwich by placing another piece of filter paper on the membrane.

(g) Using a glass tube gently roll all air bubbles out.

(h) Add the last sponge pad.

4. Close the cassette firmly and carefully so as not to move the gel and filter paper sandwich. Lock the cassette closed with the white latch.
5. Place the cassette in the electrophoresis module.
6. Add the frozen Bio-Ice cooling unit. Place in tank, and completely fill the tank with buffer. Add standard stir bar to help maintain even buffer temperature and ion distribution in the tank.
7. Put on the lid, and plug the cable into the power supply.
8. Transfer the protein at 200 mA for 60 min.
9. The time and voltage may require some optimization. We recommend following the manufacturer's instructions. Transfer to the membrane can be checked using Ponceau Red staining before the blocking step.
10. After the transfer, unclamp the blot sandwich and remove the sheets of blotting paper, exposing the blot membrane.

3.2.4 Antibody Staining

1. Block the membrane for 2 h at room temperature or overnight at 4 °C using 5 % nonfat milk blocking solution (*see* **Notes 14** and **15**).
2. Incubate membrane with appropriate dilutions of primary anti-HMGB1 antibody in 5 % or 2 % blocking solution overnight at 4 °C or for 2 h at room temperature (*see* **Notes 16** and **17**).
3. Wash the membrane three times for 10 min each in washing buffer to remove unbound antibody.
4. Incubate the membrane with the recommended dilution of labeled secondary antibody in 5 % blocking buffer at room temperature for 1 h.
5. Wash the membrane three times for 10 min each in washing buffer containing 0.05 % Tween-20. Rinse the membrane with washing buffer without Tween-20 (*see* **Note 15**).
6. To prepare the substrate, proceed according to the kit manufacturer's recommendations. For instance, mix the black and white ECL solutions (1:1 ratio) of the Amersham ECL kit.
7. Aliquot sufficient volume of substrate solution to cover and wet the membrane, and incubate the blot with the substrate for 1 min (0.1 ml/cm^2) when using Amersham ECL or 5 min when using SuperSignal Substrates.
8. Remove excess reagent, and cover the membrane in transparent plastic wrap. A plastic sheet protector works well, although

plastic wrap also may be used. Remove all air bubbles between the blot and the surface of the membrane protector.

9. Acquire image using darkroom development techniques for chemiluminescence or normal image scanning methods for colorimetric detection.
10. The relative band intensity is quantified by using the NIH image 1.59 or other software to determine HMGB1 levels with reference to standard curves generated with purified HMGB1 at various dilutions (*see* **Notes 18–22**).

4 Notes

1. According to the manufacturer, the Shino-Test Corporation capture anti-HMGB1 antibodies are highly specific to HMGB1, but not to HMGB2. However, their cross-reactivities with other plasma or serum protein are not yet known (Table 1) and should be a subject of future investigation.
2. As a highly charged molecule, HMGB1 can interact with various plasma or serum proteins such as immunoglobulins (IgGs) and thrombomodulin [14, 15]. It is not yet known how these and as-yet-unidentified HMGB1-binding molecules affect the detection of HMGB1 by using the Shino-Test or other HMGB1 ELISA kits [14].
3. Prolonged storage of blood samples before centrifugation at room temperature often leads to higher HMGB1 levels in sera [17], possibly due to leakage of HMGB1 from stressed/damaged blood cells. However, storage of serum samples after centrifugation for up to 7 days does not affect HMGB1 levels [17].
4. The minimal measurable HMGB1 concentrations are calculated by adding two standard deviations to the mean optical density value of several zero standard replicates. According to the Shino-Test Corporation, the limit of detection of HMGB1 ELISA is approximately 0.3–1 ng/ml, making it suitable for measuring plasma or serum HMGB1 in patients with sepsis or other inflammatory diseases. If serum HMGB1 levels are relatively low, consider using the recommended sensitive HMGB1 ELISA method by loading more serum samples and less HMGB1 standard on the ELISA plates (Fig. 2a).
5. In agreement with previous report [16], the measurement of HMGB1 in healthy individuals showed low levels by using the Shino-Test Corporation HMGB1 ELISA kit (Fig. 2a).
6. Research in Dr. Stoetzer's laboratory has shown that the measurement of HMGB1 in EDTA plasma samples yields considerable lower values than in sera with a mean recovery <30 % [17].

Table 1
Commercial available anti-HMGB1 antibodies for Western blotting analysis

Company	Sigma-Aldrich		Abcam		Cell signaling	OncoImmune Inc
Antibody type (Clone #)	M-MAb (Clone 2F6)	R-PAb	M-MAb	R-PAb	R-PAb	M-MAb (Clone 3B1 or 3E8)
Catalog no.	WH0003146M8	SAB2101049	ab77302	ab18256	3935	OI0001A or OI0001B
Species cross-reactivity	Human	Human Mouse Rat Bovine Canine Chicken Pig	Human	Human Mouse Rat	Human Mouse Rat Monkey	Human Mouse Rat Monkey Canine
Immunogen	Human HMGB1 residues 1–91	Human HMGB1, residues 1–91	Recombinant HMGB1, residues 1–216	Human HMGB1 residues 150–216	Synthetic peptide	Recombinant HMGB1 residues 1–216
Specificity to whole-cell lysate[a]	Hela	Hela	HMGB1-transfected 293T cell	NIH/3T3 MEF1 e PC12	NIH/3T3 DLD-1 C6 COS	Mouse spleen cells
Cross-reactivity to serum proteins	Unknown	Unknown	Unknown	Unknown	Unknown	Unknown

Note: M-MAb murine monoclonal antibodies, *R-PAb* rabbit polyclonal antibodies
[a]One specific band was detected in lysates of indicated cells

7. The ELISA method has been used by many investigators to measure HMGB1 levels in serum or plasma samples [17–26].
8. Almost all serum or plasma proteins can be readily solubilized by sodium dodecyl sulfate (SDS), making SDS-PAGE the most widely used method for determining the molecular mass of serum or plasma proteins.
9. As an ionic detergent, SDS denatures proteins by wrapping around the polypeptide backbone at a fixed ratio (1.0 g of SDS:0.7 g of polypeptide). Thus, more SDS can be added in the sampling buffer if excessive amount of plasma or serum proteins (up to 50–100 μg) are loaded onto SDS-PAGE gels in order to detect relative lower HMGB1 levels.
10. To investigate the possibility whether HMGB1 interacts with itself (aggregation) or other proteins in plasma and serum

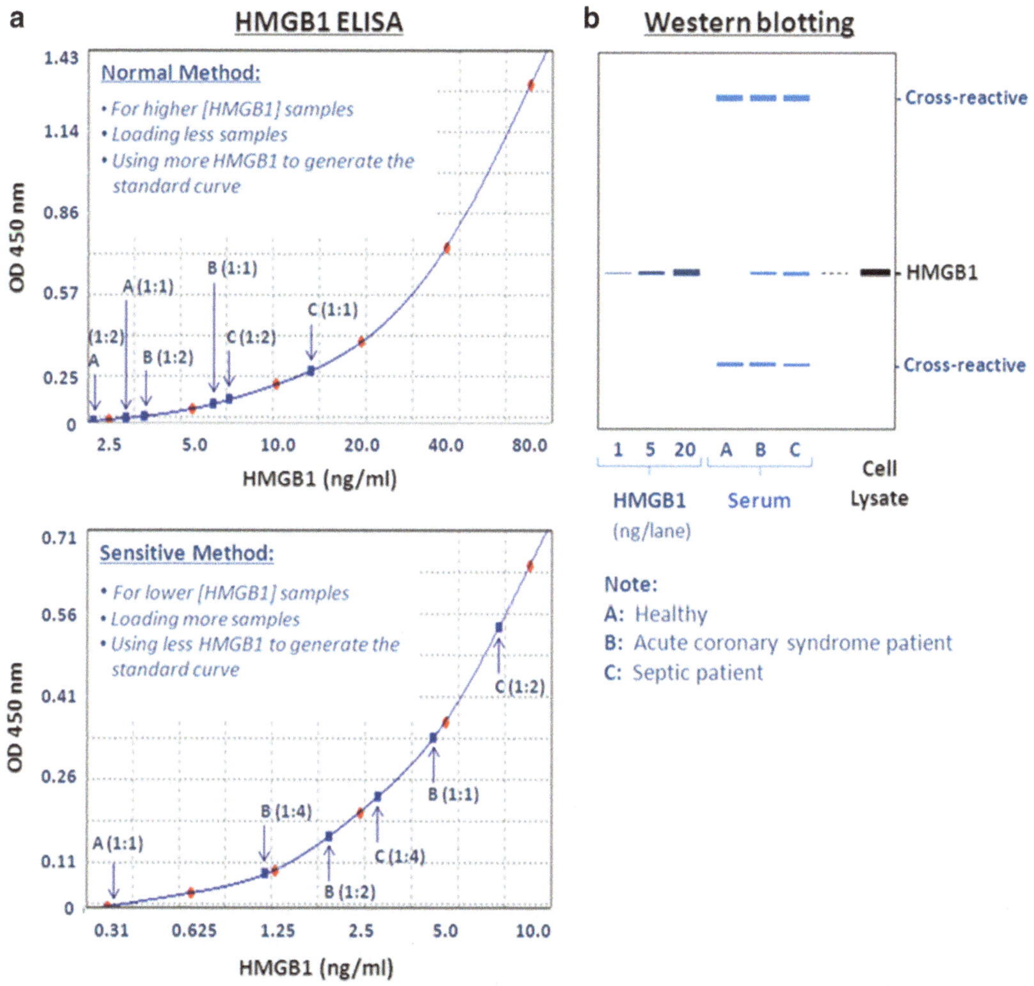

Fig. 2 Representative HMGB1 ELISA and Western blotting results of human serum samples. (**a**) Measurement of human serum HMGB1 levels using the Shino-Test Corporation ELISA kit. As per the manufacturer's recommendation, two methods with different sensitivities were used to measure HMGB1 in human serum at various dilutions. (**b**) Measurement of human serum HMGB1 by Western blotting analysis. Polyclonal antibodies were generated against recombinant HMGB1 in the authors' laboratory and used in the Western blotting analysis of human serum HMGB1 levels

samples, non-denaturing native conditions can be employed, where HMGB1 protein is electrophoresed in its native form based on charge-to-mass ratio.

11. When using different SDS-PAGE buffer systems (e.g., different pH), the charges and SDS-binding capacities of chemically modified proteins (e.g., pre-stained molecular weight standards) might be slightly affected. Consequently, there might be a slight deviation from the calculated molecular weight (based on the amino acid sequence). To confirm the identity of antibody-reactive band, whole-cell lysate can be used in parallel lanes as a reference size marker.

12. As an alternative of the classical Western blotting protocol using enzyme-conjugated secondary antibodies, direct detection using a labeled primary antibody can be tried. The direct detection takes less time and has less background signal (from the secondary antibody cross-reactivity) than a classical Western blot. However, it is generally less sensitive than the indirect detection, because a labeled primary antibody cannot provide signal amplification and occasionally loses immunoreactivity to the targeted antigen. One alternative option is biotinylating the primary antibody, which not only amplifies the signal but also eliminates the secondary antibody cross-reactivity.
13. Several types of blotting membranes are commonly used, including nitrocellulose and PVDF membranes. Whereas nitrocellulose binds proteins better and often gives better band signals, PVDF is physically stronger and easier to handle. For best results, determine empirically which membrane type, manufacturer, and lot are optimal for each Western blotting system.
14. Many different blocking reagents are available for Western blotting. Because milk contains variable amounts of endogenous biotin, it may produce higher background when using nonfat milk as a blocking reagent in the avidin/biotin systems.
15. Some systems may benefit from adding a surfactant, such as Tween®-20, to the blocking solution. Surfactants can minimize background by preventing the blocking reagent from nonspecifically binding to the target. Adding too much detergent, however, can prevent adequate blocking. Typically, a final concentration of 0.05 % detergent is used; however, for best results, determine if detergents enhance a specific system and at what optimal concentrations. Always use a high-quality detergent that is low in contaminants.
16. HMGB1 is a highly conserved protein, making it difficult to generate highly reactive antibodies in many animal species. Currently, there are several commercial sources for HMGB1-reactive polyclonal or monoclonal antibodies. These antibodies normally recognize one single band on Western blots of lysates of various types of cells (Table 1), but their cross-reactivity with serum components remains largely unknown.
17. Based on our experience, most anti-HMGB1 antibodies tested cross-reacted with several other proteins in serum or plasma samples under denaturing conditions (Fig. 2b).
18. Using highly reactive and specific polyclonal antibodies, we found that Western blotting often gave rise to higher values in serum HMGB1 levels as compared to commercially available HMGB1 ELISA (Table 2). This observation was consistent with previous report by others that sandwich HMGB1 ELISA

Table 2
Comparison of human serum HMGB1 levels measured by ELISA and Western blotting analysis

	ELISA				
	Normal method		Sensitive method[a]		Western blotting[b]
Human serum	Dilution	[HMGB1] (ng/ml)	Dilution	[HMGB1] (ng/ml)	[HMGB1] (ng/ml)
A	1:1	ND	1:1	ND	ND
	1:2	ND	1:2	ND	
	1:4	ND	1:4	ND	
B	1:1	5.9	1:1	4.5	581
	1:2	5.7	1:2	3.3	
	1:4	8.8	1:4	4.5	
C	1:1	13.2	1:1	20.9	950
	1:2	13.5	1:2	15.3	
	1:4	13.4	1:4	11.5	

Note: ND not detectable
[a]The recommended sensitive method was achieved by increasing sample loading and decreasing HMGB1 standard concentrations for the HMGB1 ELISA assay (*see* Fig. 2a)
[b]Using highly specific and reactive polyclonal antibodies generated in the authors' laboratory

often gave false low or negative results as compared to Western blots [14].

19. Although HMGB1 has been suggested as a feasible therapeutic target for experimental sepsis [3, 11, 27], its levels in unfractionated crude serum of septic patients did not correlate well with their disease severity [28].

20. Following ultrafiltration of serum proteins through membrane with 100 kDa cutoff, a 30 kDa HMGB1 band was detected by Western blotting in both the filtrate (<100 kDa) and retentate (>100 kDa) fractions of some septic patients [1]. Interestingly, HMGB1 levels in the filtrate fraction correlated well with the outcome of sepsis [1]. It supports the possibility that HMGB1 may interact with other serum components to form large (> 100 kDa) complexes.

21. In addition, chemical modification may similarly affect the immuno-detection of HMGB1. For instance, a recent study indicated that reactive oxygen species (ROS) may oxidize HMGB1 to form intramolecular disulfide bond between thiol group of Cys^{106} and Cys^{23} or Cys^{45} [29]. It will be important to investigate whether oxidization affects the immunodetection of HMGB1 in future studies.

22. We and others are still routinely using the Western blotting method to measure HMGB1 in serum samples [22, 30–41].

Acknowledgments and Funding

Work in the authors' laboratory was supported by grants from the National Institutes of Health (R01GM063075 and R01AT05076).

References

1. Wang H, Bloom O, Zhang M et al (1999) HMG-1 as a late mediator of endotoxin lethality in mice. Science 285:248–251
2. Bonaldi T, Talamo F, Scaffidi P et al (2003) Monocytic cells hyperacetylate chromatin protein HMGB1 to redirect it towards secretion. EMBO J 22:5551–5560
3. Lotze MT, Tracey KJ (2005) High-mobility group box 1 protein (HMGB1): nuclear weapon in the immune arsenal. Nat Rev Immunol 5:331–342
4. Ivanov S, Dragoi AM, Wang X et al (2007) A novel role for HMGB1 in TLR9-mediated inflammatory responses to CpG-DNA. Blood 110:1970–1981
5. Rendon-Mitchell B, Ochani M, Li J et al (2003) IFN-gamma induces high mobility group box 1 protein release partly through a TNF-dependent mechanism. J Immunol 170: 3890–3897
6. Tang D, Shi Y, Kang R et al (2007) Hydrogen peroxide stimulates macrophages and monocytes to actively release HMGB1. J Leukoc Biol 81:741–747
7. Scaffidi P, Misteli T, Bianchi ME (2002) Release of chromatin protein HMGB1 by necrotic cells triggers inflammation. Nature 418:191–195
8. Wang H, Ward MF, Fan XG et al (2006) Potential role of high mobility group box 1 in viral infectious diseases. Viral Immunol 19:3–9
9. Zhu S, Li W, Ward MF et al (2010) High mobility group box 1 protein as a potential drug target for infection- and injury-elicited inflammation. Inflamm Allergy Drug Targets 9:60–72
10. Wang H, Ward MF, Sama AE (2009) Novel HMGB1-inhibiting therapeutic agents for experimental sepsis. Shock 32:348–357
11. Wang H, Yang H, Tracey KJ (2004) Extracellular role of HMGB1 in inflammation and sepsis. J Intern Med 255:320–331
12. Wang H, Li W, Goldstein R et al (2007) HMGB1 as a potential therapeutic target. Novartis Found Symp 280:73–85
13. Wang H, Zhu S, Zhou R et al (2008) Therapeutic potential of HMGB1-targeting agents in sepsis. Expert Rev Mol Med 10:e32
14. Urbonaviciute V, Furnrohr BG, Weber C et al (2007) Factors masking HMGB1 in human serum and plasma. J Leukoc Biol 81:67–74
15. Sha Y, Zmijewski J, Xu Z et al (2008) HMGB1 develops enhanced proinflammatory activity by binding to cytokines. J Immunol 180: 2531–2537
16. He RL, Zhou J, Hanson CZ et al (2009) Serum amyloid A induces G-CSF expression and neutrophilia via Toll-like receptor 2. Blood 113:429–437
17. Lehner J, Wittwer C, Fersching D et al (2012) Methodological and preanalytical evaluation of an HMGB1 immunoassay. Anticancer Res 32:2059–2062
18. Yasuda T, Ueda T, Takeyama Y et al (2006) Significant increase of serum high-mobility group box chromosomal protein 1 levels in patients with severe acute pancreatitis. Pancreas 33:359–363
19. Gaini S, Pedersen SS, Koldkjaer OG et al (2007) High mobility group box-1 protein in patients with suspected community-acquired infections and sepsis: a prospective study. Crit Care 11:R32
20. Okazaki K, Kondo M, Kato M et al (2008) Elevation of high-mobility group box 1 concentration in asphyxiated neonates. Neonatology 94:105–109
21. Karlsson S, Pettila V, Tenhunen J et al (2008) HMGB1 as a predictor of organ dysfunction and outcome in patients with severe sepsis. Intensive Care Med 34:1046–1053
22. Rowe SM, Jackson PL, Liu G et al (2008) Potential role of high-mobility group box 1 in cystic fibrosis airway disease. Am J Respir Crit Care Med 178:822–831
23. Yoshizaki A, Komura K, Iwata Y et al (2009) Clinical significance of serum HMGB-1 and sRAGE levels in systemic sclerosis: association with disease severity. J Clin Immunol 29: 180–189
24. Chung HW, Lee SG, Kim H et al (2009) Serum high mobility group box-1 (HMGB1) is closely associated with the clinical and pathologic features of gastric cancer. J Transl Med 7:38. doi:10.1186/1479-5876-7-38

25. Dupire G, Nicaise C, Gangji V et al (2012) Increased serum levels of high-mobility group box 1 (HMGB1) in primary Sjogren's syndrome. Scand J Rheumatol 41:120–123
26. Lee H, Song M, Shin N et al (2012) Diagnostic significance of serum HMGB1 in colorectal carcinomas. PLoS One 7:e34318
27. Wang H, Yang H, Czura CJ et al (2001) HMGB1 as a late mediator of lethal systemic inflammation. Am J Respir Crit Care Med 164:1768–1773
28. Angus DC, Yang L, Kong L et al (2007) Circulating high-mobility group box 1 (HMGB1) concentrations are elevated in both uncomplicated pneumonia and pneumonia with severe sepsis. Crit Care Med 35: 1061–1067
29. Kazama H, Ricci JE, Herndon JM et al (2008) Induction of immunological tolerance by apoptotic cells requires caspase-dependent oxidation of high-mobility group box-1 protein. Immunity 29:21–32
30. Park JS, Arcaroli J, Yum HK et al (2003) Activation of gene expression in human neutrophils by high mobility group box 1 protein. Am J Physiol Cell Physiol 284:C870–C879
31. Li W, Li J, Sama AE et al (2013) Carbenoxolone blocks endotoxin-induced PKR activation and HMGB1 release. Mol Med 19:203–211
32. Li W, Li J, Ashok M et al (2007) A cardiovascular drug rescues mice from lethal sepsis by selectively attenuating a late-acting proinflammatory mediator, high mobility group box 1. J Immunol 178:3856–3864
33. Wang H, Liao H, Ochani M et al (2004) Cholinergic agonists inhibit HMGB1 release and improve survival in experimental sepsis. Nat Med 10:1216–1221
34. Li W, Zhu S, Li J et al (2011) A hepatic protein, fetuin-A, occupies a protective role in lethal systemic inflammation. PLoS One 6: e16945
35. Li W, Ashok M, Li J et al (2007) A major ingredient of green tea rescues mice from lethal sepsis partly by inhibiting HMGB1. PLoS One 2:e1153
36. Yang H, Ochani M, Li J et al (2004) Reversing established sepsis with antagonists of endogenous high-mobility group box 1. Proc Natl Acad Sci U S A 101:296–301
37. Manganelli V, Signore M, Pacini I et al (2010) Increased HMGB1 expression and release by mononuclear cells following surgical/anesthesia trauma. Crit Care 14:R197
38. Zhou Y, Xiong KL, Lin S et al (2010) Elevation of high-mobility group protein box-1 in serum correlates with severity of acute intracerebral hemorrhage. Mediators Inflamm pii: 142458. doi: 10.1155/2010/142458
39. Bruchfeld A, Wendt M, Bratt J et al (2011) High-mobility group box-1 protein (HMGB1) is increased in antineutrophilic cytoplasmatic antibody (ANCA)-associated vasculitis with renal manifestations. Mol Med 17:29–35
40. Shang GH, Jia CQ, Tian H et al (2009) Serum high mobility group box protein 1 as a clinical marker for non-small cell lung cancer. Respir Med 103:1949–1953
41. Goldstein RS, Gallowitsch-Puerta M, Yang L et al (2006) Elevated high-mobility group box 1 levels in patients with cerebral and myocardial ischemia. Shock 25:571–574

Chapter 3

ELISA-Based Assay for IP-10 Detection from Filter Paper Samples

Camilla Heldbjerg Drabe, Thomas Blauenfeldt, and Morten Ruhwald

Abstract

IP-10 is a small pro-inflammatory chemokine secreted primarily from monocytes and fibroblasts. Alterations in IP-10 levels have been associated with inflammatory conditions including viral and bacterial infections, immune dysfunction, and tumor development. IP-10 is increasingly recognized as a biomarker that predicts severity of various diseases and can be used in the immunodiagnostics of *Mycobacterium tuberculosis* and cytomegalovirus infection. Here, we describe an ELISA-based method to detect IP-10 from dried blood and plasma spot samples.

Key words ELISA, Cytokine detection, IP-10, CXCL10, Filter paper, Dried blood spots, Dried plasma spots, IGRA

1 Introduction

Interferon-γ (IFN-γ)-induced protein of 10 kDa (IP-10), also known as CXCL-10, is a small pro-inflammatory chemokine secreted from monocytes, endothelial cells, and fibroblasts as part of both innate and adaptive immune responses. IP-10 is expressed in response to multiple signals, mainly T cell-derived IFN-γ and tumor necrosis factor (TNF)-α but also interleukin (IL)-2, type II IFNs, IL-27, IL-17/IL-23, and IL-1β [1, 2]. IP-10 chemo-attracts CXCR3-positive cells, including macrophages, dendritic cells, natural killer cells, and activated T lymphocytes, and is considered a key mediator in pro-inflammatory immune responses [3]. Increased levels of IP-10 have been associated with inflammatory conditions including both viral and infectious diseases, immune dysfunction, and tumor development, and IP-10 is recognized as a biomarker that predicts severity of various diseases [3].

Ivana Vancurova (ed.), *Cytokine Bioassays: Methods and Protocols*, Methods in Molecular Biology, vol. 1172, DOI 10.1007/978-1-4939-0928-5_3, © Springer Science+Business Media New York 2014

For more than a decade, interferon-gamma release assays (IGRAs) have been commercially available for diagnosing infections with *Mycobacterium tuberculosis* (*Mtb*). The IGRAs quantify the T-cell response to peptide antigens specific for *Mtb* by measuring the level of IFN-γ released upon antigen stimulation in vitro [4, 5]. IP-10 is similarly released upon antigen stimulation and is increasingly recognized as an alternative immunodiagnostic biomarker [5, 6]. The diagnostic accuracy of IP-10 for active and latent tuberculosis seems at par with IGRAs [6], but as IP-10 is released in 100-fold higher levels than IFN-γ, it allows for simplified detection methods including lateral flow and dried blood and plasma spot (DBS/DPS) samples [1, 6–8].

In 1963, Robert Guthrie described a simple method to detect phenylketonuria in newborns using DBS [9]. Since then, DBS has been routinely used to screen newborns for metabolic and heritable disorders worldwide [10]. DPS and DBS have several advantages compared to conventional blood samples including ease of sample collection, transportation, and storage. Numerous analytes, including cytokines and chemokines, are quantifiable from samples stored in filter paper [1, 11–13], although the small volume of blood or plasma available from DBS/DPS samples can make detection of biomarkers expressed in low levels difficult [11].

Recently, Miller and McDade introduced a filter bottom plate and centrifugation-based ELISA assay for interleukin-6 detection from DBS [11]. Analyte extraction from DBS/DPS samples is done in filter bottom microtiter plate, which hereafter is stacked on top of the ELISA microtiter plate. By centrifugation of the stacked plates, the extraction buffer is forced through the filter directly into the corresponding ELISA plate well. The centrifugation step improves protein extraction from the filter paper discs leaving almost dry discs. This method has solved two drawbacks with DBS/DPS-based protein detection: optimal analyte extraction from the DBS/DPS samples and removal of filter paper discs from the extraction buffer before analysis. On top of improvements in analyte recovery, the method significantly reduces assay time and numbers of manual steps.

We have recently developed an ELISA-based assay for IP-10 detection from DBS/DPS samples using passive diffusion-based extraction from DBS/DPS samples directly in the ELISA plate [1]. This assay can be used for immunodiagnostics of *Mtb* and cytomegalovirus (CMV) infection [1, 7] and for monitoring liver fibrosis in hepatitis C virally infected patients [14]. Based on the centrifugation method developed by Miller and McDade, we have refined our assay for IP-10 detection from DBS/DPS samples. Compared to the previous method, the centrifugation method improves signal recovery by 18 % per disc (Drabe et al. in preparation). This report is a detailed description of the steps involved in sample preparation and analysis.

2 Materials

2.1 Collection of Dried Blood Spot Samples

1. Filter paper: Whatman 903 Protein Saver Card.
2. Lancet.
3. Disinfectant.
4. Gloves.

2.2 ELISA Plate-Coating Components

1. Nunc Maxisorb plates (Nunc).
2. Coating antibody (mouse anti-human IP-10 monoclonal antibody (mAb) (clone IM2) (*see* **Note 1**).
3. Coating buffer (0.05 M sodium carbonate, pH 9.6): Dissolve 4.28 g $Na_2CO_3 \times 10H_2O$ and 2.92 g $NaHCO_3$ in 900 mL of sterile H_2O, adjust pH to 9.6 with HCl/NaOH, and add sterile H_2O to a final volume of 1,000 mL. Store at 2–8 °C for up to 14 days.
4. Blocking buffer (PBS, 0.1 % BSA, 5 % trehalose): Add 0.2 g bovine serum albumin (BSA, Fraction V) and 5 g trehalose (D-(+)-trehalose dehydrate) to 100 mL of phosphate-buffered saline (PBS), pH 7.4.
5. Casein–PBS: PBS, 1 % casein.
6. Wash buffer: Add 1 mL of Tween-20 to 1,000 mL of PBS. Store for up to 1 month at room temperature.
7. Sealing tape (Nunc).
8. Plastic bags with ziplock (15 × 25 cm).
9. Desiccant bags (5 g silica gel).

2.3 Standards, Reagents, and Buffers

1. IP-10 protein standards: Recombinant human IP-10 (PeproTech) diluted in 1 % casein–PBS (pH 7.4) aliquot in 200 µL vials of 5,500 pg/mL. Store below −20 °C.
2. Dilution buffer: Add 20 g BSA to 1,000 mL of PBS. Allow to dissolve by occasional gentle shaking. Add 1 g dissolved Bronidox (*see* **Note 2**) and 1 mL of Tween-20. Adjust pH to 7.4. Keep stable at +5 °C for up to 2 months or 12 months at −20 °C.
3. Conjugate buffer: Mix IR1-HRP mAb in 3 mL of dilution buffer to a final concentration of 0.2 µg/mL.
4. Detection antibody: Vortex the horseradish peroxidase-conjugated rat anti human IP-10 mAb (Clone IR-1) (IR1-HRP mAb) (*see* **Note 1**).
5. Substrate: 3,3′,5,5′-tetramethylbenzidine (<0.02 %), hydrogen peroxide (TMB One) (Kem-En-Tec Diagnostics A/S, Taastrup, Denmark).
6. Stop solution: 1 M H_2SO_4.
7. Filter plate: HTS Filter Plate (MSHVN4510, EMD Millipore).

3 Methods

3.1 Preparing the Samples

For sample collection on filter paper two methods can be approached: Either blood is collected directly on the filter paper from the patient after a finger prick, or the sample is obtained by collecting the blood by standard venipuncture followed by pipetting sample material onto the filter paper. Always wear gloves when working with the filter paper.

3.1.1 Finger Prick Blood

1. Clearly label the Whatman Protein Saver 903 Card with patient ID and date.
2. Clean the incision area with an alcohol swab and allow to air-dry. Do not allow the incision site to come into contact with any non-sterile item or surface.
3. Prick using a lancet/needle (Fig. 1).
4. Hold the Whatman Protein Saver 903 Card without touching the filter paper (*see* **Note 3**).
5. Uniformly saturate the filter paper by quickly and gently touching the drop of blood to the Whatman Protein Saver 903 Card (*see* Figs. 2 and 3). Do not press the puncture site to the filter paper or touch the Whatman Protein Saver 903 Card at any stage of

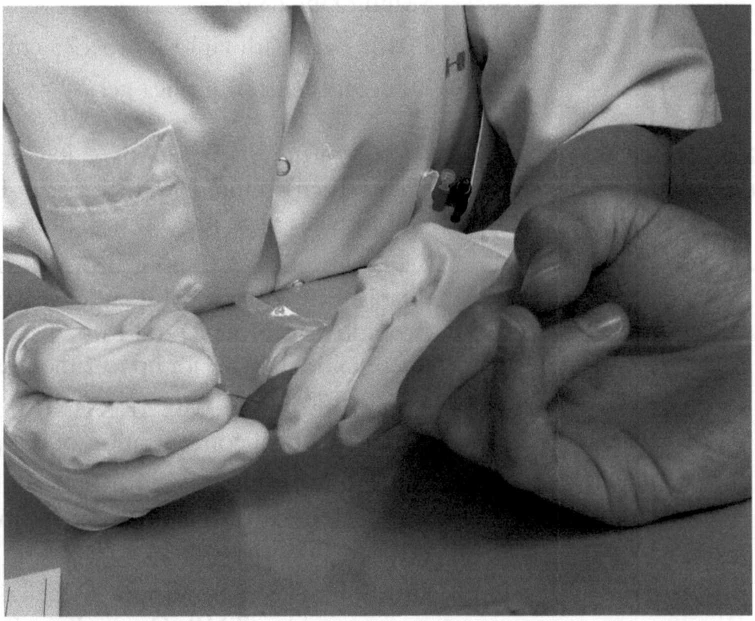

Fig. 1 Finger prick using a lancet: Start by disinfecting the area at the fingertip. Prick using a lancet. If the blood is not running immediately, a small squeeze on the sides of the finger will help the blood drops form

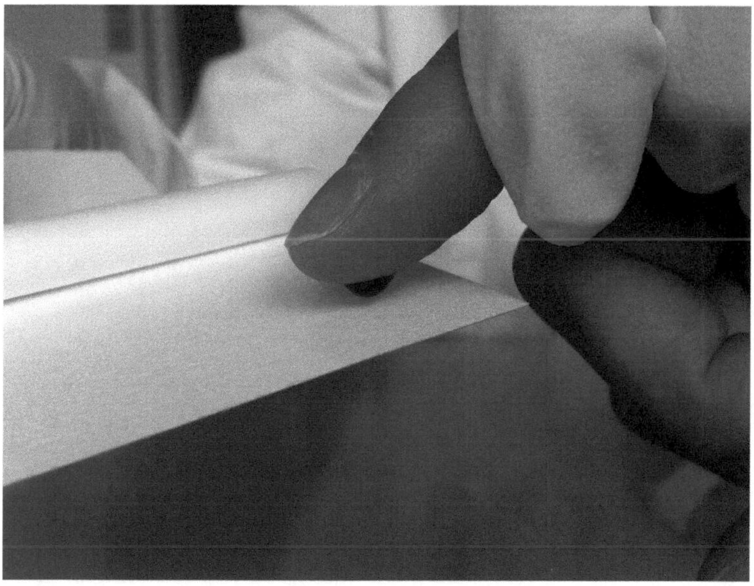

Fig. 2 Applying blood from a finger prick to filter paper: Place the filter paper horizontally under the finger

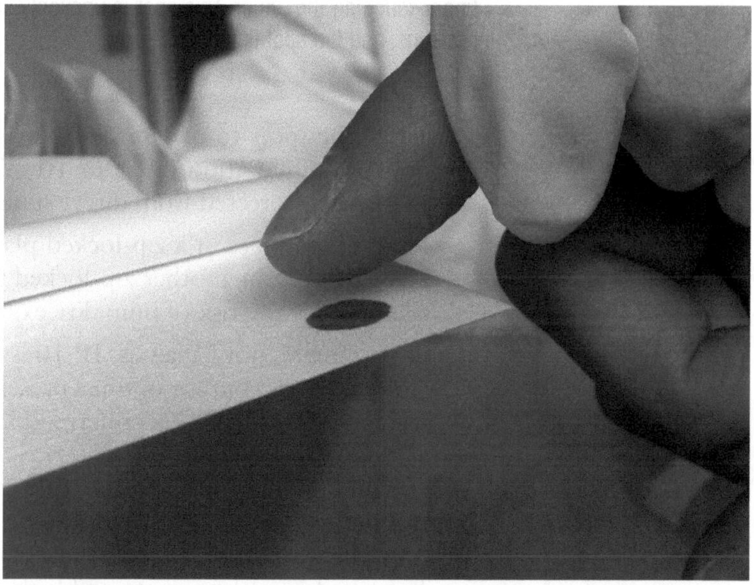

Fig. 3 Applying blood from a finger prick to filter paper: Raise the paper carefully to the drop of blood, and the blood will uniformly saturate the paper in a circle. It is very important that the finger is not directly touching the paper

collection. Do not touch the DBS circle once blood is applied (see examples of acceptable and unacceptable spots: Fig. 4).

6. After collecting five blood spots, clean the site and leave it unbandaged.

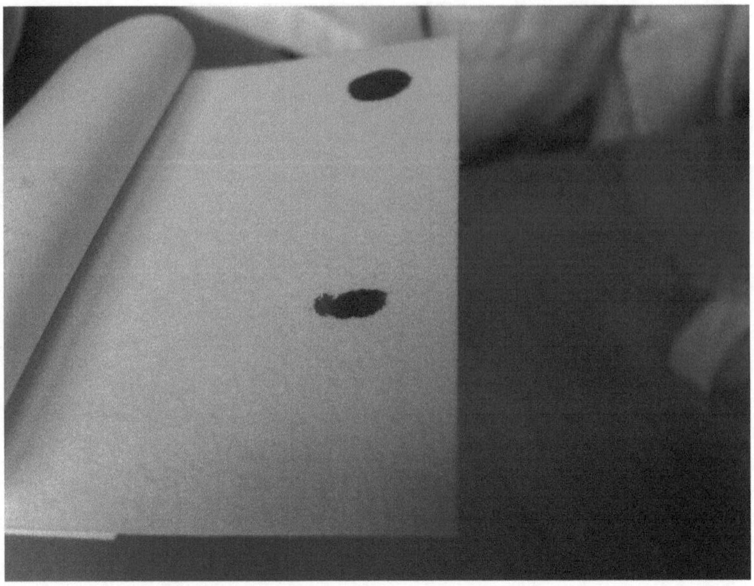

Fig. 4 Examples of an acceptable and an unacceptable blood spot. The first spot is made as shown in Figs. 2 and 3. The spot is homogenous, borders are sharply defined, and the size is acceptable (>8 mm). The second spot is made with the finger touching the paper. This spot is unevenly saturated, borders are inhomogeneous, and the spot is too little

7. Allow the spots to dry for 3–4 h at ambient temperature in a not-too-humid atmosphere or 10 min at 50 °C. The samples should be protected from direct sunlight while drying.
8. Place the samples in a zip-locked plastic bag with a desiccant. Place this bag in another zip-locked plastic bag with desiccant to minimize the risk of humidity exposure.
9. Store samples until analysis. IP-10 is stable for at least 4 weeks at up to 37 °C, but for optimal preservation for longer periods storage below −20 °C is preferred [1].

3.1.2 Sample Material Obtained by Venipuncture

This method allows for greater sample material quantities, higher homogeneity of spots, and samples to be manipulated before analysis, e.g., incubated with specific antigens (*see* **Note 4**). Samples can be stored as plasma spots or blood spots according to one's preferences (*see* **Note 5**).

1. Prepare filter paper with patient ID and date, and place the filter paper horizontally at a non-absorbent surface.
2. Blood sample preparation for DBS samples. Use heparin-, EDTA-, or citrate-stabilized blood. Ensure that cells and plasma are completely mixed; this can be done with gentle shaking of the vacutainer tube for at least 1 min. If blood clots are present in the tube the blood is insufficiently stabilized and should be discarded.

Fig. 5 Blood spots from full blood obtained from venipuncture: Apply the blood to the filter paper by holding the pipette at a 90° angle from the filter paper 1–2 mm above the filter paper. Gently spot 30–50 μL blood (or 25–40 μL plasma) on the filter paper. To assure that the sample distributes evenly on the paper, it is important that emptying the pipette is done at low speed

3. Blood sample preparation for DPS samples: Use heparin-, EDTA-, or citrate-stabilized blood (*see* **Note 3**). Ensure that cells and plasma are completely separated by centrifugation of the sample 2,000 × *g* for 10 min (no brake on the centrifuge).
4. Apply by holding the pipette at a 90° angle from the filter paper 1–2 mm above the filter paper (*see* Fig. 5). Gently spot 30–50 μL of blood or 25–40 μL of plasma on the filter paper. To assure that the sample distribute evenly on the paper, it is important that emptying the pipette is done at low speed.
5. Store as described in 3.1.1 paragraph 7–9.

3.2 Coating the ELISA Plates

1. Mix 30 μL of IM2 mAb in 10 mL of coating buffer for a final concentration of 3 μg/mL. Add 100 μL/well on the maxisorb plate with a multichannel pipette. Cover the plates with sealing tape and incubate overnight at 2–8 °C or for 2 h at 37 °C.
2. Wash 1 × 300 μL/well with wash buffer, and add 150 μL/well blocking buffer to all wells using the multichannel pipette. Cover the plates with sealing tape and incubate for 2 h at 20 °C.
3. Pour the blocking buffer out (*see* **Note 6**).
4. Dry the plates at 37 °C overnight in a ventilated incubator.
5. Store the plates in individual ziplock bags with a desiccant bag at 4 °C until use. Store under these conditions for up to 12 months.

3.3 Preparing the Plate and Samples

All reagents and plates must have room temperature. Design your plate layout.

1. Defrost one premade IP-10 protein standard (5,500 pg/mL). Resuspend vial by gentle pipetting. Serially dilute the standard to make a seven-point standard curve: Add 100 μL of dilution buffer in six 200 μL tubes, add 100 μL of premade IP-10 protein standard to the first tube of dilution buffer, and mix using the pipette. Of this suspension, take 100 μL and mix to the next tube of dilution buffer. This step is repeated until five tubes contain a mix of dilution buffer and IP-10 protein. The sixth tube is left with only dilution buffer and serves as a blank. The first standard curve point is the premade un-diluted IP-10 protein (*see* **Note 7**).
2. Place the filter bottom plate on a non-wicking surface with space between well bottom and surface (e.g., using an empty 96-well plate) (*see* **Note 8**). Note the date and number of the assay on the plate.
3. Punch one or two 5.5 mm disc from the filter sample using a standard office hole puncher or suitable other DBS punching equipment (e.g., Harris Uni-Core™ punchers (Whatman, USA)) (*see* **Note 9**). When using a classic office hole puncher, remove the bottom of the hole puncher and reverse the puncher upside down. Place the filter paper so that the sample is visible through the hole of the puncher. Make sure that the sample is centered before punching to avoid any part of the disc not containing sample material.
4. Use forceps to place the disc in the appropriate well in the filter microtiter plate according to the plate layout. Make sure that all discs are placed horizontally at the bottom of the well (*see* **Note 10**).
5. After all discs have been placed in the plate, add 80 μL of dilution buffer to all wells using a multichannel pipette.
6. Add 10 μL of standards to the appropriate wells, and mix using the pipette.
7. Avoid air bubbles under the discs by applying a gentle pressure of a pipette tip on discs that are resting on air bubbles. Use a clean tip for every well to avoid cross contamination (*see* **Note 11**).
8. Cover the plate with a lid and incubate for a total time of 1 h at room temperature (*see* **Note 12**). The plate must be protected from light during incubation.

3.4 ELISA Plate

1. ELISA plate: Note the date and number of the assay on the plate.
2. Add 20 μL of conjugate buffer to all wells of the pre-coated ELISA plate using a multichannel pipette (*see* **Note 13**).
3. Stack the filter plate on top of the pre-coated ELISA plate with conjugate buffer. Spin the plates at 4,000/min for 5 min,

and turn off the brake. The incubation time begins at the start of the centrifuge (*see* **Note 14**).

4. After completion of centrifugation, cover the ELISA plate with a lid or a sealing tape, or simply leave the filter plate on top of the ELISA plate, and incubate for a total time of 1 h at room temperature in the dark (*see* **Note 15**).
5. After incubation, wash the plate by adding 300 μL of wash buffer per well. Discard the wash buffer after 30 s. Repeat this step four times. Knock the plate dry on clean paper towels.

3.5 IP-10 Detection, Analysis, and Calculation

1. Add 100 μL of TMB substrate to all wells using a multichannel pipette, and incubate the plate for an additional 30 min at room temperature in the dark.
2. Stop the reaction by adding 100 μL of H_2SO_4 to all wells using a multichannel pipette (*see* **Note 16**).
3. The assay is now complete, and the optical densities of samples can be read with an ELISA reader at 450 nm with 630 nm reference.
4. Export the 450 nm readings after subtracting 630 nm. Use standard curve to calculate IP-10 concentration in the DBS discs.

4 Notes

1. As an alternative to IM2/IR1 antibodies, we recommend Human Quantikine ELISA kit (cat. Nr. DIP100) or the more cost-effective Human CXCL10 DuoSet (cat. Nr. DY266), both by R&D systems (Minneapolis, MN, USA). Especially when analyzing signals from unstimulated samples where IP-10 content is low, these ELISAs should be run with the highest possible volume of DBS/DPS buffer (e.g., 75 μL) to ensure high signals [11].
2. Dissolve 5 g Bronidox powder (Sigma-Aldrich) in 25 mL of 97 % ethanol under biohazard hood using protective gear, and add 5 mL of the Bronidox preservative to the solution.
3. If this method proves difficult or higher homogeneity of samples is required, blood drops can be collected in a capillary tube before rapid transfer to the filter paper.
4. If stimulating whole blood before DBS/DPS sampling, it is important only to use heparinized blood. If using chelate-stabilized blood (e.g., EDTA) the Ca^{2+} ions required for T cell signaling and activation are bound and immune responses become artificially low.
5. Plasma is less viscous and makes larger spots compared to whole blood. The less viscous plasma is easier to control when preparing spots from stimulated heparinized blood as the anticoagulant ensures that cells have formed a pellet during the

incubation period and plasma can be easily sampled. In contrast preparation of a DBS sample from stimulated blood requires shaking of the blood to resuspend the cells. This procedure can be difficult to master as too little shaking makes uneven and unacceptable spots and too much agitation can cause cell lysis and undesired results.

6. Avoid knocking, as a residual volume of trehalose is needed to stabilize the coating antibody in dried form.
7. The standard points are 500, 250, 125, 62.5, 31.3, 15.6, 7.8, and 0 pg/mL.
8. If the tip below the filter plate touches a wicking surface it will empty which affects the results.
9. When analyzing samples with low IP-10 concentration, e.g., unstimulated samples of healthy persons, the use of two discs per sample well improves signal strength and improves technical precision.
10. By gently moving the plate back and forth on the table, discs will easily fall to the bottom of the well. If there are any discs that are still misplaced after this maneuver, gently press them in place with the forceps or a pipette tip.
11. Make sure that you press the disc very gently with the pipette tip. If the pressure is too forceful, there is a risk of penetrating the bottom of the filter plate or otherwise compromising the filter barrier and cause loss of sample material from the well.
12. Applying sealing tape can lead to leaking; a lid is preferred.
13. Preparing of the conjugate buffer must be done in immediate relation to its use. The conjugate must be kept at 5 °C until use. Start preparing the conjugate buffer 5–10 min before the 1-h incubation is complete to have the ELISA plate ready for centrifugation.
14. Use only good-quality plates as contra weight in the centrifuge as some plates can break at the high g-forces causing undesired results.
15. The incubation step can take place in the centrifuge.
16. The volume of H_2SO_4 will create enough motion in the well to mix contents.

Acknowledgement

This work was supported by research grants from Hillerød Hospital and the Danish National Advanced Technology Foundation. The IP-10 ELISA was developed with support from the Danish Lung Association and the Lundbeck Foundation (Denmark). The funders had no role in study design, data collection and analysis, decision to publish, or preparation of the manuscript.

References

1. Aabye MG, Eugen-Olsen J, Werlinrud AM et al (2012) A simple method to quantitate IP-10 in dried blood and plasma spots. PLoS One 7:e39228
2. Burke SJ, Goff MR, Lu D et al (2013) Synergistic expression of the CXCL10 gene in response to IL-1β and IFN-γ involves NF-κB, phosphorylation of STAT1 at Tyr701, and acetylation of histones H3 and H4. J Immunol 191:323–336
3. Liu M, Guo S, Hibbert JM et al (2011) CXCL10/IP-10 in infectious diseases pathogenesis and potential therapeutic implications. Cytokine Growth Factor Rev 22:121–130
4. Andersen P, Munk ME, Pollock JM et al (2000) Specific immune-based diagnosis of tuberculosis. Lancet 356:1099–1104
5. Whitworth HS, Scott M, Connell DW et al (2013) IGRAs – the gateway to T cell based TB diagnosis. Methods 61:52–62
6. Ruhwald M, Aabye MG, Ravn P (2012) IP-10 release assays in the diagnosis of tuberculosis infection: current status and future directions. Expert Rev Mol Diagn 12:175–187
7. Tuuminen T, Salo E, Kotilainen H et al (2012) Evaluation of the filter paper IP-10 tests in school children after exposure to tuberculosis: a prospective cohort study with a 4-year follow-up. BMJ Open 2:6
8. Aabye MG, Ruhwald M, Praygod G et al (2010) Potential of interferon-gamma-inducible protein 10 in improving tuberculosis diagnosis in HIV-infected patients. Eur Respir J 36:1488–1490
9. Guthrie R, Susi A (1963) A simple phenylalanine method for detecting phenylketonuria in large populations of newborn infants. Pediatrics 32:338–343
10. Mei JV, Alexander JR, Adam BW et al (2001) Use of filter paper for the collection and analysis of human whole blood specimens. J Nutr 131:1631S–1636S
11. Miller EM, McDade TW (2012) A highly sensitive immunoassay for interleukin-6 in dried blood spots. Am J Hum Biol Off J Hum Biol Counc 24:863–865
12. Skogstrand K, Thorsen P, Nørgaard-Pedersen B et al (2005) Simultaneous measurement of 25 inflammatory markers and neurotrophins in neonatal dried blood spots by immunoassay with xMAP technology. Clin Chem 51:1854–1866
13. Skogstrand K, Thysen AH, Jørgensen CS et al (2012) Antigen-induced cytokine and chemokine release test for tuberculosis infection using adsorption of stimulated whole blood on filter paper and multiplex analysis. Scand J Clin Lab Invest 72:204–211
14. Ruhwald M, Andersen ES, Christensen PB et al (2012) IP-10 can be measured in dried plasma spots in patients with chronic hepatitis C infection. PLoS One 7(9):e45181

Chapter 4

Enhanced ELISA Based on Carboxymethylated Dextran Coatings

Benoît Liberelle, Charles Fortier, and Gregory De Crescenzo

Abstract

In a "sandwich" enzyme-linked immunosorbent assay (ELISA) designed to detect an antigen in a complex protein mixture, the antigen is usually captured *via* an antibody adsorbed to the wells of a microplate. Plate preparation for standard assay involves a passive adsorption of capture antibodies followed by the incubation of blocking agents. Here, we describe a new strategy that replaces these two time consuming adsorption steps (up to 15 h) by a unique step corresponding to the covalent grafting of the capture antibody on a carboxymethylated dextran (CMD) layer, a single step completed in 15 min. Taking advantage of the CMD low-fouling properties, blocking agent-free buffer solutions can be used as diluent in the improved approach.

Key words Carboxymethylated dextran, Carbodiimide chemistry, Covalent attachment, ELISA, Low-fouling

1 Introduction

Enzyme-linked immunosorbent assay (ELISA) is the most commonly used tool for antigen detection and quantification in complex solutions. In its original setup, the entire process is based on the physical adsorption of capture antibodies and blocking agents (Fig. 1a). The weak attachment between these biomolecules and the plate may however be disrupted during the washing steps involving detergent molecules [1] or under shaking [2]. Assay buffer is therefore supplemented with blocking agents to fill open spaces created by the detachment of biomolecules [3]. Bovine serum albumin (BSA), low-fat milk, albumin, or gelatin are routinely used to block the microplates and usually improve ELISA specificity [3, 4]. Benefits of blocking agents may be nevertheless limited when undesired interactions between ELISA reagents and blocking molecules induce a decrease in ELISA sensitivity [5, 6].

Based on the previously addressed limitations of classical ELISA setup, we here report the development of a new generation of ELISA allowing for faster thus more convenient ELISA (Fig. 1b).

Ivana Vancurova (ed.), *Cytokine Bioassays: Methods and Protocols*, Methods in Molecular Biology, vol. 1172,
DOI 10.1007/978-1-4939-0928-5_4, © Springer Science+Business Media New York 2014

Fig. 1 Schematic illustration of capture antibody immobilization. (**a**) Standard ELISA based on the physical adsorption of capture antibodies (up to 14 h) and blocking agents (1 h) on polystyrene wells, and (**b**) modified ELISA involving the covalent attachment of capture antibodies (15 min) onto a carboxymethylated dextran (CMD) layer grafted on aminated wells

More specifically, we propose to chemically graft a thin layer of carboxymethylated dextran (CMD) on aminated ELISA microplates. Taking advantage of its low-fouling properties and the availability of free carboxylic groups homogeneously distributed on the polymer, CMD coating may efficiently allow for the covalent immobilization of capture antibodies while acting as a protein adsorption blocking layer per se [7]. The potential of our approach was highlighted with a commercially available ELISA kit developed for the detection of human epidermal growth factor (hEGF) using a protein-free buffer as reagent diluent.

2 Materials

Prepare all solutions using ultrapure water (resistivity of 18.2 MΩ cm; total organic compounds (TOC) ≤ 5 ppb) and ACS-grade chemical reagents. Store all reagents at 4 °C (unless indicated otherwise). Glassware should be carefully cleansed by overnight immersion in a bath of potassium hydroxide-saturated isopropyl alcohol followed by intensive rinsing with ultrapure water (*see* **Note 1**).

2.1 Dextran Carboxymethylation

1. Technical grade dextran powder (*see* **Note 2**).
2. Sodium hydroxide (NaOH) pellets.
3. Bromoacetic acid.
4. Hydrochloride solution (HCl, 37 % v/v).
5. Reagent for ^{1}H NMR characterization: Deuterium oxide (D_2O).
6. Rubber/latex-free syringes (1 and 20 mL).
7. 25 mm Syringe filters (0.2 μm nylon membrane).
8. Single-use stainless steel needles.
9. Amicon Ultra-15 centrifugal filter unit with Ultracel-10 membrane.
10. 1.5 mL Tubes.

2.2 ELISA Plate Functionalization

1. Commercially available aminated microplates.
2. 0.1 M NHS aliquots: *N*-Hydroxysuccinimide (NHS) in water. Weigh 0.74 g NHS. Dissolve in 65 mL of ultrapure water. Aliquot in 750 and 150 μL volumes. Store below −18 °C until use.
3. 0.4 M EDC aliquots: Ethyl-*N'*-(3-dimethylaminopropyl) carbodiimide hydrochloride (EDC) in water. Weigh 5 g EDC. Dissolve in 65 mL of ultrapure water. Aliquot in 750 and 150 μL volumes. Store below −18 °C until use.
4. DPBS: Dulbecco's phosphate-buffered saline (8 mM sodium phosphate, 2 mM potassium phosphate, 140 mM sodium chloride, 10 mM potassium chloride, pH 7.4; *see* **Note 3**). Undissolved materials are removed by passing the solution through a 0.2 μm filter unit. Store at 4 °C.
5. 10 mM Acetate buffer, pH 5.0: Mix 10 mM sodium acetate and 10 mM acetic acid solutions. Dissolve 41 mg of sodium acetate in 50 mL of ultrapure water. Adjust the pH to 5.0 by adding a 10 mM acetic acid solution, *i.e.*, 28.6 μL of glacial acetic acid in 50 mL of ultrapure water. Undissolved materials are removed by passing the solution through a 0.2 μm filter unit. Store at 4 °C.
6. 1 M EtA, pH 8.5: Ethanolamine (EtA) in water: In a flask containing 86 mL of ultrapure water, add 9.7 mL of ethanolamine hydrochloride, EtA-HCl (99 + % purity). Prepare a fresh 1 M NaOH solution by dissolving 0.8 g of NaOH pellets in 20 mL of ultrapure water. Adjust the pH of the diluted EtA-HCl solution to pH 8.5 (*ca.* 4 mL of the 1 M NaOH solution is required). Undissolved materials are removed by passing the solution through a 0.2 μm filter unit. Store at 4 °C.

2.3 ELISA Reagents

Detection and quantification of recombinant human epidermal growth factor (rhEGF) was used in this chapter as an example. Commercially available DuoSet ELISA kit containing mouse anti-human EGF antibody (capture antibody), biotinylated goat anti-human EGF antibody (detection antibody), streptavidin–horseradish peroxidase (streptavidin–HRP, 200×), rhEGF, BSA, and substrate solution (hydrogen peroxide/tetramethylbenzidine) were purchased from R&D Systems (Minneapolis, MN).

1. DPBS-T: Wash buffer containing 0.05 % Tween 20 in DPBS. Dilute 250 μL of Tween 20 in 500 mL of DPBS (*see* **Note 4**). Store at 4 °C.
2. DPBS-BSA: Blocking solution containing 1 % BSA in DPBS. Store at 4 °C.
3. Capture antibody aliquots (200×): Dissolve 720 μg of lyophilized antibodies in 0.9 mL of DPBS. Aliquot in 30 μL volumes. Store below −18 °C until use.

4. Standard rhEGF aliquots (1.65 μM): dissolve 0.1 mg in 10 mL of PBS–BSA.
5. Detection antibody aliquots (200×): Dissolve 9 μg of lyophilized antibodies in 0.9 mL of DPBS–BSA. Aliquot in 30 μL volumes. Store below −18 °C until use.
6. Stop solution: 2 N sulfuric acid solution: In a glass bottle, dilute 2.8 mL of sulfuric acid (99 + % purity, 35.6 N) in 47.2 mL of ultrapure water. Store in a well-ventilated place.

3 Methods

3.1 Production of Carboxymethylated Dextran

1. Add 1.19 g of NaOH pellets, 10 mL of ultrapure water, and a large magnetic stirrer in a 100 mL conical flask. Slowly add 400 mg of dextran under stirring (*see* **Note 5**). Once dissolved, add 1.4 g of bromoacetic acid, close the conical flask using aluminum foil, and seal using parafilm. Keep the reaction mixture under mild stirring for 15 h at room temperature (RT).
2. Remove the aluminum foil from the flask. Adjust the pH to 5 by adding HCl dropwise with a 1 mL syringe (*see* **Note 6**). Stop stirring.
3. Collect the solution using new disposable 20 mL syringe and needle. Discard the needle, and mount a 0.2 μm filter to the syringe. Inject 2.5 mL in a centrifugal filter (*see* **Note 7**).
4. Top the centrifugal filter to 15 mL with ultrapure water. Centrifuge for 30 min at 4,000 × *g* in a swinging bucket rotor. Discard the filtrate.
5. Repeat **step 4** another five times.
6. Homogenize the retentate by pipetting up and down several times with a 200 μL pipette. Collect the retentate, and aliquot it by up to 750 μL in 1.5 mL tubes. Let the aliquots freeze in a −20 °C freezer. Lyophilize. Store the powder at 4 °C until use.
7. Rinse and store the centrifugal filter at 4 °C in a 1 % sodium benzoate solution until use.

3.2 Characterization of Carboxymethylated Dextran

1. Sample preparation for ^{1}H NMR characterization: Weigh 5 mg of dextran. Dissolve each powder in 1 mL of D_2O. Sonicate until complete solubilization.
2. Collect ^{1}H NMR spectra (*see* ref. [7]) at 50 °C, and suppress the water signal by pulse sequences (72.4°). Representative spectra are shown in Fig. 2. The degree of carboxymethylation (or *dcm*) can be evaluated from the ^{1}H NMR peak integration ratio of the newly added methyl protons (peak *b*) to the anomeric protons (peaks *a* and *a′*):

$$dcm = \frac{b}{2\left(a + a'\right)}$$

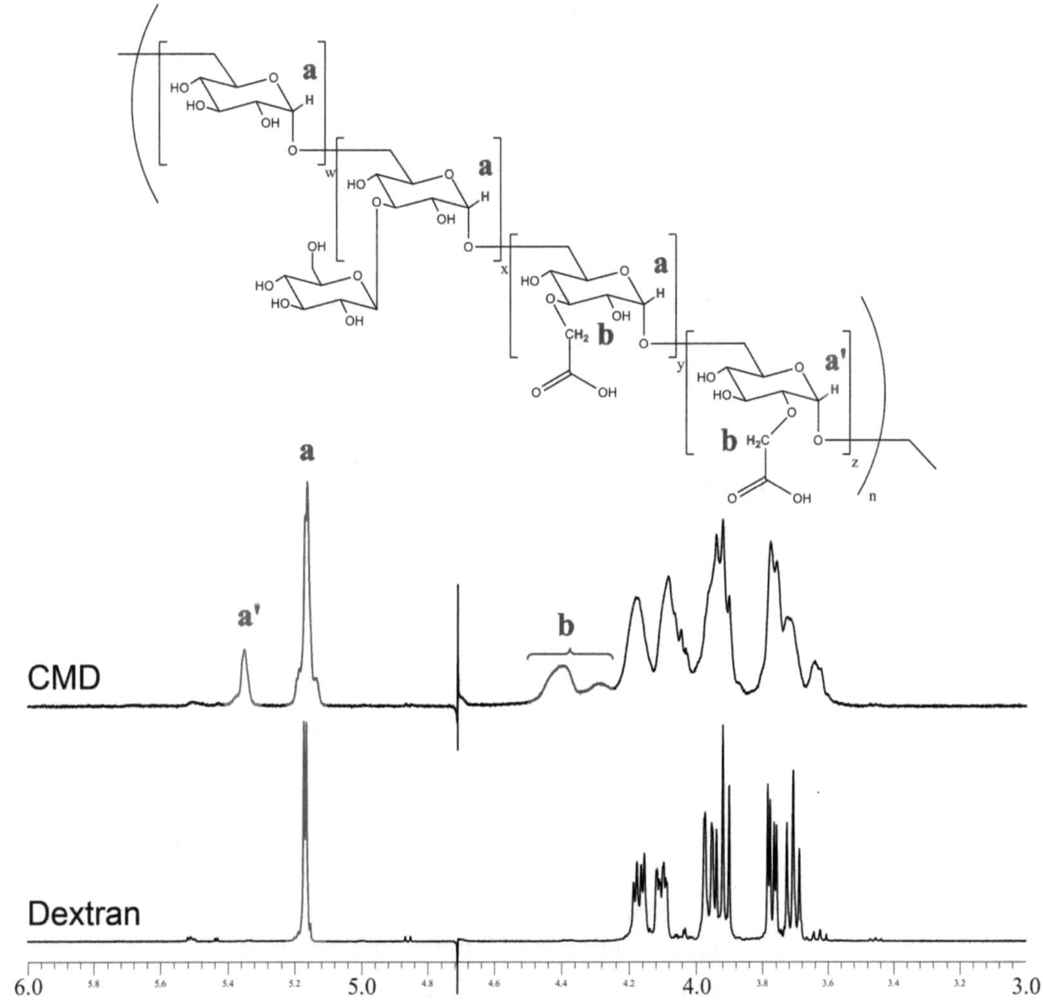

Fig. 2 ^{1}H NMR spectra and corresponding peak identification for dextran and carboxymethylated dextran (CMD). Here, dextran was carboxymethylated for 1 h to yield a CMD with a degree of carboxymethylation (*dcm*) of *ca.* 33 %. The water signal (located at *ca.* 4.7 ppm at 50 °C) was suppressed by pulse sequences

3.3 Carboxymethylated Dextran Layer Grafting

1. Dissolve 20 mg of CMD in 9 mL of ultrapure water. Thaw and add two aliquots of 0.1 M NHS (2 × 750 μL) and two aliquots of 0.4 M EDC (2 × 750 μL). Mix and immediately inject 110 μL of the NHS-activated CMD solution into each aminated well. Cover the ELISA plate with an adhesive protective film and incubate on an orbital plate shaker for 30 min at RT.
2. Rinse the wells three times (3 × 200 μL) with DPBS and five times (5 × 200 μL) with ultrapure water (*see* **Note 8**).

3.4 Capture Antibody Grafting

For the following steps, do not allow the plate to dry at any point. During the incubation times, the plates are covered with an adhesive film and placed on an orbital plate shaker.

1. Thaw and mix eight aliquots of 0.1 M NHS (8×750 μL) and eight aliquots of 0.4 M EDC (8×750 μL). Immediately inject 100 μL of the carbodiimide solution into each aminated well. Incubate for 5 min at RT.
2. Rinse the wells one time with 200 μL of ultrapure water (*see* **Note 9**), and immediately inject 100 μL of the capture antibody solution (2×30 μL of capture antibody dissolved in 12 mL of 10 mM acetate buffer, pH 5.0) into each well for 15 min at RT.
3. Rinse the wells three times with 200 μL of DPBS-T.
4. Incubate 100 μL of 1 M EtA (pH 8.5) in each well for 10 min at RT (*see* **Note 10**).
5. Rinse the wells three times with 200 μL of DPBS-T.

3.5 Assay Procedure

For the following steps, do not allow the plate to dry at any point. During the incubation times, the plates are covered with an adhesive film and placed on an orbital plate shaker.

1. Incubate 100 μL of standard rhEGF solutions and samples for 1 h at RT.
2. Rinse the wells three times with 200 μL of DPBS-T.
3. Dilute two aliquots of detection antibody (2×30 μL) in 12 mL of DPBS-T. Incubate 100 μL of the diluted detection antibody solution for 30 min at RT.
4. Rinse the wells three times with 200 μL of DPBS-T.
5. Dilute 60 μL of streptavidin–HRP (200×) in 12 mL of DPBS-T (*see* **Note 11**). Incubate 100 μL of the diluted streptavidin–HRP solution for 20 min in the dark at RT.
6. Rinse the wells three times with 200 μL of DPBS-T.
7. Mix 6 mL of color reagent A (hydrogen peroxide) and 6 mL of color reagent B (tetramethylbenzidine) (*see* **Note 12**). Incubate 100 μL of the substrate solution for 30 min in the dark at RT.
8. Add 50 μL of the stop solution in each well. Gently tap the plate to ensure thorough mixing.
9. Read the optical density of each well using a microplate reader set at 450 nm (*see* **Note 13**). If wavelength correction is available, subtract readings at 540 nm (or 570 nm) from the readings at 450 nm (*see* **Note 14**). Standard ELISA dose–response curve is proposed in Fig. 3.

4 Notes

1. "Base bath" (saturated potassium hydroxide (KOH) in isopropanol) is commonly used to remove organic contaminants that are physically adsorbed or chemically attached to the laboratory

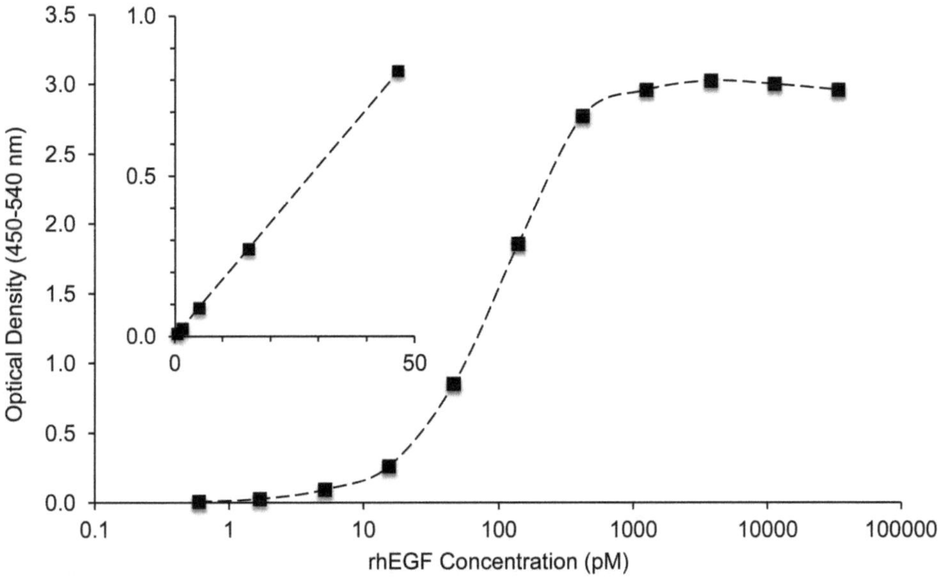

Fig. 3 Typical standard curve generated for human EGF detection with DuoSet capture antibodies chemically grafted on a CMD layer

glassware. Overnight soaking in the cleaning bath can indeed dissolve a thin outer layer of the glass. Safety precautions are required when this highly caustic solution is used; that is, wear apron, eye protection, and thick butyl gloves; keep ignition sources away from thc base bath; protect the work area from the caustic solution with aluminum foil. Place a large plastic container in a fume hood; add approximately 200 g of solid KOH pellets and 4 L of isopropyl alcohol. Dry and oil-free glassware is left to soak. Replace the cover onto the plastic container. After overnight immersion, the glassware is rinsed thoroughly with ultrapure water and dried in air. Note that KOH pellets should be added occasionally to maintain a KOH saturation and the base bath efficiency.

2. The impact of the dextran molecular weight (Mw = 10, 70, and 500 kDa) on ELISA response was evaluated in a previous study (*see* ref. [8]). We recommend 70 and 500 kDa dextrans to maximize both ELISA sensitivity and amplitude of standard dose–response curve response.
3. Dulbecco's phosphate-buffered saline (modified PBS, without calcium chloride and magnesium chloride) is a physiological buffer commonly used for ELISA. Presence of divalent ions may indeed induce some nonspecific electrostatic interactions between ELISA reagents. Dry-blend powders of Dulbecco's PBS are commercially provided to eliminate weighing time, pH adjustments, and concerns about long-term stability.

4. Tween 20 is a viscous liquid. Cutting off a small amount of the pipette tip is helpful to aspirate the required volume. Once added to DPBS, gently mix the solution to prevent any excessive foaming.
5. Slowly add dextran to the aqueous solution to avoid formation of dextran aggregates. Note that ultrasonic bath can be used to dissolve any remaining dextran aggregates and obtain a homogeneous mixture.
6. HCl is injected in the alkaline reactive solution to stop dextran carboxymethylation process. 600–1,000 μL HCl are required to reach pH 5 for 10 mL of reagent. Acidity can be evaluated using pH paper.
7. We recommend that the total amount of CMD processed in such centrifugal filter do not exceed 100 mg in order to avoid viscosity issues that may slow down purification and limit each centrifugation yield. The centrifugal filter membrane cutoff should be at least three times smaller than the dextran molecular weight. For the 70 kDa and 500 kDa dextrans, we recommend cutoffs of 10 kDa and 100 kDa, respectively.

   As an alternative to centrifugal filters (or when CMD is produced in larger quantity), dialysis membranes (standard regenerated cellulose membranes, molecular weight cutoff = 8 kDa) can be used. Rinse the pre-wetted membranes with ultrapure water. Inject the syringe content into the dialysis tubing passing through a 0.2 μm filter. Dialyze at least five times against ultrapure water (use a large water container; slowly increase the dialysis time from 15 min to overnight). Filter (0.2 μm) and lyophilize the retentate. Rinse and store the dialysis membrane at 4 °C in a 1 % sodium benzoate solution until use.
8. For long-term storage, CMD-coated microplates may be air-dried (preferably in a laminar-flow cabinet to prevent dust deposition), sealed with an adhesive protective film, and stored at 4 °C until use. Note that comparable ELISA responses were obtained using freshly made CMD-covered plates and air-dried CMD plates that had been stored for 8 weeks at 4 °C.
9. The stability of NHS-activated carboxyl groups decreases in the presence of salt (*see* ref. [7]). Ultrapure water is preferred for the rinsing step following a carbodiimide activation.
10. EtA is used to deactivate the remaining NHS-reactive groups in order to prevent any undesired attachment of amine-containing molecules. Note that if the capture antibody concentration is high (as in the chosen example, *i.e.*, 4 μg/mL), the EtA blocking step is not required (*see* ref. [8]).
11. HRP is light sensitive. Once prepared, wrap the labware containing diluted streptavidin–HRP solution with aluminum foil. During incubation time, protect the plate from light with a box.

12. Substrate solution is unstable and light sensitive. Prepare the solution only a few minutes before injection, and wrap the labware containing the substrate solution with aluminum foil. During incubation time, protect the plate from light with a box.
13. Clean the bottom surface of the plate with a lint-free wipe to remove dust and dried liquid before reading the plate.
14. Readings made directly at 450 nm without correction may be higher and less accurate.

Acknowledgments

This work was supported by the Canada Research Chair on Protein-enhanced Biomaterials (GDC).

References

1. Prusaksochaczewski E, Luong JHT (1989) An improved elisa method for the detection of salmonella-typhimurium. J Appl Bacteriol 66:127–135
2. Steiner JM, Teague SR, Williams DA (2003) Development and analytic validation of an enzyme-linked immunosorbent assay for the measurement of canine pancreatic lipase immunoreactivity in serum. Can J Vet Res 67: 175–182
3. Sentandreu MA, Aubry L, Toldra F et al (2007) Blocking agents for ELISA quantification of compounds coming from bovine muscle crude extracts. Eur Food Res Technol 224:623–628
4. Steinitz M (2000) Quantitation of the blocking effect of Tween 20 and bovine serum albumin in ELISA microwells. Anal Biochem 282:232–238
5. Jeyachandran YL, Mielczarski JA, Mielczarski E et al (2010) Efficiency of blocking of non-specific interaction of different proteins by BSA adsorbed on hydrophobic and hydrophilic surfaces. J Colloid Interf Sci 341:136–142
6. Vogt RFJ, Phillips DL, Henderson LO et al (1987) Quantitative differences among various proteins as blocking agents for ELISA microtiter plates. J Immunol Methods 101:43–50
7. Liberelle B, Boucher C, Chen JK et al (2010) Impact of epidermal growth factor tethering strategy on cellular response. Bioconjugate Chem 21:2257–2266
8. Liberelle B, Merzouki A, De Crescenzo G (2013) Immobilized carboxymethylated dextran coatings for enhanced ELISA. J Immunol Methods 389:38–44

Chapter 5

A Novel Three-Dimensional Biosensor Based on Aluminum Oxide: Application for Early-Stage Detection of Human Interleukin-10

Michael Lee, Abdoullatif Baraket, Nadia Zine, Miguel Zabala, Francesca Campabadal, Raffaele Caruso, Maria Giovanna Trivella, Nicole Jaffrezic-Renault, and Abdelhamid Errachid

Abstract

Immunosensors based on electrolyte-oxide-semiconductors (EOS) have been extensively researched over the last few decades. By electrochemical impedance spectroscopy (EIS) the specific molecular biorecognition of the antibody–antigen (Ab–Ag) can be detected providing an alternative quantitative system to immunoassay techniques. The electrochemical variations from a fabricated immunosensor can provide quantitative values for the analyte of interest at reduced costs and analysis time. In this context, a novel EOS substrate based on aluminum oxide (Al_2O_3) grown by atomic layer deposition on silicon was applied. The interaction between recombinant human (rh) interleukin-10 (IL-10) with the corresponding monoclonal antibody (mAb) for early cytokine detection of an anti-inflammatory response due to left ventricular assisted device implantation was studied. For this purpose, a 3D biosensor was composed of multi-walled carbon nanotubes with carboxylic acid functionalities (multi-walled carbon nanotubes–COOH) to increase the surface area for the range of human IL-10 detection. These were activated with *N*-hydroxysuccinimide and *N*-(3-dimethylaminopropyl)-*N′*-ethyl-carbodiimide hydrochloride for the immobilization of the anti-human IL-10 mAb. First, the interaction between the Ab and Ag was observed by fluorescence patterning to ensure that the biorecognition event was achievable. Then, EIS is explained for the quantification of commercial human IL-10 on this capacitance-based EOS macroimmuno-FET sensor.

Key words Immunosensor, Aluminum oxide, Interleukin-10, Silane, Carbon nanotubes, Fluorescence, Electrochemical impedance spectroscopy

1 Introduction

Biosensors are widely regarded with interest due to the potential that they can be applied as an alternative to immunoassay techniques (e.g., enzyme-linked immunosorbent assay (ELISA)) when utilized within clinical diagnostics. Biosensors based on electrical measurement are devices that employ biochemical molecular recognition for desired selectivity with a specific biomarker of interest,

Ivana Vancurova (ed.), *Cytokine Bioassays: Methods and Protocols*, Methods in Molecular Biology, vol. 1172,
DOI 10.1007/978-1-4939-0928-5_5,

e.g., the antibody–antigen (Ab–Ag) complex. The bioreceptor (Ab) is incorporated onto the transducer, and the biosensitive material is converted into a quantitative signal that can be measured in the form of an electrochemical response. One of these techniques is known as electrochemical impedance spectroscopy (EIS) where upon biorecognition a change can be observed in the interfacial charge, resistance, capacitance, mass, and thickness at the electrode surface. These biosensors require no or little sample preparation [1–5], though nonspecific binding (NSB) is an issue. There are no discriminatory processes that are capable of differentiating between the measured signal from specific and nonspecific interactions [6]. Therefore, the surface of the biotransducer has to be developed for sensitivity towards only the binding of the desired interaction whilst retaining its high specificity against NSB.

Over the past few decades, biosensor fabrication has been based upon the semiconducting properties of silicon with thermally grown silicon dioxide (SiO_2) and silicon nitride (Si_3N_4) being most favorable, when used as transistor gates within field effect transistors (FETs) [7–11]. In today's climate, thickness reduction of SiO_2 complementary metal oxide semiconductor (CMOS) devices has introduced a fundamental challenge to downscaling as high gate oxide leakage becomes apparent when the quality of the SiO_2 layer is jeopardized. Capacitance can be enhanced by increasing the dielectric constant (k), and, therefore, many materials have been considered as potential alternatives for gate-k materials instead of SiO_2. These metal oxides can present the necessary capacitance due to their physical thickness and, thus, a reduction of the gate leakage current [12]. Previously, we have demonstrated the application of a novel immunosensor based on hafnium oxide (HfO_2) that was functionalized with an aldehyde-silane ((11-(triethoxysilyl) undecanal (TESUD)) for the direct immobilization of the anti-human interleukin-10 (IL-10) Ab. This was followed by the EIS detection of human IL-10. This biosensor was sensitive between 0.1 pg/mL and 20 pg/mL, respectively [13]. Therefore, to improve the sensitivity of a biosensor, carbon nanotubes (CNTs) have been intensively researched due to their unique properties and promising applications in nanotechnology. This interest reflects upon their extraordinary chemical, electrical, mechanical, and structural properties [14, 15]. For electrochemical sensors, CNTs are advantageous due to their large surface area that can produce high sensitivity, improved chemical stability, and good biocompatibility for electrochemical biosensors that require electron transfer of redox proteins and enzymes [16].

To ensure the functionality of a biosensor, fluorescent imaging is a rapid tool for analyzing biolayers, which ensures that detection can be made by formulating biorecognition processes due to the high affinity an Ab has for its corresponding Ag. The soft-lithographical technique, micro-contact printing (μCP), facilitates the printing of the required pattern by applying a structured poly(dimethylsiloxane) (PDMS) stamp [17]. The functionality of the immobilized receptor by the fluorescent pattern is, therefore, firstly required before EIS measurements can be undertaken.

Here, we present a novel substrate of Al_2O_3 where the bio-recognition process of human IL-10 on the immobilized multi-walled carbon nanotubes (MWCNTs) was first of all characterized by fluorescence patterning using μCP. After ensuring that the bio-recognition process worked, an immuno-FET was fabricated and analyzed using EIS to evaluate this novel biosensor for early-stage detection of human IL-10 inflammation for left ventricular assisted device (LVAD) recipients.

2 Materials

1. Anti-human IL-10 monoclonal antibody (R&D Systems): Dilute the antibody to 10 μg/mL in PBS.
2. Recombinant human IL-10 protein (R&D Systems): Dilute to 25 μg/mL in PBS.
3. Carboxyfluorescein-conjugated anti-human IL-10 monoclonal antibody (R&D Systems): Dilute to 2.5 μg/mL in PBS.
4. Carboxylic acid-functionalized MWCNT solution (DROPSENS).
5. PDMS (Sylgard 184).
6. Piranha solution: Mix three volumes of H_2SO_4 and one volume of H_2O_2.
7. Base piranha solution: Mix three volumes of NH_4OH and one volume of H_2O_2.
8. Octadecyltrichlorosilane (OTS) solution: 160 μL of OTS and 10 mL of CCl_4 in 40 mL of heptane.
9. 3-(Aminopropyl)triethoxysilane (APTES) 1 % solution: Mix 50 μL of APTES and 4,950 μL of DI water.
10. MWCNT–COOH mixture: Disperse 0.17 μL from 0.25 mg/mL MWCNT in 183 μL of DI H_2O containing sodium 5.77 mg (0.1 M) dodecyl sulphate (SDS). This produces a final volume of 200 μL.
11. 0.8 M Solution of *N*-(3-dimethylaminopropyl)-*N′*-ethyl-carbodiimide hydrochloride (EDC).
12. *N*-hydroxysuccinimide (NHS) solution: Prepare 0.2 M solution of NHS (2.30 mg) and 0.8 M solution of EDC (12.42 mg) in PBS, and then add together to produce a final volume of 200 μL.
13. 0.1 % Ethanolamine (ETA): Mix 1 μL of ETA in 999 μL of PBS in Eppendorf tube; centrifuge.

14. Sulphuric acid (H_2SO_4).
15. Ammonium hydroxide (NH_4OH).
16. Hydrogen peroxide (H_2O_2).
17. Carbon tetrachloride (CCl_4).
18. Heptane.
19. Ethanol.
20. Acetone.
21. UV/Ozone cleaner (BioForce).

3 Methods

3.1 Preparation of Al_2O_3

The macro-FET structures were developed at the Centro Nacional de Microelectrónica (CNM) in Barcelona using 100 mm diameter p-type silicon wafers (<100>), oriented and with a resistivity 4–40 Ω cm. After standard cleaning, the high-*k* dielectric is deposited by atomic layer deposition (ALD) (Cambridge NanoTech Savannah 200 ALD system). The system uses deionized water (DI H_2O) as the oxygen precursor, together with trimethylaluminum (TMA) for Al_2O_3 deposition and N_2 as the carrier/purging gas. Deposition of the Al_2O_3 layer is carried out at a temperature of 200 °C and at a base pressure of 300 mTorr using 90 ALD cycles. A first estimation of the deposited Al_2O_3 layer thickness is carried out by means of ellipsometry, obtaining a thickness of 10.8 nm having fixed the refractive index to 1.64.

Finally, a 500 nm thick aluminum (Al) layer is deposited on the back of the wafers for electrically contacting the silicon substrate (Fig. 1).

3.2 Biorecognition by Micro-contact Printing

3.2.1 Preparation of Micro-contact Printing Tools

1. Fabricate a positive 10 μm² square patterned silicon mold. Define microstructures into the surface of the silicon to produce a master with positive superficial structures. Here, a silicon wafer (<100> orientation) with a photoresist is patterned using a mask containing the required squared structures by established photolithography techniques. Etch the surface of the photoresist using deep reactive ion etching (DRIE) (601 DRIE, Alcatel) to the required depth. The process for the etching conditions is optimized to machine whole-silicon wafers [18].

Fig. 1 Schematic illustrations of the macro-FET fabricated at the CNM, Barcelona

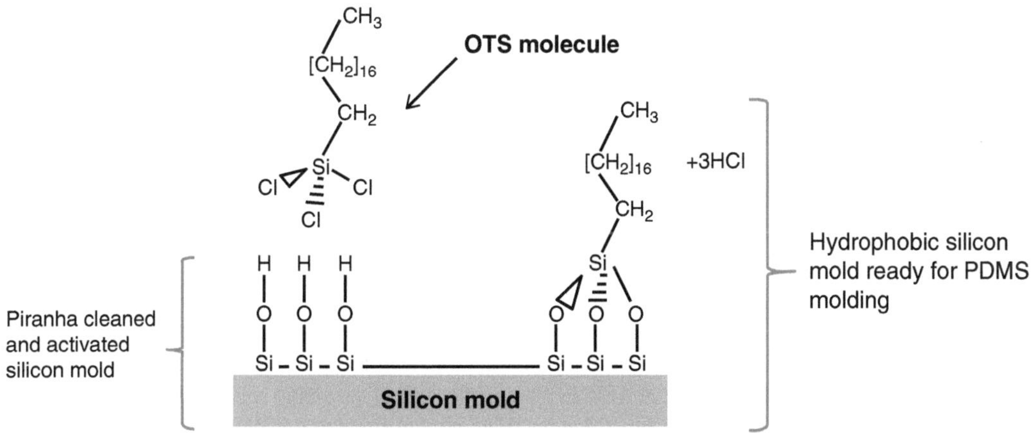

Fig. 2 Hydrophobic monolayer to enable PDMS peeling from the silicon master mold

2. Silanize the mold to ensure that PDMS replicas can be easily demolded by creating a hydrophobic monolayer onto the silicon surface (Fig. 2). To do this, clean and activate the silicon mold in freshly prepared piranha solution. Rinse the mold comprehensively with DI H_2O and dried with nitrogen. (Attention: Piranha solution is highly corrosive and a powerful oxidizer. Wear suitable laboratory clothing and activate within a fume cupboard.)
3. Incubate the silicon mold in OTS solution for 20 min at room temperature (RT) [19]. Here, the OTS molecules are chemically adsorbed onto the silicon surface.
4. Rinse the mold in heptane, and place it in the oven at 100 °C for 1 h. This will silanize the OTS molecules to the silicon substrate.
5. Rinse the mold again in heptane to ensure that all non-bound OTS is removed from the surface.
6. The silicon mold is now hydrophobic (*see* **Note 1**) and can be applied to produce a polymeric replica.

3.2.2 Soft Lithography and Replica Molding

1. Prepare PDMS by weighing a 10:1 (w/w) mix of the elastomer:cross-linking agent within a plastic and disposable container.
2. Mix the polymer thoroughly with a plastic stirrer, and ensure that the two mixtures have comprehensively mixed.
3. Degas the PDMS by desiccation to remove all air bubbles that are generated by mixing.
4. Place the silanized silicon mold with its structures top side up within a small petri dish (60 × 15 mm).
5. Pour the PDMS mixture on top of the silicon mold at a depth of 3–5 mm, and degas once again until all air bubbles have been removed.

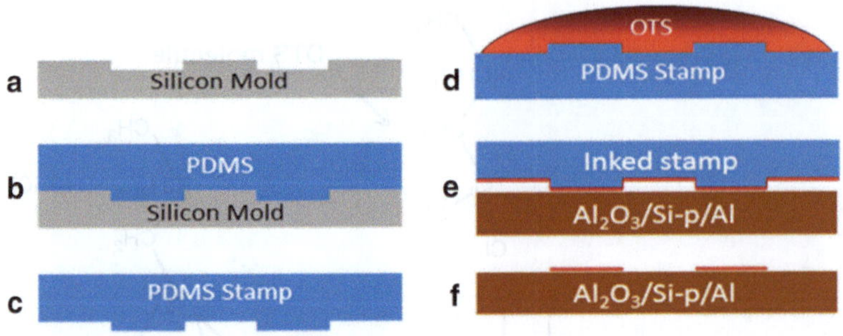

Fig. 3 Scheme for the fabrication of a structured PDMS stamp by soft lithography: (**a**) Silanized silicon master mold, (**b**) PDMS (10:1 w/w) poured onto the silicon mold, (**c**) cured PDMS stamp, (**d**) inking of the stamp with physically adsorbed OTS molecules, (**e**) μCP of the OTS from the structured PDMS stamp onto the activated Al_2O_3 substrate, and (**f**) transferred OTS pattern onto the Al_2O_3 substrate

6. Place the petri dish containing the PDMS and silicon mold into a preheated oven at 90 °C for 1 h. This enables the PDMS to polymerize and cure.
7. After hardening, remove the PDMS from the petri dish using a scalpel and separate the structured PDMS from the silicon mold using a scalpel and tweezers (*see* **Notes 2** and **3**).
8. This produces a PDMS stamp that now obtains negative features (*see* **Note 4**).

3.2.3 Preparation of Al_2O_3

1. Clean the bare Al_2O_3 substrates by sonication in ethanol for 10 min, followed by thorough rinsing in DI H_2O. Dry the substrate with nitrogen.
2. Activate the surface of the Al_2O_3 substrate using a UV/Ozone cleaner for 20 min, followed by 15 min in base piranha solution at RT (*see* **Note 5**).
3. Rinse the substrate thoroughly in DI H_2O and dry with nitrogen. (Attention: Base piranha is highly corrosive and a powerful oxidizer. Wear suitable laboratory clothing, and activate within a fume cupboard.)

3.2.4 Micro-contact Printing

1. Clean the negative 10 μm square patterned PDMS stamp by sonication in heptane for 1 min (*see* **Note 6**). Afterwards, dry the PDMS stamp with nitrogen.
2. Immerse the PDMS stamp in 5 mM OTS solution made up in heptane for 1 min (*see* **Note 7**). Here, the stamp adsorbs the OTS molecules.
3. Dry the PDMS stamp with nitrogen, and micro-contact print the OTS onto the hydroxyl activated Al_2O_3 substrate. Maintain printing for 10 s. This will produce a negative pattern of OTS on the Al_2O_3 surface (Fig. 3) (*see* **Note 8**).

3.2.5 Amino Functionalities and Multi-walled Carbon Nanotubes Immobilization

1. After μCP, immerse the Al_2O_3 substrate in APTES solution for 45 min (*see* **Note 7**).
2. Afterwards, gently rinse the Al_2O_3 substrate in DI H_2O and dry by nitrogen.
3. Place the Al_2O_3 substrate into a preheated oven at 100 °C for 1 h and the monolayer will silanize by the double-silanization process.
4. Prepare MWCNT and NHS/EDS solutions.
5. Mix 200 μL of MWCNT solution and 200 μL of NHS/EDC solution to formulate the active EDC/NHS ester groups on the MWCNTs. This produces a final volume of 400 μL of MWCNT/NHS/EDS solution.
6. Immerse Al_2O_3 in MWCNT/NHS/EDC solution for 2 h at RT, and follow this by rinsing in PBS to remove all non-bound MWCNTs (*see* **Note 9**).
7. This produces a 0.11 μg/mL concentration of functionalized MWCNTs.

3.2.6 Biorecognition of Human IL-10

1. Micropipette diluted anti-human IL-10 antibody onto the surface of the Al_2O_3-MWCNTs and incubate for 1 h at 4 °C. Afterwards, rinse the substrate with PBS (*see* **Note 10**).
2. Micropipette the diluted human IL-10 protein onto the surface of the Al_2O_3-MWCNTs-anti-IL-10 and incubate for 1 h at 4 °C. Then, rinse the substrate with PBS (*see* **Note 10**).
3. Micropipette the diluted carboxyfluorescein-conjugated anti-human IL-10 antibody onto the surface and incubate for 1 h at 4 °C. Finally, rinse the substrate with PBS and place in a darkened container.
4. Analyze the substrate for fluorescence using a fluorescence microscopy. For fluorescein, the excitation is made with a 470 (±40) nm band-pass filter and the fluorescence is observed with a 525 (±50) nm band-pass filter [20]. The schematic for the complete blocking, functionalization, and biodetection steps for human IL-10 can be observed in Fig. 4.
5. In Fig. 5, the fluorescence image of the biodetection for human IL-10 is shown with the positive squared regions fluorescing due to the biorecognition of human IL-10. The darkened negative regions are consistent to the blocking of the surface with hydrophobic OTS molecules.
6. After ensuring the functionality of the biorecognition event with human IL-10, the Al_2O_3 substrate can now be prepared to formulate a macro-immunoFET for the detection of human IL-10 by EIS measurements.

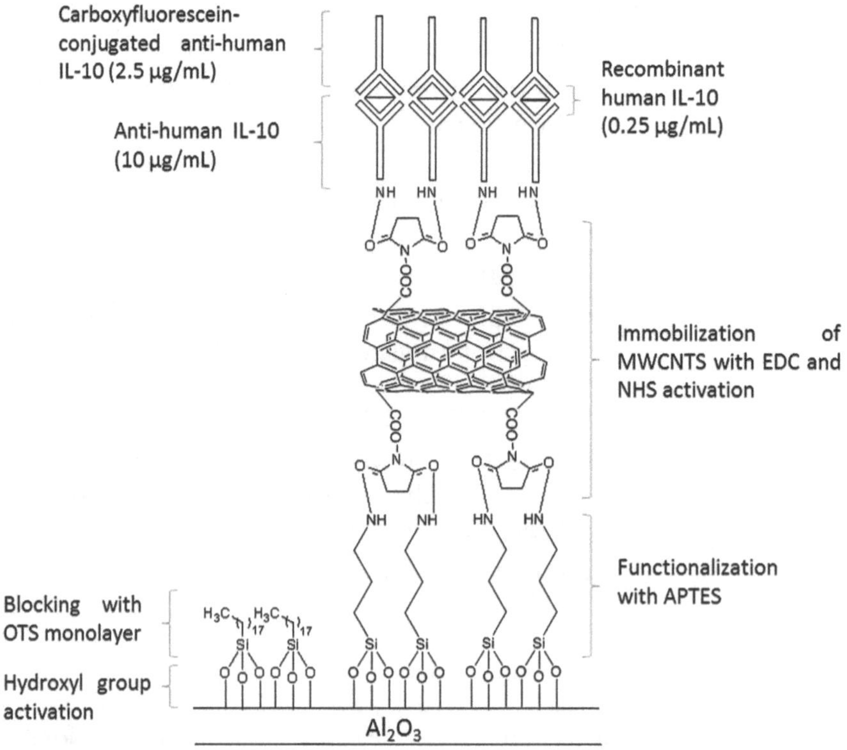

Fig. 4 Al_2O_3 oxidation followed by μCP of OTS to block the surface using a positive PDMS (10 μm) stamp. The positive regions were functionalized with APTES for the immobilization of MWCNTs with NHS and EDC activation. Then, the anti-human IL-10 was immobilized onto the MWCNT followed by the biorecognition with the recombinant human IL-10 and the carboxyfluorescein-conjugated anti-human IL-10 antibody

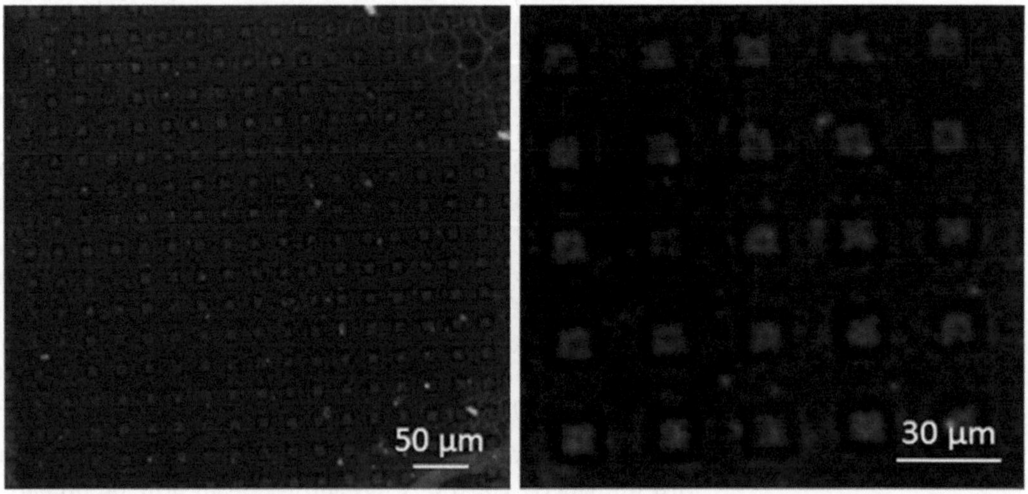

Fig. 5 Fluorescence microscope image of the Ab–Ag–Ab biorecognition for the detection of human IL-10 with positive square patterns demonstrating the detection of human IL-10 using a fluorescent microscope (Zeiss Axio Scope.A1, France)

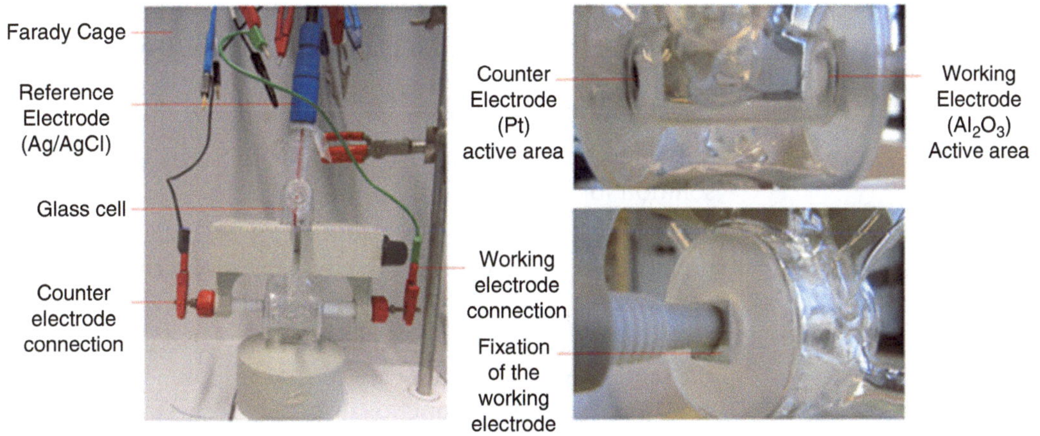

Fig. 6 Impedance setup for the detection of human IL-10. All measurements were made inside a Faraday cage using a specialized glass cell with the working electrode (Al_2O_3), counter electrode (Pt), and reference electrode (Ag/AgCl)

3.3 Electrochemical Impedance Spectroscopy

1. The setup for EIS measurements can be observed in Fig. 6. Here, the Al_2O_3 substrate is connected (after all immobilization steps) to the active area for the working electrode (WE).
2. A platinum (Pt) plate electrode is used as the counter electrode (CE). Clean the CE by sonication in acetone for 10 min, rinse thoroughly with DI H_2O, and dry with nitrogen. Connect the Pt electrode to the perpendicular active area towards the WE.
3. A calomel reference electrode (RE) made from silver/silver chloride (Ag/AgCl) is then inserted through the top to complete the three-electrode system.
4. Here, the electrolyte solution is freshly prepared PBS at pH 7.4. Fill the glass cell with PBS to ~8 mL.
5. To reduce the signal:noise ratio, carry out all EIS measurements inside a Faraday cage.

3.3.1 Calculation of the Monolayer of APTES by Electrochemical Impedance Spectroscopy

1. The percentage coverage of the amino silane can be calculated by the estimation

$$\theta = 1 - \frac{R_{before\ silanization}}{R_{after\ silanization}}$$

where R is the calculated resistance value.

2. Therefore, clean a bare Al_2O_3 substrate by sonication in ethanol for 10 min, followed by thorough rinsing in DI H_2O.
3. An impedance measurement of the bare Al_2O_3 substrate is made by the connection shown in Fig. 6. Apply with the following conditions (potential, amplitude, and frequency at −2 V, 200 mV, and 200 kHz to 100 mHz, respectively (56 s/scan)) (*see* **Note 11**).

4. After the measurement, clean the same Al_2O_3 substrate in ethanol and activate using UV/Ozone cleaner for 20 min, followed by 15 min in base piranha at RT. Then, rinse the substrate thoroughly in DI H_2O and dry with nitrogen.
5. Directly afterwards, immerse the Al_2O_3 substrate in 1 % APTES solution for 45 min at RT.
6. Then, gently rinse the substrate with DI H_2O and place into a preheated oven at 100 °C for 1 h. This allows the monolayer to silanize onto the Al_2O_3 surface.
7. Connect the Al_2O_3-APTES substrate to the glass cell, and apply the same conditions as the bare Al_2O_3 substrate. This produces a Nyquist plot for the functionalized Al_2O_3 surface (frequency: 200 kHz to 10 mHz (5 min 2 s/scan)) (*see* **Note 12**).
8. After the formation of the two plots for the Al_2O_3 (bare) and Al_2O_3-APTES-functionalized surfaces, fit the curves to extract the component values using an equivalent circuit that was chosen on the interfaces seen in the Nyquist plot of the semicircles. Normalize the impedance data with the fitting software installed from the instrument provider. Here, we modeled with EC-Lab V10.18 modeling software (Bio-Logic Science Instrument, France).
9. Fit with the randomize + simplex method with randomization stopped on 10,000 iterations and the fit stopped on 5,000 iterations.
10. Due to the different observable interfaces, we applied two different equivalent circuits to normalize the data (Fig. 7, inset) (*see* **Note 13**). The polarization resistance value for bare Al_2O_3 was measured at 9,950 kΩ ($X^2 = 0.0436$), and the Al_2O_3-APTES functionalized was 27,380 kΩ ($X^2 = 0.0465$) (Fig. 7); therefore, the silane monolayer was equivalent to 64 %.

3.3.2 Biodetection by Electrochemical Impedance Spectroscopy

1. Activate the Al_2O_3 substrate as described in Subheading 3.2.3.
2. Afterwards, functionalize the macro-FET by all the steps described in Subheading 3.2.5 (note: no μCP is made here).
3. Connect the Al_2O_3-MWCNTs substrate (WE) and the Pt plate electrode (CE) to the glass cell as shown in Fig. 6.
4. Turn the glass cell onto its side (*see* **Note 14**), and micropipette the diluted anti-human IL-10 antibody onto the WE active area of the Al_2O_3-MWCNTs. Incubate for 2 h at 4 °C (*see* **Note 15**).
5. After this time, rinse the glass cell thoroughly with PBS to ensure that the excess anti-human IL-10 antibody has been removed (*see* **Note 16**).
6. Block the surface of the MWCNTs with the mAb in 0.1 % ETA solution for 20 min. This deactivates all binding sites and prevents NSB from occurring.

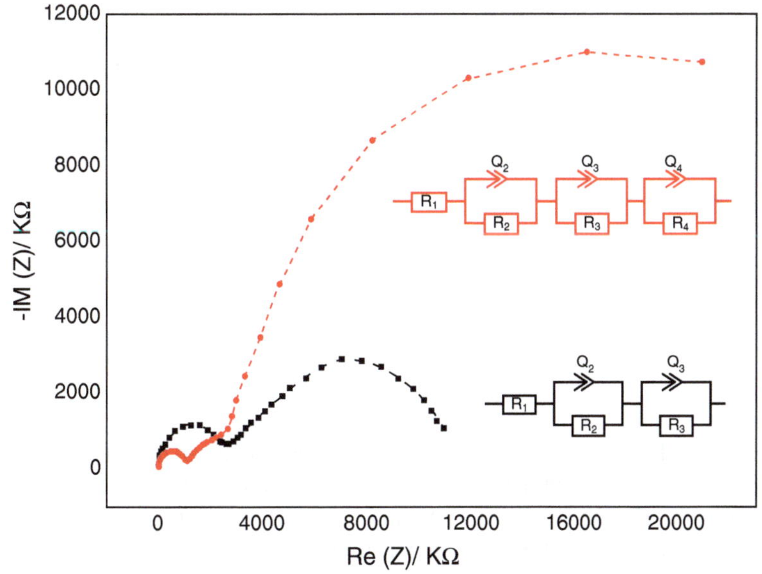

Fig. 7 Nyquist plot of the bare Al_2O_3 (*black square*) and the Al_2O_3-APTES (*red circle*) functionalized substrate with a 64 % of the silane monolayer on the Al_2O_3 surface. The inset shows the applied equivalent circuits to normalize the Nyquist plots

7. Fill the glass cell with PBS (~8 mL), and make all electrical connections (WE, RE, and CE) to the potentiostat. Keep all conditions (potential (−2 V), amplitude (−200 mV), and frequency (200 kHz to 600 mHz) (23 s/scan)) constant to those applied in Subheading 3.3 (*see* **Note 11**).
8. By impedance, firstly, measure the anti-human IL-10 mAb. (This is already immobilized on the Al_2O_3 substrate.)
9. After stabilization of the semicircle, stop the scan and disconnect all electrical connections.
10. Then, follow with the detection of the human IL-10. Starting with the weakest concentration, inject 50 μL of the human IL-10 directly on top of the Al_2O_3 active area. Incubate for 30 min at 4 °C (*see* **Notes 14–16**).
11. After the incubation period, rinse the glass cell with PBS and refill the cell with fresh PBS. Continue the impedance measurements until the stability of each human IL-10 concentration is observable.
12. Continue with the EIS measurements with increased concentrations of human IL-10 until the saturation of the immunosensor can be observed. This is achieved by superimposing the Nyquist plots together. The resulting Nyquist plot can be seen in Fig. 8.
13. After the formation of the Nyquist plots, normalize the impedance data with the software provided from the instrument provider. Here, we applied EC-Lab V10.18 modeling software.

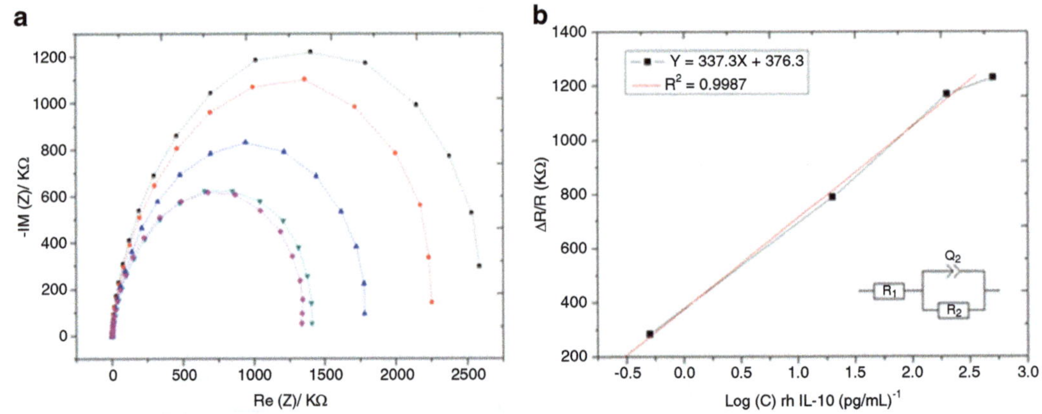

Fig. 8 (**a**) Nyquist impedance plot for the detection of human IL-10 using the following conditions: frequency between 200 kHz and 600 mHz, a sinus amplitude of 200 mV, and a polarization potential of −2 V. *Black square*, IL-10 mAb (10 μg/mL); *red circle*, rh IL-10 (0.5 pg/mL); *blue triangle*, rh IL-10 (20 pg/mL); *inverted green triangle*, rh IL-10 (200 pg/mL); and *pink diamond*, rh IL-10 (500 pg/mL). (**b**) Normalized curve by EIS detection of the logarithm of rh IL-10 with concentrations ranging from 0.5 to 500 pg/mL against the variation of the R_p calculated by $\Delta R/R$. Inset: Applied equivalent circuit for calculation of the fitting parameters

Table 1
Fitting parameters obtained from the equivalent circuit for the detection of human IL-10

Ag conc. (pg/mL)	R_1 (Ω)	Q_2 (nF·s$^{(a-1)}$)	R_2 (kΩ)	X^2
0	442.7 ± 0.2	$9.9 \pm 27.7 \times 10^{-6}$	$2{,}797 \pm 0.5 \times 10^{-3}$	0.0805
0.5	479.9 ± 0.2	$10.5 \pm 31.1 \times 10^{-6}$	$2{,}312 \pm 1.2 \times 10^{-3}$	0.0789
20	571.5 ± 0.2	$11.6 \pm 44.0 \times 10^{-6}$	$1{,}808 \pm 1.5 \times 10^{-3}$	0.0776
200	748.7 ± 0.2	$13.2 \pm 70.6 \times 10^{-6}$	$1{,}429 \pm 1.2 \times 10^{-3}$	0.1106
500	662.2 ± 0.2	$12.4 \pm 72.7 \times 10^{-6}$	$1{,}367 \pm 0.4 \times 10^{-3}$	0.0482

The R_2 values are the polarization resistance values used to normalize the data (*note*: 0 pg/mL is the mAb (10 μg/mL))

14. Apply an equivalent circuit to calculate the best fit of the data. The selection of the equivalent circuit is dependent upon the interfaces of the Nyquist plot to produce the smallest error (X^2) (*see* **Note 17**).
15. Here, we used $R_1 + Q_2/R_2$, where R_1 is the resistance of the electrolytic solution and Q_2 is the constant-phase element (CPE) and in parallel is R_2, which is the polarization resistance (R_p) and the value required to normalize the data (Table 1) (*see* **Note 18**).
16. For the extraction of the data, apply the equation $\Delta R/R$. The resistance for the real component decreases, and then Ab–Ag = resistance of the immunosensor. (The concentrations of human IL-10 are converted to logarithm values (Table 2).)

Table 2
Calculation of the logarithm values for human IL-10 concentrations and the plotted Δ*R*/*R* values shown in the normalized plot

Ag conc. (pg/mL)	Log conc. (pg/mL)	Δ*R*/*R* (kΩ)
0.5	-0.30	285
20	1.30	789
200	2.30	1,168
500	2.70	1,230

17. Plot these values onto a graph (Fig. 8b) to demonstrate the linearity for human IL-10 detection. Here, the sensitivity of the rh IL-10 detection was linear between 0.5 and 200 pg/mL at a linearity of $R^2 = 0.9987$. At 500 pg/mL the anti-human IL-10 antibodies on the immunosensor surface were saturated. This is observed by its nonlinearity towards the other human IL-10 concentrations that were measured.

4 Notes

1. To ensure that the silicon mold for μCP has silanized and formed correctly, a droplet of water will provide a very high hydrophobic surface on the OTS monolayer of silicon. (Silicon is a hydrophilic material.)
2. Upon removal of the PDMS replica stamp from the silicon master mold, remove all excess PDMS from the petri dish leaving an easier and smaller dimension of the PDMS for removal. Plastic tweezers and a scalpel are easier to apply when peeling the two apart; however, pay attention not to scrape the silicon mold from the backside as this will remove the OTS monolayer.
3. Also, ensure that excess PDMS that has covered the backside of the silicon mold has been removed as the silicon mold can break upon trying to remove it from the PDMS stamp.
4. As the Al_2O_3 substrate is first of all blocked with OTS, a negative PDMS stamp is preferential to leave the positive regions available for the observation of the biorecognition event. Here, we applied square patterns though other designed relief patterns that can demonstrate blocking and biorecognition are just as useful.
5. For the activation of the Al_2O_3 substrate, base piranha was preferred due to its less aggressive nature compared to piranha solution (H_2SO_4:H_2O_2—3:1 v/v). A piranha solution is extremely exothermic; however, it will remove the metal oxide

and aluminum layer. Base piranha will also act in a similar fashion if it is heated up to 60 °C, and, here, NH_4OH at 4 °C was applied. The Al_2O_3 substrate was placed top side down at the liquid interface with air to prevent the aluminum contact from activating. Activation by oxygen plasma or hydrogen peroxide can also be applied to oxidize the surface.

6. Heptane will swell PDMS and alter the PDMS stamp structures on the top surface; therefore, cleaning of the PDMS stamp is recommended 10–15 min before incubation with the OTS solution.
7. Prepare the silane solution just before the transfer of the monolayer. The silane reagent will silanize quickly, and this reaction can be sped up at higher temperatures.
8. To reduce issues with swelling and nonconformal μCP, the PDMS stamp can be supported on a homogeneous glass substrate. By oxygen plasma treatment, both surfaces can be activated to produce strong chemical bonds (Si–O–Si). Here, the backside of the PDMS stamp (20 s) and glass (2 min) can be oxygen activated at 90 W. This bonding will be irreversible though prolonged periods in heptane (~30–1 h) will disrupt this bonding. By bonding the PDMS to glass, this will also help with future printing as the structured PDMS surface will be known.
9. For the functionalization of the Al_2O_3 substrate with MWCNTs this was made with a total volume of 400 μL. This was achievable using a small glass beaker. However, depending upon your glassware these volumes may need to be increased. If these are increased pay attention to the required mass of EDC as the chemical is a fine powder and an increased mass will be required.
10. For the mAb and human IL-10 concentrations, the solutions were previously made up in Eppendorf tubes and stored at −20 °C. Prior to analysis, these were allowed to defrost for 20–30 min. After defrosting, the solutions were mixed by centrifugation.
11. For our applied substrates, setup, and potentiostat these were our optimized conditions. Other ALD–Al_2O_3 substrates and applied instruments may observe different required conditions and, therefore, require optimization by the user.
12. For EIS measurements, the percentage coverage of the amino silane from the bare Al_2O_3 to Al_2O_3 functionalized with APTES will require a lower frequency range, and, therefore, an increased scan time will be observed with an increased resistance value.
13. For the choice of the equivalent circuit, this is dependent upon how many interfaces are seen, e.g., one, two, or three semicircles within the Nyquist plot. Therefore, the equivalent circuit has to be adapted to measure all parts of the observed Nyquist plot.

14. For each incubation step of the mAb to the human IL-10 concentrations, the glass cells were turned sideways and balanced on a glass beaker when placed at 4 °C. Due to the inlets on the glass cells, these were closed during the incubation step using parafilm.
15. The incubation of the anti-human mAb and the increasing human IL-10 concentrations was made at a fixed volume of 50 μL. By micropipette, the volume can be observed as the droplet falls within the concave active area of the WE. This ensures that the Al_2O_3 was in direct contact with the solutions, that a specific volume for each incubation step was maintained, and that the analysis remained consistent.
16. After incubation at 4 °C the glass cell and the connections condense with water molecules. Therefore, the cell and connections were dried with a stream of nitrogen. Leaving the cell for 5–10 min will also ensure that the glass cell returns to ambient temperature.
17. For the normalization of the fitting data ensure that X^2 is always small. This ensures a better fitting.
18. For the fitting, depending upon the form of the semicircle pure capacitive components cannot be applied as the semicircles appear compressed or elongated. Therefore, the imperfect capacitors of CPE will function on improving the fitting values.

Acknowledgements

We acknowledge the funding through the Ministère de L'enseignement Supérieur et de la Recherche (MESR) and from the SensorART project funded by the European Communities Seventh Framework Programme (FP7/2007–2013) under the grant agreement No. 248763, the InFuLOC project under grant agreement No 230749, the SMARTCANCERSENS project (FP7-PEOPLE-2012-IRSES) under the grant agreement No. 318053, and NATO project, SPS (NUKP.SFPP984173).

References

1. Hnaien M, Lagarde F, Bausells J et al (2011) A new bacterial biosensor for trichloroethylene detection based on a three-dimensional carbon nanotubes bioarchitecture. Anal Bioanal Chem 400:1083–1092
2. Vidic J, Pla-Roca M, Grosclaude J et al (2007) Gold surface functionalization and patterning for specific immobilization of olfactory receptors carried by nanosomes. Anal Chem 79:3280–3290
3. Bourigua S, Hnaien M, Bessueille F et al (2010) Impedimetric immunosensor based on SWCNT-COOH modified gold microelectrodes for label-free detection of deep venous thrombosis biomarker. Biosens Bioelectron 26:1278–1282
4. Chebil S, Hafaiedh I, Sauriat-Dorizon H et al (2010) Electrochemical detection of d-dimer as deep vein thrombosis marker using single-chain

d-dimer antibody immobilized on functionalized polypyrrole. Biosens Bioelectron 26:736–742
5. Hleli S, Martelet C, Abdelghani A et al (2006) An Immunosensor for haemoglobin based on impedimetric properties of a new mixed self-assembled mono layer. Mat Sci Eng R C 26:322–327
6. Berggren C, Bjarnason B, Johansson G (2001) Capacitive biosensors. Electroanalytical 13: 173–180
7. Wang J (2006) Electrochemical biosensors: towards point-of-care cancer diagnostics. Biosens Bioelectron 21:1887–1892
8. Castellarnau M, Zine N, Bausells J et al (2007) Integrated cell positioning and cell-based ISFET biosensors. Sens Actuat B 120:615–620
9. Gustavsson J, Altankov G, Errachid A et al (2008) Surface modifications of silicon nitride for cellular biosensor applications. J Mater Sci Mater Med 19:1839–1850
10. Marques de Oliveira IA, Caballero D, Baccar ZM et al (2007) Immobilization of DNA on silicon nitride by microcontact printing: characterisation of DNA-IS sensors by atomic force microscopy and impedance spectroscopy methods, paper presented to XIXth international symposium on bioelectrochemistry and bioenergetics, 1–4 April 2007, Toulouse, France
11. Amari A, El Bari N, Zine N et al (2008) Nanocharacterization of a novel biosensor sensitive to beta casein using AFM and impedance spectroscopy, paper presented to Eurosensors XXII, 7–10 September 2008, Dresden, Germany
12. Wilk GD, Wallace RM, Anthony JM (2001) High-k gate dielectrics: current status and materials properties considerations. J Appl Phys 89:5243–5275
13. Lee M, Zine N, Baraket A et al (2012) A novel biosensor based on hafnium oxide: application for early stage detection of human interleukin-10. Sens Actuat B 175: 201–207
14. Wang J (2005) Carbon-nanotube based electrochemical biosensors: a review. Electroanalytical 17:7–14
15. Pulickel M, Ajayan PM, Tour JM (2007) Materials science: nanotube composites. Nature 447:1066–1068
16. Putzbach W, Ronkainen NJ (2013) Immobilization techniques in the fabrication of nanomaterial-based electrochemical biosensors: a review. Sensors 13:4811–4840
17. Xia Y, Whitesides GM (1998) Soft lithography. Angew Chem Int Ed 37:550–575
18. Mills CA, Martinez E, Bessueille F et al (2005) Production of structures for microfluidics using polymer imprint techniques. Microelectron Eng 78–79:695–700
19. Caballero D, Plâ-Roca M, Bessueille F et al (2006) Atomic force microscopy characterization of a microcontact printed, self-assembled thiol monolayer for use in biosensors. Anal Lett 39:1721–1734
20. Baraket A, Lee M, Zine N et al (2013) Diazonium modified gold microelectrodes onto polyimide substrates for impedimetric cytokine detection with an integrated Ag/AgCl reference electrode. Sens Actuat B Chem 189:165–172

Chapter 6

Quantification of Multiple Cytokines and Chemokines Using Cytometric Bead Arrays

Gemma Moncunill, Joseph J. Campo, and Carlota Dobaño

Abstract

Quantitative suspension array technology allows the simultaneous measurement of different cytokines and chemokines in small sample volumes. The possibility of measuring multiple variables is important for discovery of biomarkers of pathogenesis or protection in complex diseases as well as measurement of antigen-specific cellular responses. Measurements can be made in biological specimens, such as plasma or serum, cell culture supernatants, and others. This technology is based on a capture-detection sandwich-type assay using fluorescent microspheres analyzable by Luminex instruments or flow cytometers. The complexity and cost of producing highly multiplexed cytokine/chemokine in-house assays make them especially apt for commercial production. There are several commercial kits available that vary in absolute cytokine concentration, sensitivity, reproducibility, and cost. This chapter gives an overview of cytometric bead array technology, introduces some of the kits, and provides detailed information about the one that performed well in a comparative study (Cytokine Human Magnetic 30-Plex Panel from Life Technologies™).

Key words Multiplex bead assay, Cytokines, Chemokines, Flow cytometry, Luminex

1 Introduction

The introduction of fluorescent bead-based technology represents a significant advancement for measuring cytokine and chemokine responses. It allows the simultaneous measurement of multiple analytes, or multiplexing, in a reduced sample volume with time-saving advantages and less cost compared to other immunoassays such as ELIspot, intracellular cytokine staining, and flow cytometry or ELISA, which are limited in the number of variables that can be measured simultaneously. The advantage of having more endpoints for less input is greater analytical power for complex problems such as pattern recognition and correlations within big data. In fact, multiplexing technologies have been a boon to biomarker

Electronic supplementary material The online version of this chapter (doi:10.1007/978-1-4939-0928-5_6) contains supplementary material, which is available to authorized users.

Ivana Vancurova (ed.), *Cytokine Bioassays: Methods and Protocols*, Methods in Molecular Biology, vol. 1172,
DOI 10.1007/978-1-4939-0928-5_6, © Springer Science+Business Media New York 2014

discovery, providing insight into pathology and immunity in a number of studies [1–5], and may be a key driver in identifying immune correlates of protection for malaria, HIV/AIDS, tuberculosis, and other major infectious diseases in which none has yet been identified [6].

There are a variety of commercial kits available that allow quantification of a wide range of cytokines and chemokines from serum, plasma, body fluids, and cell/tissue culture supernatant samples. All vendors offer kits with premixed panels as well as singleplex and customized kits (*see* **Note 1**). Kits are based on fluorescent microspheres, but the detection method varies, and beads may be acquired by a flow cytometer or by a Luminex apparatus. The assays based on Luminex instruments use xMAP® technology, which was originally developed using the principles of flow cytometry, and Luminex instruments (Luminex 100/200™ and FLEXMAP 3D® systems) share common components with general flow cytometry instruments, such as lasers, fluidics, and optics but have been optimized to analyze xMAP microspheres.

In the Luminex-based assays, the beads are internally labeled with two fluorescent dyes (red and infrared fluorophores). Fluorescence intensities are made to vary, which allows for discerning different bead sets. Each bead set is given a unique bead "region" (number) and is conjugated to a specific capture antibody that recognizes a single analyte. When added along with samples, the analytes are captured by the antibodies conjugated to the beads. Analyte-specific biotinylated detection antibodies are added and bind to the analytes immobilized on the surface of the microspheres. Streptavidin conjugated to a fluorescent protein, most often phycoerythrin (PE), is added and binds to the biotinylated detector antibodies bound to the immune complexes on the beads. These complexes are measured using a Luminex instrument that has two lasers to identify particles with fluorescence characteristics of both the bead and the detector. The intensity of PE fluorescence for all sandwich complexes from a specific bead set reveals the concentration of that particular analyte (Fig. 1).

The concentration of cytokines is calculated from the median fluorescent intensity (MFI) detected by the instrument using a reference standard curve. The standards that comprise the curve are a collection of samples of known concentration of the analyte titrated in a dilution series. The standard curve is then created with an equation that relates the MFI to sample concentration and is estimated from the responses of the standards. Each kit provides all specific reagents needed, including lyophilized standards for the standard curves. Premixed kits usually have the standard curves lyophilized together or in two sets, whereas the flexible and customized kits have the standard of each analyte separately and must be mixed. There are several software programs from different vendors for analysis of Luminex data, each of which calculates and creates graphs of the standard curves and calculates cytokine concentrations of the test samples. The acquisition software

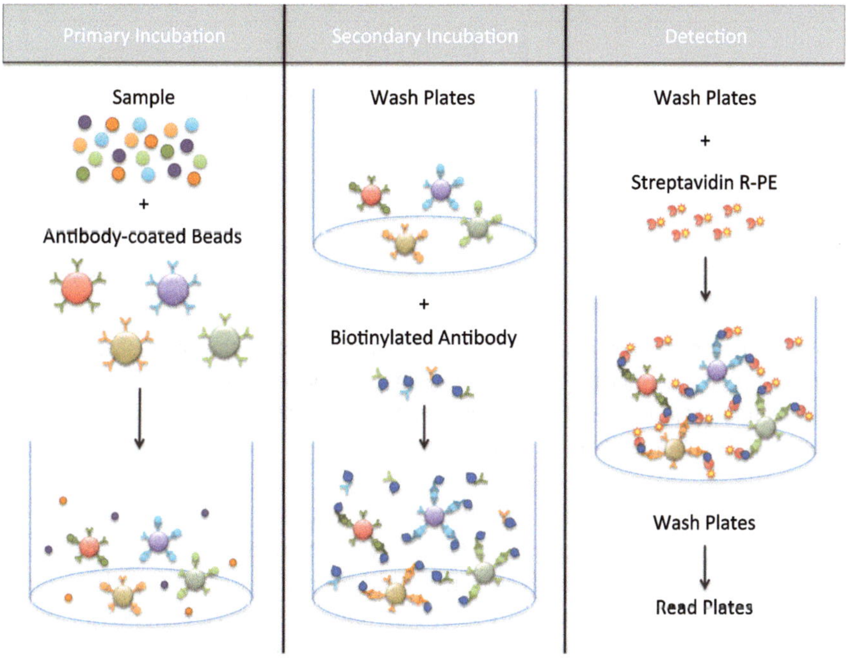

Fig. 1 Cytometric bead array assay. During primary incubation, biological specimens are mixed with multiplex, antibody-labeled microspheres, which bind specifically to target cytokines/chemokines. After washing to remove primary incubation matrix and excess molecules, a cocktail of biotinylated secondary antibodies bind to captured analytes. Excess is again washed away, and detection occurs through union of streptavidin–PE to bound secondary antibody. Fluorescence intensity is used as a measurement of quantity of analyte

accompanying Luminex instruments also permits analysis, and it can be installed on additional computers, or data can be exported and analyzed in external analysis software.

The xMAP beads used in the Luminex platform can be of two types: MicroPlex® Microspheres or the more recently developed MagPlex® Microspheres, which are magnetic beads. MicroPlex Microspheres are carboxylated polystyrene particles of 5.6 μm that have been color coded into 100 spectrally distinct sets, allowing multiplexing up to 100 different proteins for simultaneous detection (Fig. 2). Similar to MicroPlex® Microspheres, MagPlex Microspheres are carboxylated polystyrene beads but of higher diameter (6.5 μm) and contain magnetite. The assay is performed and read in a 96-well plate (provided usually by vendors). MicroPlex Microspheres are best used with filter-bottom plates and washed using vacuum filtration equipment. Instead, magnetic beads can be washed more easily with the use of a magnetic separator, improving percent recovery during handling and wash steps.

Flow cytometer-based assays, unlike Luminex assays, can use a combination of beads of different sizes and color intensities to identify individual microspheres, although the amount of analytes simultaneously detected is usually lower. For example the eBioscience® FlowCytomix™ kit uses two sizes of beads (5 and 4 μm) and

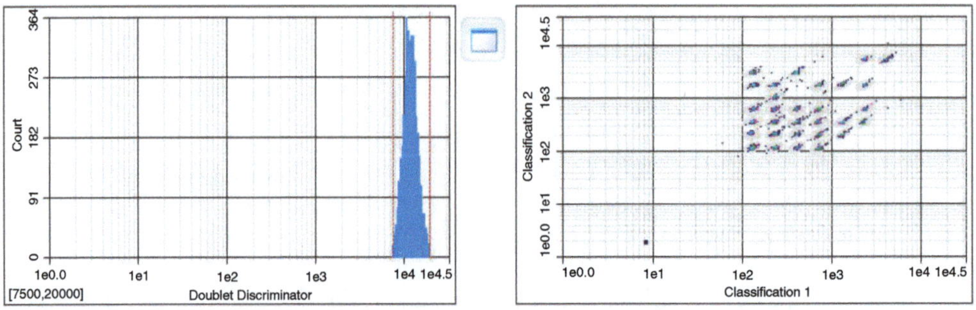

Fig. 2 Multiplex bead classification of a 30-plex kit (Luminex based). Microsphere-sized particles are gated (*left panel, histogram*) in order to limit readings, or events, to only beads. Each event is measured by a dual-fluorescence signature to classify bead identity or region (*right panel, dot plot*). Each event that appears in a designated region is then measured for PE fluorescence

differential intensity of internal dye to detect simultaneously up to 20 markers. The BD™ Cytometric Bead Array (CBA) kits are based on beads dyed with a single red fluorophore that is used to discriminate bead sets, but only up to seven markers can be detected at a time. Direct PE conjugates are used as detection reagents instead of biotin and streptavidin with a fluorochrome. However, BD CBA Flex Set assays have dual-color fluorescent beads that allow assessment of up to 30 markers. The BD CBA Enhanced Sensitivity Flex Set system, based on the established BD CBA Flex Set system, has a higher sensitivity but a longer protocol and up to ten markers multiplexed. No kits with magnetic beads are currently available for the flow cytometry platform. Acquisition of bead samples by a flow cytometer may be more time consuming, but the flow cytometry assays also support the utilization of 96-well plates by the high-throughput sampler (HTS) platform available on some BD instruments. For the analysis of data acquired by flow cytometry, each vendor provides its own analysis software.

Choosing the most appropriate kit within the diversity of options available is not straightforward, as they vary in reproducibility and sensitivity. Although it is advisable to test different kits before starting any study and ideally qualify or validate them for reproducibility and precision, some studies that have focused on the performances and characteristics of kits may help the decision-making process [7–14]. Although cytokines/chemokines may follow similar patterns between different kits, absolute concentrations are not comparable [14], and it is recommended to use the same kit, even the same lot, if comparisons are to be made between different data sets. Several reasons may account for the variation in performance of multiplex kits from different manufacturers in addition to the heterogeneity in analytes within each kit. Probably the most important factor is the different antibody used for capture and detection of individual analytes by each manufacturer. In addition, there may be differences among purified recombinant proteins used to generate the standard curves as well as in the assay buffers supplied by the manufacturer.

We compared the performance of six human cytokine/chemokine multiplex kits currently available using culture supernatants from peripheral blood mononuclear cells (PBMC) stimulated with *Plasmodium falciparum* malaria parasite lysates [8]. The following human cytokine kits were tested: a FlowCytomix™ kit from eBiosciences® (formerly Bender MedSystems®), a Bio-Plex Pro™ Plex Assay from Bio-Rad®, a MILLIPLEX® MAP Kit from Millipore™, a nonmagnetic bead kit and a magnetic bead kit from Life Technologies™ (formerly Invitrogen), and a BD™ CBA Human Enhanced Sensitivity Plex kit. Differences in the number of samples detected within the accurate range and reproducibility were observed depending on the method used and even the cytokine detected, although Luminex-based kits with magnetic beads proved to be better. In addition to the individual cytokine performance in each kit, other factors may be important in the final selection of a multiplex cytokine profiling kit as they vary widely: the number of cytokines measured within each kit, the volume of sample needed, and the time and the cost of the assays. Other commercial kits for the Luminex platform available from other vendors that we did not test include Fluorokine® MAP Multiplex kits, VersaMAP™ Multiplex Development kits and VersaMAP™ Magnetic Multiplex kits from R&D Systems, and Procarta kits from Affymetrix.

Cytokine Human Magnetic 30-Plex from Life Technologies was the one we selected for our following studies, although the MILLIPLEX® MAP kit also performed well. Below, we provide a detailed protocol for this kit, based on the one provided by the vendor (LHC6003M Hu Cytokine Magnetic 30-Plex Panel), that includes the following cytokines and chemokines: EGF, eotaxin, FGF basic, G-CSF, GM-CSF, HGF, IFN-α, IFN-γ, IL-1ra, IL-1β, IL-2, IL-2r, IL-4, IL-5, IL-6, IL-7, IL-8, IL-10, IL-12(p40/p70), IL-13, IL-15, IL-17, IP-10, MCP-1, MIG, MIP-1α, MIP-1β, RANTES, TNF, and VEGF. In addition, we outline the overall workflow (Fig. 3) and provide notes that can be applied to any of the cytometric bead array kits.

2 Materials

2.1 Materials Provided

The Human Cytokine 30-Plex kit includes both antibody bead reagents and buffer reagents (*see* **Notes 2** and **3**) for 96-well plate(s) (100 tests):

1. One vial of 2.5 mL of Human Cytokine Magnetic 30-Plex Antibody bead concentrate (1×) (contains 0.05 % sodium azide).
2. One vial of 1 mL of Human Cytokine Magnetic 30-Plex Biotinylated Antibody concentrate (10×) (contains 0.1 % sodium azide).

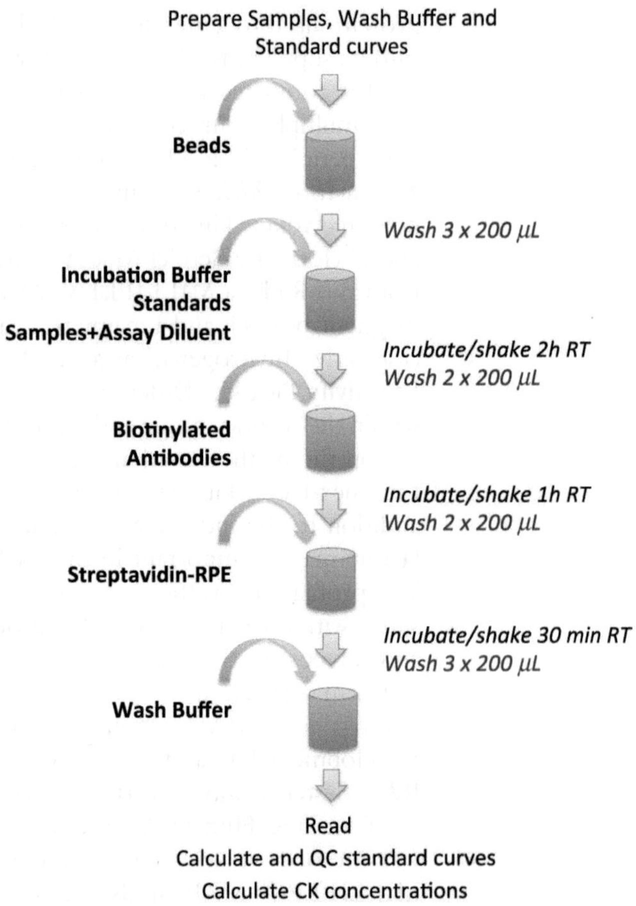

Fig. 3 Overview of assay workflow. Each well (standard, control, *or* test wells) of a 96-well plate follows a common procedure. CK: Cytokines and chemokines

3. Two vials of Hu 16-Plex Standard (contains 0.1 % sodium azide).
4. Two vials of Hu 14-Plex Standard (contains 0.1 % sodium azide).
5. Two bottles of 15 mL wash solution concentrate (20×) (contains 0.1 % sodium azide).
6. One bottle of 15 mL of assay diluent (contains 0.1 % sodium azide).
7. One bottle of 12 mL of incubation buffer (contains 0.05 % sodium azide).
8. One bottle of 12 mL of biotin diluent (contains 3.3 mM thymol).
9. One vial of 1 mL of streptavidin–RPE concentrate (10×) (contains 0.1 % sodium azide).
10. One bottle of 12 mL of streptavidin–RPE diluent (contains 3.3 mM thymol).
11. One 96-well filter plate.
12. One 96-well Greiner black mylar bottom plate.

2.2 Materials Required but Not Provided

1. Polypropylene microfuge tubes.
2. Aluminum foil.
3. Reagent reservoirs.
4. Absorbent pads.
5. Plate sealers.
6. Plate stand.

2.3 Equipment

1. Luminex® xMAP® system (100 or 200) or FLEXMAP 3D® system with data acquisition and analysis software.
2. Magnetic or vacuum filtration washing equipment (manual or automatic).
3. Sonicating water bath.
4. Vortex mixer.
5. Orbital shaker (small-diameter rotation recommended).

3 Methods

3.1 Preparing Reagents

Bring all reagents and samples to room temperature before use (*see* **Note 4**).

3.1.1 Wash Solution Preparation

1. Prepare a 1× working wash solution by transferring the entire contents of the wash solution concentrate bottle to a container, and then add 285 mL of deionized water. Mix well. Upon storage at 2–8 °C, a precipitate may form in the 20× wash solution concentrate. If this occurs, warm the 20× wash solution concentrate to 37 °C and mix until the precipitate is dissolved.
2. To prepare smaller volumes of 1× working wash solution, mix 1 part of 20× concentrate with 19 parts of deionized water.
3. The 1× working wash solution is stable for up to 2 weeks when stored at 2–8 °C.

3.1.2 Standard Curve Preparation

1. Reconstitute the protein standard within 1 h of performing the assay (*see* **Note 5**).
2. The preparation of the Human 30-Plex standard curves requires one vial of Human 16-Plex Standard plus one vial of Human 14-Plex Standard.
3. Reconstitute the lyophilized standards. When using serum or plasma samples, reconstitute the standard with the assay diluent provided. If using other sample types (e.g., cell culture supernatant), reconstitute the standard with a mixture, composed of 50 % assay diluent and 50 % of the matrix, which closely resembles the sample type (50 %/50 % mixture). For example, when the sample type is RPMI medium containing

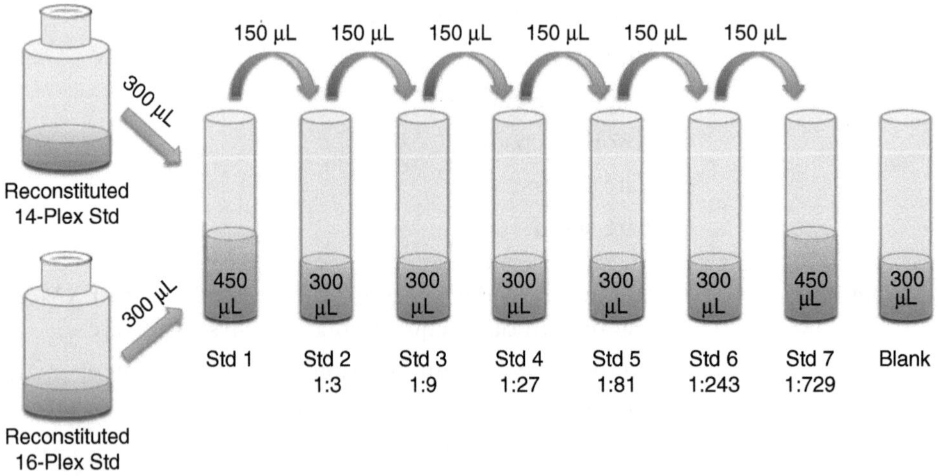

Fig. 4 Preparation of the standards. Both reconstituted multiplex standard (Std) solutions are mixed first and then diluted in a 1:3 series for seven dilutions (or alternatively ten for a longer curve). Always include a pair of wells with no standard (blanks) to adjust for background reporter signal

10 % fetal bovine serum (FBS), penicillin, and streptomycin, the standards should be reconstituted in a mixture composed of 50 % assay diluent and 50 % RPMI containing 10 % FBS, penicillin, and streptomycin.

4. To prepare standard 1, first reconstitute each vial with 0.5 mL of appropriate diluents.
5. Do not vortex. When mixing or reconstituting protein solutions, always avoid foaming.
6. Replace the vial stopper, and allow the vial to stand undisturbed for 10 min.
7. Gently swirl and invert the vial 2–3 times to ensure complete reconstitution, and allow the vial to sit at room temperature for an additional 5 min.
8. While waiting for the reconstitution of the standards, prepare eight polypropylene tubes to do the seven serial dilutions of the standard curves (Fig. 4). Add 300 µL of assay diluent in the Std2 to Std7 and blank for serum or plasma samples or 300 µL of a 50 %/50 % mixture of assay diluent and the matrix for other types of samples (e.g., cell culture media).
9. Then combine 300 µL from each standard vial, and mix by gently pipetting up and down 5–10 times to prepare the Std1.
10. Perform a 1:3 dilution series of the Std1 as shown in Fig. 4 to prepare the seven-point standard curve. Use a fresh pipette tip for each dilution step, and mix by pipetting up and down 5–10 times. Do not vortex (*see* **Note 6** on recommended reverse pipetting).

11. This volume is appropriate to do duplicates of the standard curve.
12. It may be beneficial to expand the standard curve up to ten serial dilution points to reach functional sensitivity for some analytes.

3.1.3 Prepare 1× Antibody Beads

The antibody bead concentrate is already supplied as a 1× concentrate in this kit (*see* **Note 7**). The fluorescent beads are light sensitive (*see* **Note 8**).

3.1.4 Biotinylated Antibody Preparation

1. The biotinylated antibody is supplied as a 10× concentrate and must be diluted prior to use with the provided diluent.
2. To prepare a 1× biotinylated antibody stock, dilute 10 μL of 10× biotinylated antibody in 100 μL of biotin diluent per assay well. Each well requires 100 μL of the diluted biotinylated antibody.
3. Use a dilution factor of 1:11 for extra pipetting volume. For example, for 96 wells, 9.6 mL of biotin diluent is needed and 0.96 mL of 10× biotinylated antibody.

3.1.5 Streptavidin–RPE Preparation

1. The streptavidin–RPE is supplied as a 10× concentrate and must be diluted prior to use with the provided diluent. Protect streptavidin–RPE from light during handling.
2. To prepare a 1× streptavidin–RPE stock, dilute 10 μL of 10× streptavidin–RPE in 100 μL of streptavidin–RPE diluent per assay well. Each well requires 100 μL of the diluted streptavidin–RPE.
3. Use a dilution factor of 1:11 for extra pipetting volume. For example, for 96 wells, 9.6 mL of streptavidin diluent is needed and 0.96 mL of 10× streptavidin–RPE.

3.2 Washing Methods

This assay may be washed using a vacuum manifold (requires the filter-bottom plate provided) or may be washed using the aid of a magnetic separator (requires the flat-bottom plate provided). Incomplete washing adversely affects assay results. Perform all wash steps with the wash solution supplied with the kit.

3.2.1 Manual Magnetic Separator Method (Recommended)

1. Place a Greiner flat-bottom plate on the magnet.
2. Allow the plate to soak for 60–90 s (90 s after primary incubation) to allow immobilization of the magnetic beads.
3. Grab the magnet from the bottom while holding the plate on it tightly.
4. Gently flick off the liquid by inverting the magnet with the plate (held tightly) over a sink or another disposal container in

one smooth downward motion followed by abrupt stop (*see* **Note 9**).

5. Keep the magnet and plate inverted, and very gently tap them onto a short stack of paper towels to absorb residual drops.
6. After the first tap, allow the paper towels to absorb the excess fluid, followed by moving over to dry paper towels for several additional gentle taps to ensure equal evacuation from all wells.
7. Return the magnet and plate upright, and remove the plate from the magnet.
8. Repeat for any additional washes as needed.

3.2.2 Filtration Washing Method with Manual Vacuum Manifold (See **Note 10**)

1. To wash beads, place the filter plate on the vacuum manifold and aspirate the liquid with gentle vacuum pressure (*see* **Note 11**).
2. Stop the vacuum pressure as soon as the wells are empty. Do not attempt to pull the plate off the vacuum manifold while the vacuum is still on or filter plate damage may occur. Release the vacuum prior to removing the plate.
3. After all wells are empty, lightly tap or press the filter plate onto clean paper towels (hold the plate in the center for tapping) to remove excess fluid from the bottom of the filter plate. Do not invert plate. Following the last aspiration and plate taps, use a clean absorbent towel to blot the bottom of the filter plate before addition of the next buffer or reagent.

3.2.3 Automatic Wash Methods

Some optimization of the automated plate washer setup may be required, and we do not recommend automatic methods.

3.3 Sample Preparation

Bring all samples to room temperature and invert or vortex briefly to mix. Centrifuge samples in a microfuge for <30 s to pull liquid from tube caps (*see* **Notes 12–16** for advice on sample specimens and **Note 16** for plate design).

3.4 Analyte Capture

1. The filter-bottom plate requires pre-wetting before use in the assay. Add 200 μL of 1× wash solution. Incubate the plate for 30 s at room temperature, and aspirate the wash solution from the wells using the vacuum manifold. The solid-bottom plates do not require this step.
2. Vortex (30 s) and sonicate (30 s) the antibody bead solution. Immediately add 25 μL to each assay well (*see* **Note 17**).
3. Add 200 μL of 1× wash solution, and wash the beads following the washing steps explained above, depending on the method. Repeat the washing step two more times (total three times).
4. Add 50 μL of incubation buffer to each test sample well.
5. Add 100 μL of standard into designated wells. For wells designated for samples, add 50 μL of assay diluent followed by

	1	2	3	4	5	6	7	8	9	10	11	12
A	Standard 1	Standard 1	Unknown Sample	Unknown Sample	Unknown Sample	Unknown Sample	Unknown Sample	Unknown Sample	Unknown Sample	Unknown Sample	Unknown Sample	Unknown Sample
B	Standard 2	Standard 2	Unknown Sample	Unknown Sample	Unknown Sample	Unknown Sample	Unknown Sample	Unknown Sample	Unknown Sample	Unknown Sample	Unknown Sample	Unknown Sample
C	Standard 3	Standard 3	Unknown Sample	Unknown Sample	Unknown Sample	Unknown Sample	Unknown Sample	Unknown Sample	Unknown Sample	Unknown Sample	Unknown Sample	Unknown Sample
D	Standard 4	Standard 4	Unknown Sample	Unknown Sample	Unknown Sample	Unknown Sample	Unknown Sample	Unknown Sample	Unknown Sample	Unknown Sample	Unknown Sample	Unknown Sample
E	Standard 5	Standard 5	Unknown Sample	Unknown Sample	Unknown Sample	Unknown Sample	Unknown Sample	Unknown Sample	Unknown Sample	Unknown Sample	Unknown Sample	Unknown Sample
F	Standard 6	Standard 6	Unknown Sample	Unknown Sample	Unknown Sample	Unknown Sample	Unknown Sample	Unknown Sample	Unknown Sample	Unknown Sample	Unknown Sample	Unknown Sample
G	Standard 7	Standard 7	Unknown Sample	Unknown Sample	Unknown Sample	Unknown Sample	Unknown Sample	Unknown Sample	Unknown Sample	Unknown Sample	Unknown Sample	Control
H	Blank	Blank	Unknown Sample	Unknown Sample	Unknown Sample	Unknown Sample	Unknown Sample	Unknown Sample	Unknown Sample	Unknown Sample	Unknown Sample	Control

Fig. 5 Typical plate design. A typical plate design includes two columns of diluted standards (or rows if ten standard points are used) in duplicates and test samples. It is good practice to include at least one duplicate positive control sample per assay plate

50 μL of the sample (*see* **Notes 17** and **18**). Figure 5 shows a typical plate design.

6. Seal the plate with an adhesive plate seal (*see* **Note 19**), and cover the plate with an opaque plate cover or aluminum foil. Incubate the plate for 2 h at room temperature on an orbital plate shaker at 500–600 rpm (*see* **Note 20**).

3.5 Analyte Detection

1. Prepare 1× biotinylated detector antibody 10 min before finishing the incubation.
2. Remove the liquid of the wells using a magnet or a vacuum equipment, and wash the wells twice with 200 μL of 1× wash solution.
3. Add 100 μL of diluted biotinylated detector antibody to each assay well (*see* **Note 17**).
4. Seal the plate with a new adhesive plate seal, cover, and incubate the plate for 1 h on a plate shaker (500–600 rpm).
5. During the incubation step, the Luminex® 100™, 200™, or FlexMAP 3D® instrument can be prepared (Subheading 3.7) if the plate is going to be read on the same day (*see* **Note 21**).
6. Prepare 1× streptavidin–RPE solution 10 min before finishing the incubation.
7. Remove the liquid of the wells using a magnet or a vacuum equipment, and wash the wells two times with 200 μL of 1× wash solution.
8. Add 100 μL of diluted streptavidin–RPE to each assay well (*see* **Note 17**).
9. Seal the plate with a new adhesive plate seal, cover, and incubate the plate for 30 min on a plate shaker (500–600 rpm).
10. Remove the liquid of the wells using a magnet or a vacuum equipment, and wash the wells three times with 200 μL 1× wash solution.

11. Add 125 μL of 1× wash solution to each assay well.
12. Seal with a new plate sealer, cover the plate, and place it on the plate shaker (500–600 rpm) for 2–3 min.

3.6 Plate Reading

Read the plate on any Luminex® 100™, 200™, FM3D, or MAGPIX instrument (*see* **Note 22**) or put it at 4 °C if it is going to be read the next day.

3.7 Setup for Luminex® 100™ and 200™ and FLEXMAP 3D® using xPONENT Software

1. Prepare to read the plates: Turn on the instrument, and ensure that the waste container is empty and that the sheath fluid reservoir has sufficient buffer to run the plate read.
2. System initialization: Open the xPONENT program, log in, go to the Maintenance tab, select "System Initialization," prepare the control plate according to on-screen instructions, and press Run (*see* **Note 23**). This will run for approximately 35 min.
3. Open a sample protocol: In the Protocols tab, open a new protocol, enter a name and version number, select the Bead Type (Microplex or MagPlex), enter the Volume as 75 μL, enter the DD gate (e.g., 7,500–20,000; *see* **Note 24**), set Reporter Gain to DEFAULT (*see* **Note 25**), set Analysis Type to Quantitative, enter the number of standard points (7 or 10), check the Fit of all standards option (*see* **Note 26**), and then click Next.
4. In the Analytes tab, highlight the appropriate *Bead Regions* for each analyte (Refer to the Technical Data Sheet of the kit for all bead regions), type in the name of cytokines/chemokines corresponding to each region in the table, ensure that the analysis column is set to Logistic 5PL Weighted (*see* **Notes 27–29**) by clicking the Change button next to Default Analysis, ensure that the Units column is set to pg/mL (or as noted in the Technical Data Sheet) and that the count is 100 for each region by entering these values and clicking the Apply All button, and then click Next.
5. In the Plate Layout tab, highlight the wells of the plate corresponding to the standards and click the round Standard (S) button with the replicate count set to 2, highlight the two wells corresponding to the blanks and click the Background (B) button, highlight the test sample wells and click the Unknown (U) button with the appropriate replicate numbers selected, highlight the wells corresponding to any controls present (*see* **Note 30**) and click the Control (C) button, manually enter IDs in the table if desired, and then click Save (*see* **Note 31**).
6. Open a Standards & Controls protocol: In the Stds & Ctrls tab, click Create New Std/Ctrl Lot, enter the name, Std/Ctrl Lot#, Expiration and Manufacturer, enter the Standard and

Control concentrations in the tables for each analyte according to the Technical Data Sheet, set a dilution factor of 1:3, click the Apply Dilution button, and then click Save.

7. Run a batch: In the Batch tab, click Create a New Batch from an Existing Protocol, scroll through the list and select the appropriate sample protocol, select the appropriate standards and controls protocol in the Stds & Ctrls tab, ensure that the plate layout is correct, and then click Run Batch to begin acquisition.
8. When acquisition is completed, check for any acquisition errors (*see* **Notes 32** and **33**) and analyte by analyte the standard curves (*see* **Notes 34–39**).
9. To export the data, go to the Results tab, select the completed batch run under Saved Batches, and export the data as a csv file using the Exp Results button, or go to Reports and generate a PDF report by selecting the completed batch run and pressing the Generate Report button.

4 Notes

1. Multiplexing: When several singleplex kits are multiplexed, it is possible that the background might increase and the overall assay signals of other standard points might be reduced. This can result in lower dynamic range or loss in sensitivity in some assays. This effect might be greater as more assays are added to the multiplex.
2. Do not freeze any component of the reagent kit. Store kit components at 4 °C when not in use.
3. Do not mix or substitute reagents with those from other kits or lots. However, if the whole kit is used only once (96 tests or less), there will be leftover reagents that can be mixed with others of the same lot.
4. Allow all reagents to warm to room temperature before use (air-warm all reagents at room temperature for at least 30 min or in a water bath for 20 min). In some laboratories, daily temperature fluctuations introduce significant variation in assay performance. In these cases, it is recommended to warm all reagents in a benchtop incubator set to 22 °C and maintain reagents there until use.
5. Do not reuse or store resuspended antigen standards.
6. For pipetting, we recommend using a modified reverse-pipetting technique for all pipetting during assay setup to reduce imprecision due to residual fluid volumes within pipette tips.

A multichannel pipettor is also recommended for all common reagents during the assay.

(a) To use a modified reverse pipetting technique, set the pipette to the appropriate volume needed.

(b) Press the plunger button slowly to the first stop, and then press slightly past it (just enough to aspirate a small extra volume of the reagent). Alternatively, pipette up and down slowly several times (at least three times) going to the first stop only (this allows residual fluid to build up in the tip).

(c) Immerse the tip into the liquid, just below the level of the solution.

(d) Release the plunger button slowly and smoothly to the top resting position to aspirate the set volume of liquid.

(e) Be careful not to have excess of liquid outside the tip (touch your tip against the edge of the vessel or the reservoir to remove excess liquid).

(f) Place the end of the tip at an angle against the inside wall of the receiving vessel, above the fluid level.

(g) Press the plunger button slowly and smoothly to the first stop (do not go past the first stop). A small volume of residual liquid will remain in the tip, and it should not be dispensed.

(h) Remove the tip from the recipient vessel. The liquid remaining in the tip can be pipetted back into the original solution or thrown away with the tip (when pipetting replicates of the same sample or solution the tip can be reused, but change tips if contamination is possible).

7. For the Life Technologies magnetic kits, when using the Greiner black mylar bottom plate to perform the assay with magnetic beads, Life Technologies recommends an extra dilution due to greater bead retention compared to using the filter plate. The Human 30-Plex capture beads are supplied at 1× and need an additional 1:2 dilution to 0.5× for magnetic washing (load 25 μL of 0.5× beads per well). However, in future Life Technologies magnetic kit lots, concentration of the beads may be changed to correct this dilution step.

8. Fluorescent beads are light sensitive. Protect the beads from direct light to avoid photobleaching of the embedded dye. Use aluminum foil to cover the tubes used in the assay. Cover plates containing beads with an opaque or aluminum foil-wrapped plate cover. Since the amber vial does not provide full protection, keep the vial protected from light when not in use.

9. When using a manual magnetic washer, avoid splashing back into the plate when shaking out the fluid. This can happen

with a slight backward motion after stopping the flick (either pull back hard, or do not pull back at all). We recommend an abrupt stop followed directly by blotting.

10. An alternative to filtration washing with nonmagnetic beads is to perform the assay in low-binding round-bottom 96-well plates. To perform the washing steps, centrifuge the plate(s) in a benchtop centrifuge at 500 × *g* for 5 min, followed by flicking off of the wash buffer as described above, but with no patting (simply place the inverted plate briefly on the paper towels to absorb the residual buffer). In our hands, this method has worked well with up to 11-plex assays.
11. If you are using filter plates, leaking and clogging are common problems inherent to this assay format that lead to reduced accuracy and precision. If you can choose, in our hands it is better to use magnetic manual washers.
    (a) Be careful to avoid over-vacuuming (do not exceed 5 mmHg). Excessive vacuum can cause the membrane to tear. Prevent any vacuum surge by opening and adjusting the vacuum on the manifold before placing the plate on the manifold surface.
    (b) The bottom of the filter plate should not be in direct contact with any surface during the entire assay, including plate setup or incubation times. To prevent capillary action, a plate stand or a non-flat side of a plate cover is recommended to rest the filter plate.
    (c) If you experience clogging, try to remove the clog by gently flicking the underside of the well with a pipette tip or a needle.
    (d) Following plate washing, remove excess liquid and blot from the bottom of the plate by pressing the plate on clean paper towels.
12. Cytokines/chemokines in general are quite labile and will degrade over time even when stored at −80 °C. Freeze–thaw cycles also affect the concentration of cytokines. Lower concentrations may be detected for some cytokines, whereas for others a higher concentration may be measured. Sample storage strategies should be determined empirically prior to making them standard practice.
13. For plasma samples, it is important to take into account the anticoagulant used. For some cytokines, different concentrations may be detected if the plasma has EDTA or heparin.
14. If using frozen serum or plasma samples, we recommend centrifuging the samples after thawing to remove debris as it may interfere with the beads. We have observed some bead aggregation.

15. Vortex and sonicate the beads immediately before adding to the plate to ensure an equal distribution of bead within the wells. Beads aggregate and precipitate easily.
16. The intraplate and inter-plate variability may introduce a bias in your data. To compare and pool data from different plates, we recommend randomization of the samples to balance against plate or batch effects.
17. When pipetting reagents, maintain a consistent order of addition from well to well to ensure equal incubation times for all wells.
18. For the Life Technologies magnetic kits some modifications can be done if only low sample volumes are available: If you have less than 50 μL of sample/well available, reduce all reagent volumes by equal volumes equivalent to the sample volume available (e.g., for 25 μL of sample, reduce all reagent volumes by a factor of 2: 25 μL incubation buffer, 50 μL standard, 25 μL assay diluent, 25 μL sample). Minimum total volume per well of 37.5 μL is suggested.
19. Plate sealers are not provided in some kits, but their use is recommended to avoid well-to-well contaminations. If the seal is wet from some samples be careful not to contaminate other wells when it is removed (hold the plate firmly pressed on the benchtop when removing the film to avoid splashing).
20. The speed of the orbital shaker may need to be adjusted depending upon the radius of the orbital shaker (lower speeds for larger orbital radius). We check that the solution is mixing, as in a vortex, but that the liquid level is not reaching above the top quarter of the well.
21. To save time, begin system initialization of the reading instrument during the incubation with streptavidin–RPE.
22. Ensure that the instrument has passed QC (calibration and validation) prior to starting the assay.
23. If the Luminex instrument has multiple users, be careful that the probe height is adjusted.
24. We recommend that the doublet discriminator gates be set at 7,500–20,000 for the Luminex® 100™ and 200™ or 4,800–15,000 for FLEXMAP 3D® as the initial setting when using Magplex beads. The gate should be smaller when using Microplex beads (e.g., 4,000–14,000). Adjustment of these values is required for individual instruments.
25. Use the default setting of low PMT for the Luminex® 100™ and 200™, and use "Standard PMT" setting for the Luminex FLEXMAP 3D®.

26. Some Luminex analysis software let the user choose between using individual standard points vs. mean of each standard replicate. The use of individual standard points provides the model with more points to fit leading to a better curve, as a result of having greater degrees of freedom. However, this method is more sensitive to outliers, so then it is better that you eliminate any outliers that you might have before proceeding with the standard curve calculations. If you do not want to discard outliers, you can use the mean of each standard replicate. This method is less susceptible to outliers, but it generates worse curve fits as it uses less data points.

27. Curve-fitting model for the standard curves: A number of curve-fitting model equations are available in Luminex analysis software (including xPONENT). We recommend the 5-Parameter Logistic (5PL) nonlinear regression model (Fig. 6). It is an asymmetric function that in our hands has proved the best for analyzing cytokine/chemokine Luminex data:

$$y = A_{max} + \frac{A_{min} - A_{max}}{\left(1 + \left(x / \mathrm{EC}_{50}\right)^{\mathrm{Hill}}\right)^{\mathrm{Asym}}}$$

A_{min} is the lower asymptote, A_{max} is the upper asymptote, Hill is the slope at the inflection point, EC_{50} is the concentration at the inflection point, and Asym is a factor of asymmetry.

28. Curve fitting may also be performed in external statistical program, such as STATA (*see* **Note 39** and Fig. 6), R, or SPSS using the raw data.

29. Weighting: Usually in immunoassays the variance of the dependent variable varies across the data. In cytometric bead arrays, the MFI values at higher concentrations usually have a greater variance than MFI values on the low end of the standard curve (Fig. 7). However, the 5PL model assumes equal variance. For analyzing Luminex data, there are some weighting algorithms available in the analysis software to account for this variance. The $1/Y^{\wedge}2$ is the most recommended weighting function for the 5PL model equation, which minimizes residuals (errors) based on relative MFI values. Basically, it puts less emphasis on getting a close fit to the data points at the high end of the curve than the bottom.

30. Add controls for QC. Controls are useful for knowing if your assay worked and for comparing the performance between different assays. Ideally three controls of high, medium and low concentration should be used. Depending on the values of your control you may want to discard the results of that plate. Having at least one control in duplicate in each plate also allows calculation of inter-plate variability. There are commercially

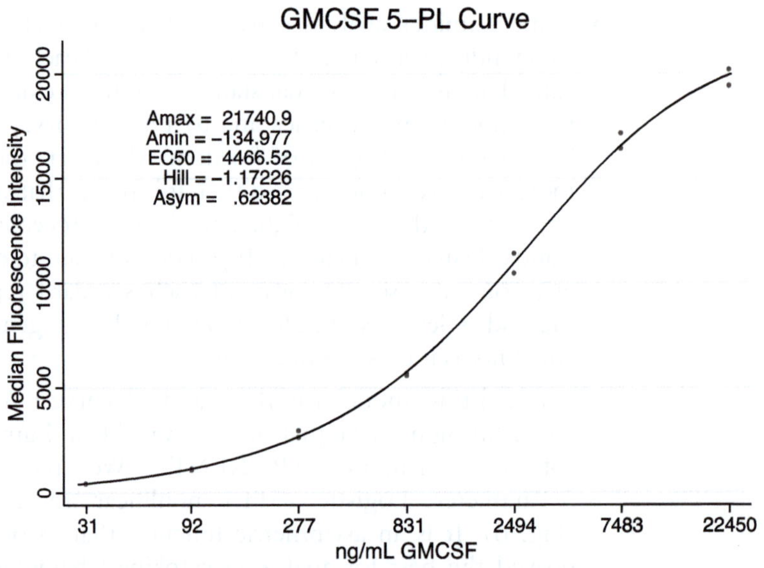

Fig. 6 5-Parameter Logistic Standard Curve. An iterative nonlinear regression model fits MFI values to expected concentration to produce a sigmoidal curve from which sample concentrations can be estimated. The function includes an *upper* (A_{max}) and a *lower* (A_{min}) asymptote. The parameters can be combined to estimate concentration for any MFI values within the asymptote range. Weighting by the inverse variance adds greater emphasis (read: closer fit) in the lower range of the curve

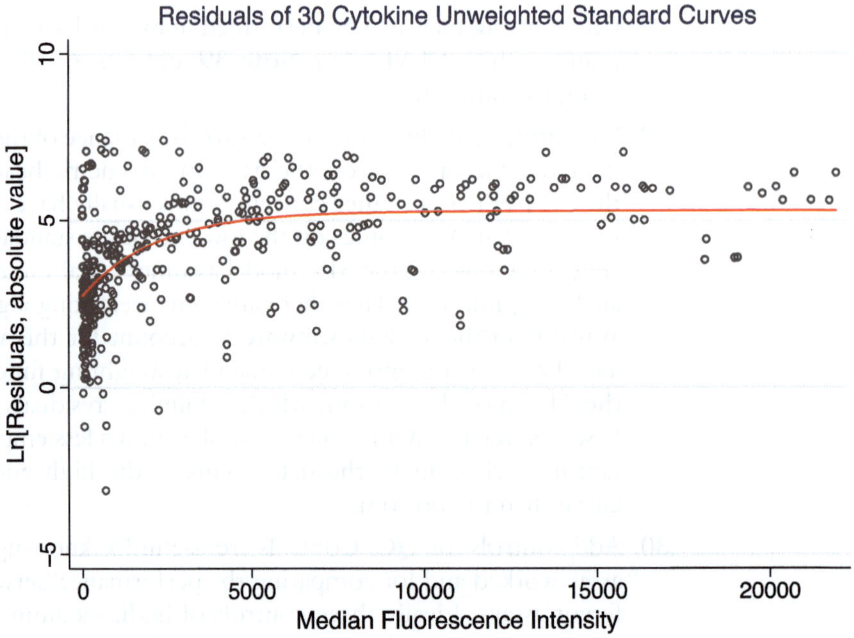

Fig. 7 Variation in standard signal intensity. The natural log of the absolute value of the residuals for 30 5PL standard curves of different cytokines/chemokines is plotted against MFI to demonstrate the increase in variability at higher signal intensities. The *red line* represents a fit of the data by an exponential curve nonlinear model

available controls, but you can use a sample with known cytokine concentration values or make your control using the same matrix that you use in your assay with spiked-in recombinant proteins (lyophilized standards remaining from any kit could be used).

31. Take into account the dilution factors and introduce them in the software. Use of incorrect dilution factors will result in incorrect final calculated concentrations. For the Life Technologies kits, in addition to dilutions performed on the sample prior to running the assay, the sample concentration calculated from the standard curve must be multiplied by an additional factor of 2 to correct for the 1:2 dilution made when samples were added to the plate (50 μL of sample + 50 μL of assay diluient vs. 100 μL of standard points).

32. Check that bead counts reached 100 per analyte, although we usually considere a reading as valid if a minimum of 30 beads is counted per well.

33. Check for slow bead acquisition or erratic event counts during acquisition, abnormal scatterplot populations or shifts in the doublet discriminator or additional peaks, and acquisition timeout or "sample ran to empty" warnings. This may mean that the probe is clogged and needs to be flushed/sonicated or that an error occurred during the assay procedures resulting in bead loss.

34. Standard outliers should be invalidated. Outliers in your standards can be identified by analyzing the % recovery (calculated concentration/expected concentration) × 100. The probability of a data point being an outlier increases as the % recovery deviates from 100 %, and we consider any standard that has <70 % recovery or >130 % recovery as outliers. % Recovery is calculated based on all the standard points that are validated. As you invalidate standard points and reanalyze the standard curve, be sure to recheck the % recovery calculations, despite being wary of over-fitting a curve by removing too many points. If strict reproducibility is wanted, a tighter range (90–110 %) should be used. Alternatively, if the software allows it, residuals can be used to detect and invalidate outliers.

35. All data with MFI value greater than 25,000 should be excluded, as higher MFI are saturated, at least for Luminex 100/200 instruments. It is very inaccurate to be interpolating or extrapolating any concentration values near the limit of the instrument saturation.

36. Precision should also be checked. Acceptance criteria for replicate % CV of concentration and MFI is generally a maximum of 30 %.

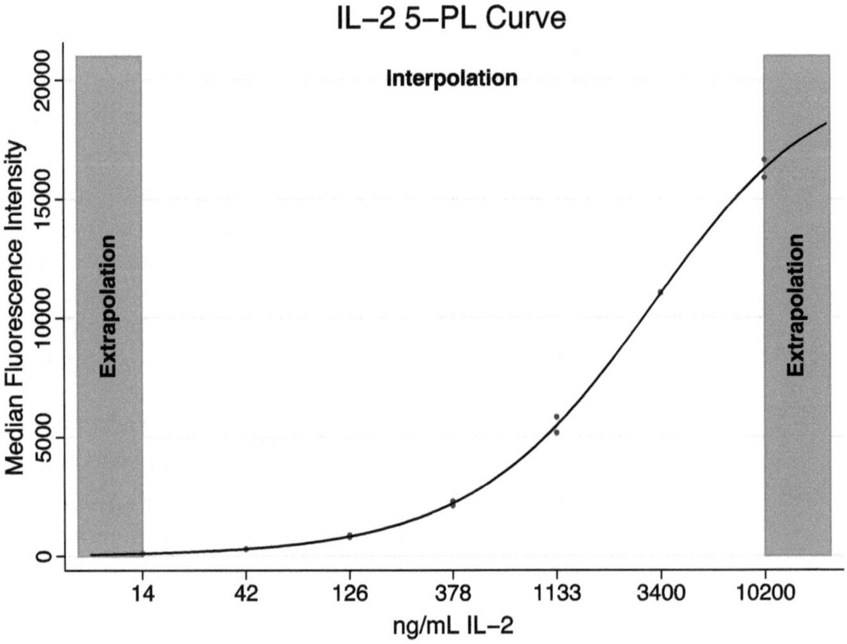

Fig. 8 Extrapolation and interpolation. Estimation within the range of the first and last standard points is interpolation, whilst estimation outside of the range of the standard points is extrapolation. Care should be taken about the range of the curve used for estimation, especially in the case of extrapolation. Precision decreases at the extremes

37. Extrapolation of data is required when concentrations must be estimated for points that are within calculable limits but are outside of the range of standards inside of the standard curve (Fig. 8). Extrapolated values are less accurate, in particular if values are on the flat parts of the curve or if a curve does not reach saturation. As the curve flattens, small changes in the MFI values may translate into big changes in concentration estimates. Likewise, standard points that end while still in the linear portion of the curve can result in unnaturally high limits of quantification. If the samples are above the standard curve range, and you have more sample volume, you can retest the sample after diluting them taking into account then the dilution factor when calculating the sample concentrations. If your sample is below the standard curve range, you can extend your serial dilutions in the following assays.

38. For values that fall out of the range of standards, we usually give a concentration value of half the detection limit for that cytokine or chemokine (described by the manufacturer) to any MFI below the lowest standard point or below the limits of detection of the experiment. For MFI over the last validated standard point we give a value of the expected concentration for this last standard point.

```
cap drop yhat
cap ereturn clear
foreach i of var cytokinelist {
    summarize mfi_`i', meanonly
    local Amin = r(min)
    local Amax = r(max)
    summarize unit_`i' if mfi_`i'!=., meanonly
    local EC50 = r(mean)
    nl (`i' = {Emax} + (({Emin} - {Emax}) / ((1 + ((unit_`i' / {EC50})^{Hill}))^{Asym})) ) ///
        if mfi_`i'!=. [w=1/(mfi_`i'^2)], ///
        iter(100) initial(Emin `Emin' Emax `Emax' EC50 `EC50' Hill 1 Asym 1)
    if e(ic)<100 {
        predict yhat
        local emax: display %8.0g _b[/Emax]
        local emin: display %8.0g _b[/Emin]
        local ec50: display %8.0g _b[/EC50]
        local hill: display %8.0g _b[/Hill]
        local asym: display %8.0g _b[/Asym]
        scatter mfi_`i' unit_`i' if unit_`i'!=., mc(gs4) msize(vsmall) || ///
            line yhat unit_`i', sort lc(black) legend(off) xsc(log) ///
            xlab(5 50 500 5000 50000) yti("MFI") ylab(0(5000)31000) xti("ng/mL") ///
            text(25000 25) "Emax = `emax'" "Emin = `emin'" "EC50 = `ec50'" ///
            "Hill = `hill'" "Asym = `asym'") title("Weighted 5-PL Fit of `i'") ///
            name(pfit, replace) graphregion(color(white) lwidth(thick) margin(small) ///
            icolor(white) ilwidth(thick))
        graph export STUDY_fit_`i'.pdf, replace
        replace unit_`i' = ((((_b[/Emin] - _b[/Emax]) / (mfi_`i' - _b[/Emax]))^(1/_b[/Asym]) ///
            -1)^(1/_b[/Hill]))*_b[/EC50]
    }
    if e(ic)>=100 {
        di "5-PL regression for cytokine `i' did not converge"
    }
}
```

Fig. 9 Script example

39. Fitting your own curves in STATA: A simple script can be used to fit standard curves using the 5PL model and apply the curve estimates to the raw data. This may require some organization of exported data in a spreadsheet program, such as Microsoft Excel. An example script is shown in Fig. 9.

References

1. Prakash D, Fesel C, Jain R et al (2006) Clusters of cytokines determine malaria severity in Plasmodium falciparum-infected patients from endemic areas of Central India. J Infect Dis 194:198–207
2. Rovira-Vallbona E, Moncunill G, Bassat Q et al (2012) Low antibodies against Plasmodium falciparum and imbalanced pro-inflammatory cytokines are associated with severe malaria in Mozambican children: a case–control study. Malar J 11:181
3. To KKW, Hung IFN, Li IWS et al (2010) Delayed clearance of viral load and marked cytokine activation in severe cases of pandemic H1N1 2009 influenza virus infection. Clin Infect Dis 50:850–859
4. Sun Y, Jin C, Zhan F et al (2012) Host cytokine storm is associated with disease severity of severe fever with thrombocytopenia syndrome. J Infect Dis 206:1085–1094
5. Stacey AR, Norris PJ, Qin L et al (2009) Induction of a striking systemic cytokine cascade prior to peak viremia in acute human immunodeficiency virus type 1 infection, in contrast to more modest and delayed responses in acute hepatitis B and C virus infections. J Virol 83:3719–3733
6. Dobaño C, Campo JJ (2009) Understanding protective immune mechanisms induced by malaria vaccines in the context of clinical trials. Hum Vaccin 5:562–565

7. Berthoud TK, Manaca MN, Quelhas D et al (2011) Comparison of commercial kits to measure cytokine responses to Plasmodium falciparum by multiplex microsphere suspension array technology. Malar J 10:115
8. Moncunill G, Aponte JJ, Nhabomba AJ et al (2013) Performance of multiplex commercial kits to quantify cytokine and chemokine responses in culture supernatants from Plasmodium falciparum stimulations. PLoS One 8:e52587
9. Richens JL, Urbanowicz RA, Metcalf R et al (2010) Quantitative validation and comparison of multiplex cytokine kits. J Biomol Screen 15:562–568
10. Djoba Siawaya JF, Roberts T, Babb C et al (2008) An evaluation of commercial fluorescent bead-based luminex cytokine assays. PLoS One 3:e2535
11. Nechansky A, Grunt S, Roitt IM et al (2008) Comparison of the calibration standards of three commercially available multiplex kits for human cytokine measurement to WHO standards reveals striking differences. Biomark Insights 3:227–235
12. Young S-H, Antonini JM, Roberts JR et al (2008) Performance evaluation of cytometric bead assays for the measurement of lung cytokines in two rodent models. J Immunol Methods 331:59–68
13. Khan IH, Krishnan VV, Ziman M et al (2009) A comparison of multiplex suspension array large-panel kits for profiling cytokines and chemokines in rheumatoid arthritis patients. Cytometry B Clin Cytom 76:159–168
14. Khan SS, Smith MS, Reda D et al (2004) Multiplex bead array assays for detection of soluble cytokines: comparisons of sensitivity and quantitative values among kits from multiple manufacturers. Cytometry B Clin Cytom 61: 35–39

Chapter 7

A Useful Guide for Analysis of Immune Markers by Fluorochrome (Luminex) Technique

Maria Faresjö

Abstract

The fluorochrome Luminex technique is a bead-based sandwich immunoassay that combines the enzyme-linked immuno sorbent assay (ELISA) with flow cytometry. The Luminex technique allows multiple cytokines to be measured simultaneously in small volumes and provides a convenient and sensitive tool for the detection of a large number of, e.g., extracellular secreted cytokines to characterize cytokine profiles.

The technique is based on the so-called microspheres (beads) that serve as a solid phase for molecular detection. These individually dyed micro-beads have monoclonal antibodies directed against the cytokines and chemokines of interest and allow simultaneous detection of up to nearly 100 cytokines and chemokines in a dual-laser flow analyzer. Immune markers can be detected in serum and plasma samples as well as in cell culture supernatants from in vitro-stimulated peripheral blood mononuclear cells (PBMC).

This chapter describes the Luminex technique for detection of multiple cytokines by magnetic bead sandwich immunoassay, with a special focus on some important pre-analytical factors, such as cell separation, cryopreservation, and PBMC thawing that may affect the detection outcome of immune markers. This method can also be easily adapted to measuring other biomarkers in biological samples.

Key words Luminex, Immune markers, Cytokines, Peripheral blood mononuclear cells, Cell separation, Cryopreservation of PBMC

1 Introduction

The fluorochrome Luminex technique is a bead-based sandwich immunoassay [1]. The technique is based on the so-called microspheres (beads) that serve as a solid phase for molecular detection. The microspheres are available in 100 different fluorescent color tones of red and infrared, each carrying its own detection reagent on the surface. As a result, the technique allows simultaneous detection of up to 100 parameters in a small single sample [2].

Each bead set is coated with a capture antibody, specific to a particular bioassay, allowing the capture and detection of a specific analyte. In a secondary step, a biotinylated antibody together with streptavidin is captured to the bead complex and, finally, a phycoerythrin (PE) fluorescent reporter is coupled to the complex (Fig. 1).

Ivana Vancurova (ed.), *Cytokine Bioassays: Methods and Protocols*, Methods in Molecular Biology, vol. 1172, DOI 10.1007/978-1-4939-0928-5_7,

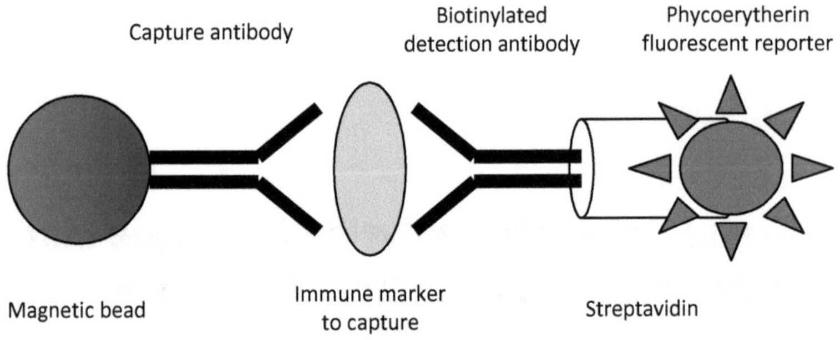

Fig. 1 Principle of magnetic bead sandwich immunoassay. Each magnetic bead set is coated with a capture antibody for detection of each specific immune marker. In a secondary step, a biotinylated antibody together with streptavidin is captured to the bead complex and, finally, a phycoerythrin fluorescent reporter is coupled to the complex

In principle, the multiplex assay is performed directly in a microtiter plate. Each bead set, defined by an individual fluorescent color tone on its surface, is mixed to a multiplex bead mix and put into the wells on the plate. The sample (and standard) is, in a secondary step, mixed with the beads, and thereafter the biotinylated antibody as well as streptavidin–phycoerytherin are attached to the bead set of interest before readout of the plate (Fig. 2).

On a dual-laser flow-based detection instrument, the amount of bound analyte correlates directly with the fluorescent intensity, allowing the quantification of each specific analyte. During the readout of the plate, the beads are analyzed by a red laser (635 nm) that excites the internal dyes to identify each microsphere particle. At the same time, a green laser (532 nm) excites the reporter dye (streptavidin–PE) in direct proportion of the fluorescence of the analyte for quantification (Fig. 3).

2 Materials

2.1 Serum, Plasma, or Whole-Blood Sample

1. Venous blood samples from healthy human volunteers or patients are drawn in tubes without supplemented (for serum) or supplemented with sodium heparin (for plasma and preparation of peripheral blood mononuclear cells, PBMC).

2.2 Isolation of Peripheral Blood Mononuclear Cells

1. RPMI-1640 without glutamine.
2. Washing buffer: RPMI-1640 without glutamine supplemented with 2 % heat-inactivated fetal calf serum (FCS).
3. Separation reagent: Ficoll-Paque density gradient centrifugation.

2.3 Cellular Number and Viability

1. Türk's solution.
2. Trypan blue solution.

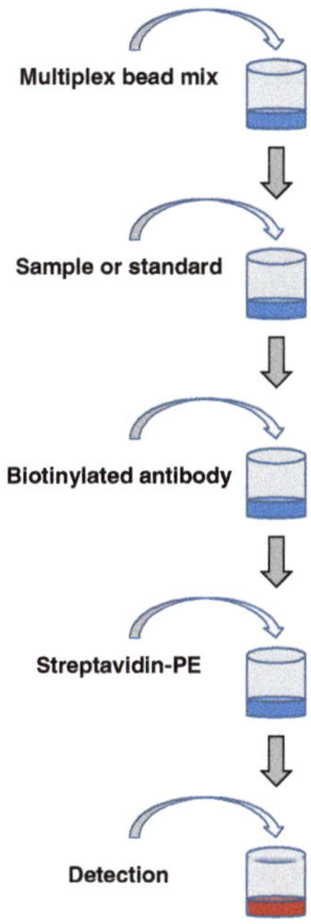

Fig. 2 Methodological principle of the Luminex technique. The multiplex assay is performed directly in a 96-well microtiter plate. Each bead set is mixed to a multiplex bead mix and put into the wells on the plate. The sample and standard are mixed with the beads, and thereafter the biotinylated antibody as well as streptavidin–phycoerythrin are attached to the bead set of interest. Finally, the samples are analyzed by reading each individual well on a dual-laser flow-based detector instrument

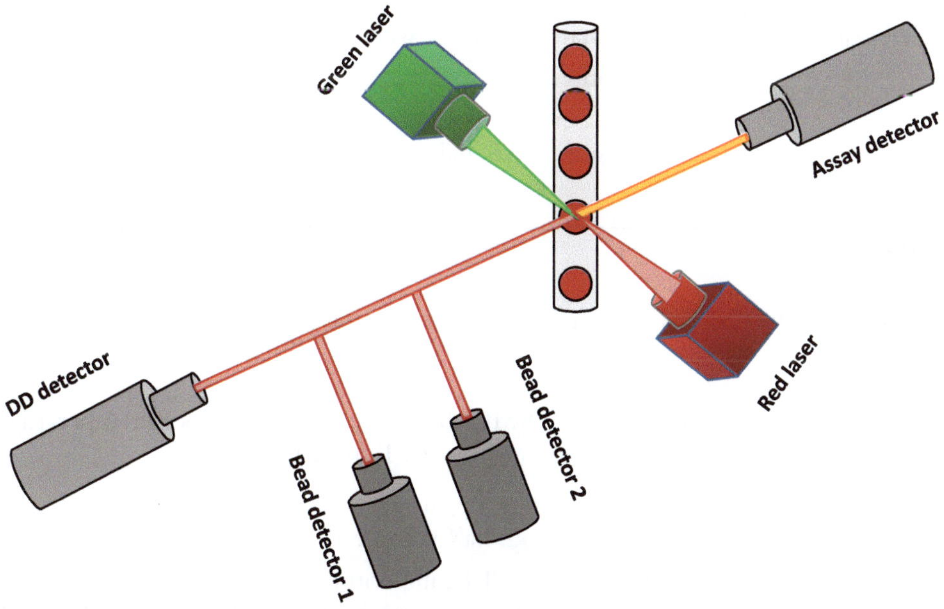

Fig. 3 Dual-laser flow-based detection. The beads are analyzed by a red laser (635 nm) that excites the internal dyes to identify each microsphere particle. The green laser (532 nm) excites the reporter dye (streptavidin–PE) to quantify the amount of bound immune marker on the beads

2.4 Cryopreservation, Thawing, and Stimulation of PBMC

1. Freezing medium: 50 % RPMI-1640 without glutamine, 40 % FCS, and 10 % dimethyl sulfoxide (Me_2SO).
2. Cryo 1 °C Freezing Container (Nalge Nunc International) containing isopropanol.
3. Vials appropriate for cryopreservation (Nalge Nunc International).
4. Washing buffer: RPMI-1640 without glutamine, supplemented with 10 % heat-inactivated FCS.
5. AIM V research-grade serum-free medium (Gibco) supplemented with 2 mM L-glutamine, 50 μg/l streptomycin sulfate, 10 μg/l gentamicin sulfate, and 2×10^{-5} M 2-mercaptoethanol.

2.5 Luminex

2.5.1 Luminex Kit Including the Following Reagents

1. Standard for Luminex assay.
2. Luminex diluent for standard (standard diluent for serum and plasma or cell culture medium for cell culture supernatant).
3. Luminex sample diluent (for serum and plasma).
4. Luminex assay buffer.
5. Luminex wash buffer.
6. Coupled magnetic beads (10×).
7. Detection antibodies (10×).
8. Detection antibody diluent.
9. Streptavidin–PE (100×).
10. Flat bottom plate (96-well).
11. Sealing tape.

2.5.2 Equipment

1. Wash station (for use with magnetic bead-based assays).
2. Microtiter plate shaker.
3. Dual-laser flow-based plate reader including software for data acquisition and analysis.

3 Methods

3.1 Serum or Plasma Sample

1. Centrifuge the blood samples at 2,000×*g* for 10 min, and transfer the serum/plasma to sterile tubes.

3.2 Isolation of Peripheral Blood Mononuclear Cells

Separation of PBMC should be handled at room temperature in a ventilated hood (*see* **Notes 1** and **2**).

1. Dilute venous heparinized blood in RPMI, 1:2 (i.e., half-volume blood, half-volume RPMI).
2. Invert the Ficoll-Paque bottle several times to ensure thorough mixing. Next, withdraw the required volume of Ficoll-Paque using aseptic technique. The Ficoll:blood ratio should be 3:5 (i.e., 12 ml of Ficoll:20 ml of diluted blood sample).

3. Transfer the required volume of Ficoll-Paque to a sterile centrifuge tube.
4. Carefully lay the diluted blood sample on the Ficoll layer to avoid mixing between Ficoll and the blood sample.
5. Centrifuge at 400 × *g* for 30 min, without brake, at 18–20 °C.
6. Carefully draw up the layer of PBMC using a clean Pasteur pipette, and transfer PBMC to a sterile centrifuge tube.
7. Resuspend PBMC in 20 ml of washing buffer.
8. Centrifuge at 400 × *g* for 10 min, with low brake, at 18–20 °C.
9. Remove the supernatant, and wash PBMC by repeating the **steps 7–8** once more.
10. Remove the supernatant.

3.3 Cellular Number and Viability

1. Dissolve the PBMC pellet in 1 ml of washing medium.
2. For determination of the number of PBMC, dilute 10 μl of PBMC suspension with 90 μl of Türk's solution (1:10).
3. Place 10 μl of PBMC suspension in a Bürker's chamber. Count the number of PBMC using a light microscope (*see* **Note 3**).
4. For determination of membrane integrity, dilute 10 μl of PBMC suspension with 10 μl of trypan blue solution (1:1).
5. Place 10 μl of PBMC suspension in a Bürker's chamber. Count 100 PBMC in a light microscope (*see* **Note 4**).

3.4 Cryopreservation of PBMC (See Note 5)

1. Centrifuge the solution of PBMC at 400 × *g* for 10 min at 18–20 °C.
2. Remove the supernatant.
3. Resuspend PBMC in freezing medium (4 °C) by adding freezing medium dropwise, while the tube is continuously agitated, to a concentration of 5–10 × 10^6 PBMC/ml.
4. Transfer 1 ml of PBMC solution to a vial suitable for cryopreservation.
5. Place the vial with PBMC in a precooled (4 °C) Cryo 1 °C Freezing Container containing isopropanol.
6. Place the container at −70 °C (the freezing rate is −1 °C/min).
7. Transfer the vial to liquid nitrogen (−196 °C) the following day.

3.5 Thawing of PBMC

1. Thaw PBMC, directly from −196 to +37 °C, in a water bath under continuous agitation.
2. Immediately thereafter, add washing buffer dropwise to the cells until a total volume of 10 ml is reached.
3. Centrifuge the solution of PBMC at 400 × *g* for 10 min at 18–20 °C.

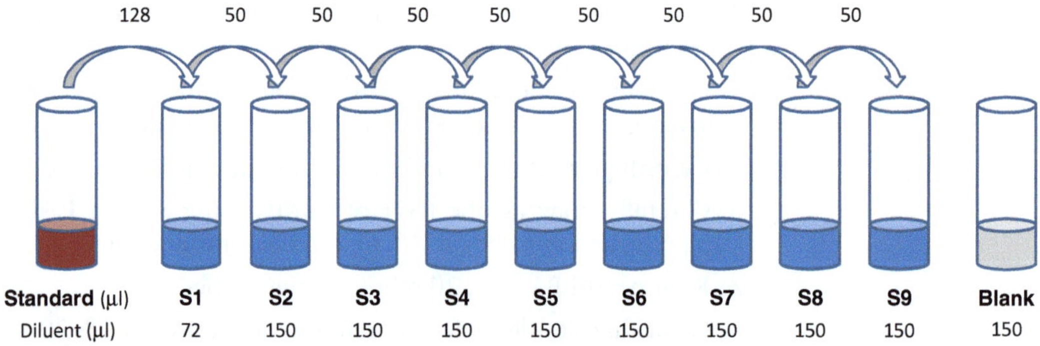

Fig. 4 Preparation of fourfold serious of standard. A nine-point standard dilution series is prepared by adding 72 μl of diluent to tube S1 (standard diluent for serum and plasma or cell culture medium for cell culture supernatant) and 150 μl of diluent to tubes S2–S9. Reconstituted standard (128 μl) is transferred into tube S1 and, thereafter, serially diluted fourfold from tubes S1 thru S9 by transferring 50 μl between each tube. To the tube marked as blank, exclusively 150 μl of diluent (for serum samples) or cell culture medium (for cell culture supernatants) is transferred

4. Remove the supernatant.
5. Resuspend PBMC in 1 ml of AIM V research-grade serum-free medium with supplements.
6. Count the cell number, and determine the cell viability (*see* Subheading 3.3).

3.6 Stimulation of PBMC and Collection of Cell Supernatant

1. Adjust the concentration of PBMC in AIM V medium with supplements to 1×10^6/ml.
2. Transfer 1 ml of the PBMC suspension to each tube for stimulation.
3. Add antigen or control to appropriate concentration.
4. Incubate PBMC between 24 and 96 h at 37 °C with 5 % CO_2.
5. After completed incubation of PBMC, centrifuge the cells for 10 min at $400 \times g$.
6. Transfer aliquots of cell supernatants to sterile tubes.

3.7 Luminex

Standard, diluents, and cell culture medium should be at room temperature, while serum, plasma, or cell culture supernatants should be kept on ice before preparation.

3.7.1 Preparation of Standard and Blank

1. Reconstitute the vial of standard in 500 μl of diluent (standard diluent for serum and plasma or cell culture medium for cell culture supernatant), vortex, and incubate on ice for 30 min.
2. Prepare the nine-point standard dilution series and blank. Label tubes S1–S9, and blank.
   Add 72 μl of diluent (standard diluent for serum and plasma or cell culture medium for cell culture supernatant) into tube S1 and 150 μl of diluent to tubes S2–S9 and blank (Fig. 4).

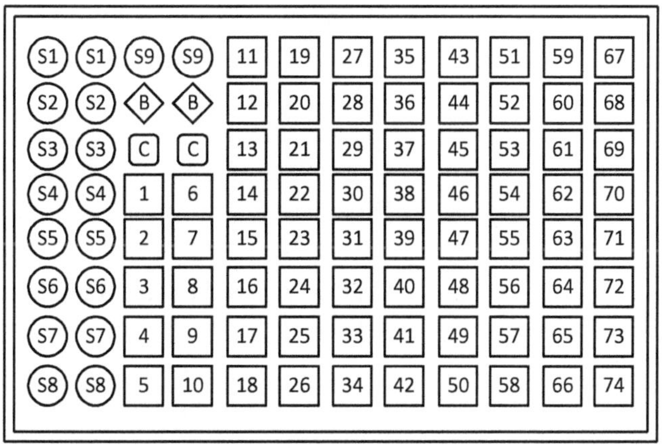

Fig. 5 Plate setup. An example of plate setup for standards (S1–S9), blank (B), control (C), and samples (serum or cell culture supernatants; positions 1–74) for the wells on the plate

3. Transfer 128 μl of reconstituted standard into S1 tube.
4. Serially dilute fourfold from tubes S1 thru S9 by transferring 50 μl between each tube (vortex between each transfer).
5. To the tube marked as blank, add 150 μl of diluent (for serum samples) or cell culture medium (for cell culture supernatants).

3.7.2 Preparation of Samples and Beads

1. Dilute serum samples in sample diluent (e.g., 1:4; 50 μl of serum and 150 μl of sample diluent).
2. Dilute cell culture supernatants in cell culture medium (if high concentration of immune markers is expected).
3. Dilute 575 μl of beads in 5,175 μl of assay buffer (1:10).

3.7.3 Running the Assay

1. Add 50 μl/well of beads (diluted 1:10) to the assay plate.
2. Wash the plate by adding 100 μl of wash buffer per well, and repeat this step once.
3. Add 50 μl of samples (serum or cell culture supernatants), standards (S1–S9), blank, and, if used, also control (Fig. 5).
4. Cover the plate with sealing tape and incubate for 30 min in the dark at room temperature with continuous shaking at 300 rpm.
5. Within 10 min remaining of the incubation step, prepare the detection antibody.

   Dilute 300 μl of detection antibody in 2,700 μl of diluent (1:10).
6. Wash the plate by adding 100 μl of wash buffer per well, and repeat this step twice.
7. Add 25 μl/well of detection antibody (diluted 1:10) to the assay plate.
8. Cover the plate with sealing tape and incubate for 30 min in the dark at room temperature with continuous shaking at 300 rpm.

9. Meanwhile, prepare software protocol, e.g., normalized standard (S1–S9) values.
10. Within 10 min remaining of the incubation step, prepare 1× streptavidin–PE in assay buffer. Dilute 60 μl of streptavidin–PE in 5,940 μl of assay buffer (1:100).
11. Wash the plate by adding 100 μl/well of wash buffer, and repeat this step twice.
12. Add 50 μl/well of streptavidin–PE (diluted 1:100) to the assay plate.
13. Cover the plate with sealing tape and incubate for 10 min in the dark at room temperature with continuous shaking at 300 rpm.
14. Wash the plate by adding 100 μl/well of wash buffer, and repeat this step twice.
15. Resuspend beads in 125 ml of assay buffer/well, cover the plate with sealing tape, and shake at 1,100 rpm for 30 s.
16. Remove the sealing tape and the plate cover, and place the plate in the reader. Read the plate by choosing the running protocol (prepared in advance). A minimum of 100 beads per region should be analyzed.
17. Raw data are thereafter analyzed using suitable software to apply a standard curve, with a cutoff for minimum detectable concentration, for each individual immune marker.

4 Notes

1. Time interval between blood sampling and sample processing: The time period between blood sampling and sample processing is very important. Our own research group has found that blood samples left at room temperature for 24 h before separation of PBMC will increase the expression of CD3/CD4 (unpublished data). It has also been shown that PBMC handled and cryopreserved within 8 h from venipuncture have better viability, higher cell recovery, and higher concentration of IFN-γ compared to PBMC handled after 8 h from blood sampling [3].
2. Enrichment of PBMC: There is a number of different ways to enrich PBMC. The most frequently used techniques for isolation of PBMC are density gradient separation by Ficoll or Lymphoprep or separation by vacutainer cell preparation tube (CPT). Ficoll is routinely used to isolate mononuclear cells from bone marrow, peripheral blood, and umbilical cord blood. In principle, PBMC are enriched from whole blood that is layered onto a density gradient. Gentle centrifugation at

room temperature results in a buffy coat of monocytes and lymphocytes under a layer of plasma, with the remaining white blood cells together with red blood cells passing through the interface and collecting at the bottom of the tube. The PBMC interface is collected and washed for several times in either phosphate-buffered saline (PBS) or cell culture medium (washing buffer) to remove any contaminating separation medium.

The vacutainer CPT is a single-tube system, supplemented with sodium citrate, for collection of whole blood and the separation of PBMC. In principle, blood is collected in the CPT, the tube is centrifuged, and the cell pellet is resuspended in PBS or cell culture medium before subsequent assay or procedure.

3. Cellular number: In the most common design, the volume of each large square of a Bürker's chamber is 0.1 mm^3. The cells in four large squares are counted, and cells over or touching the lines on top and on the left are counted, but cells over or touching the right or the bottom lines are ignored. The concentration in cells per ml will be calculated as cells in four large squares/4 × 10,000.

4. Cell recovery and viability: Viability thresholds should be used in, e.g., clinical trials in order to obtain reliable results of functional assays. The most convenient way to check the viability of PBMC is to manually count PBMC by light microscopy after the cells have been pre-colored with, e.g., trypan blue. In a viable cell, trypan blue is not absorbed; however, it traverses the membrane of dead cells. Hence, dead cells appear as distinctive blue color dots when visualized by light microscopy.

   In principle, the number of live (white) cells versus the number of dead (blue) cells is counted. The number of white cells/100 PBMC gives the viability in percentage.

   A more reliable way to ensure the viability of cells is detection by flow cytometry. In principle, PBMC should be resuspended in a flow cytometry staining buffer and thereafter cell count and viability analysis are performed. However, this requires more cells and thus may not be possible in limited sample sizes.

5. Cryopreservation of PBMC: The functional assays on cryopreserved PBMC are associated with viability of the cells. For practical reasons, it can be an advantage to collect cells during a limited period of time and perform all analyses collectively, post-cryopreservation, in order to overcome, e.g., inter-assay variation in the methodology analyzing cytokine secretion, e.g., by Luminex. Cryopreservation is a convenient way to handle PBMC. Actually, frozen cells are used in the great majority of all studies of cell-mediated immunity. We and others have been able to show that cryopreserved PBMC can be used, trustworthy, after cryopreservation for detection of

cytokines and chemokines [4] and Treg-associated markers [5] and also that cryopreserved PBMC maintain a stable expression of the T-regulatory marker FOXP3 (% and MFI) in the CD4+CD25hi cell population after cryopreservation [6]. Collectively, these studies show that cryopreserved PBMC can be used and still give trustworthy and useful information.

References

1. De Jager W, te Velthuis H, Prakken BJ et al (2003) Simultaneous detection of 15 human cytokines in a single sample of stimulated peripheral blood mononuclear cells. Clin Diagn Lab Immunol 10:133–139
2. Lagrelius M, Jones P, Franck K et al (2006) Cytokine detection by multiplex technology useful for assessing antigen specific cytokine profiles and kinetics in whole blood cultured up to seven days. Cytokine 33: 156–165
3. Bull M, Lee D, Stucky J et al (2007) Defining blood processing parameters for optimal detection of cryopreserved antigen-specific responses for HIV vaccine trials. J Immunol Methods 322:57–69
4. Axelsson S, Faresjö M, Hedman M et al (2008) Cryopreserved peripheral blood mononuclear cells are suitable for the assessment of immunological markers in type 1 diabetic children. Cryobiology 57:201–208
5. Kivling A, Nilsson L, Faresjö M (2009) How and when to pick up the best signal from markers associated with T-regulatory cells? J Immunol Methods 345:29–39
6. Rydén A, Faresjö M (2011) Efficient expansion of cryopreserved CD4+CD25+CD127lo/- cells in type 1 diabetes. Res Immunol 1:36–44

Part II

Cytokine Bioassays

Chapter 8

Control of Pro-inflammatory Cytokine Release from Human Monocytes with the Use of an Interleukin-10 Monoclonal Antibody

Hardik Patel and Dennis Davidson

Abstract

The monocytes (MONOs) can be considered as "double-edge swords"; they have both important pro-inflammatory and anti-inflammatory functions manifested in part by cytokine production and release. Although MONOs are circulating cells, they are the major precursors of a variety of tissue-specific immune cells such as the alveolar macrophage, dendritic cells, microglial cells, and Kupffer cells. Unlike the polymorphonuclear leukocyte, which produces no or very little interleukin-10 (IL-10), the monocyte can produce this potent anti-inflammatory cytokine to control inflammation. IL-10, on an equimolar basis, is a more potent inhibitor of pro-inflammatory cytokines produced by monocytes than many anti-inflammatory glucocorticoids which are used clinically. This chapter describes how to isolate monocytes from human blood and the use of IL-10 monoclonal antibody to determine the effect and timing of endogenous IL-10 release on the production and release of pro-inflammatory cytokines.

Key words Endotoxin, Interleukins, Immunoglobulin, RT-qPCR, ELISA, Antibody-labeled microbeads

1 Introduction

The predominant circulating leukocytes in the healthy human are polymorphonuclear leukocytes (PMNs) and lymphocytes. Monocytes generally represent less than 10 % of circulating leukocytes. Many studies use the term peripheral blood monocytic cells (PBMCs) to describe the combination of lymphocytes and monocytes which are mononuclear cells that can be separated from PMNs. To study the monocyte alone and to convert monocytes in vitro to other cell types in their lineage, e.g., macrophages, monocyte purity from human blood must be very high [1–3].

To study the role of cytokines produced by inflammatory cells in vitro one must choose a clinically significant stimulus in a physiologic dose and at an appropriate time after stimulation.

Ivana Vancurova (ed.), *Cytokine Bioassays: Methods and Protocols*, Methods in Molecular Biology, vol. 1172,
DOI 10.1007/978-1-4939-0928-5_8, © Springer Science+Business Media New York 2014

Endotoxin in the dose range of 1–100 ng/mL of cell culture media is a commonly used stimulant used throughout the literature [4–6]. Some cytokines are released early, such as IL-8, while others may be released after an hour or more, such as interleukin-10 (IL-10) [7]. Accordingly, the investigator should start with some pilot studies that will examine submaximal and physiologically important stimuli under their experimental conditions. Unlike PMNs that undergo apoptosis in standard culture media in the first 12–24 h, monocytes survive much longer and can be studied for at least 18 h with high viability.

The use of a monoclonal antibody to a cytokine such as IL-10 can be used in cell culture to understand the relative importance of IL-10 release on other cell functions such as pro-inflammatory cytokine release. Inhibition of IL-10 will eliminate the negative control on pro-inflammatory cytokine release. To determine the concentration of antibody, a dose–response study is needed to find a submaximal concentration of antibody that will enhance pro-inflammatory cytokine release (e.g., IL-6) without interfering with the ELISA assay. To determine that the monoclonal antibody is not having an adverse effect on ELISA measurements, a control with an equal concentration of IgG is used in all experiments.

2 Materials

2.1 Monocyte Isolation

1. Clinical supplies for venipuncture including heparinized preservative-free clinical test tubes (*see* **Note 1**).
2. Ficoll-Paque PLUS (GE Healthcare Life Sciences).
3. Phosphate-buffered saline (PBS).
4. Percoll Gradient: 24.25 mL Percoll (GE Healthcare Life Sciences), 20.75 mL cell culture water, and 5 mL 1.6 M NaCl.
5. 15 mL Conical test tubes.
6. 50 mL Tubes.
7. RBC 10× lysing solution: 1.55 M NH_4CL, 100 mM $NaHCO_3$, 1 mM EDTA, pH 7.4.
8. MACS monocyte isolation kit II (Miltenyi Biotec, Auburn, CA).
9. CD15 microbeads (Miltenyi Biotec, Auburn, CA).
10. MACS buffer: PBS with 2 mM EDTA and 1 % fetal calf serum (FCS).
11. 0.4 % Trypan blue solution.
12. Hemocytometer.
13. Light microscope.
14. Handheld cell counter.

2.2 Neutralizing Anti-IL-10 Antibody

1. IL-10 monoclonal antibody (MAB 2171, R&D Systems).
2. Mouse IgG2B isotype control (MAB004, R&D Systems).
3. Lipopolysaccharide (L4391 from *Escherichia coli* 0111:B4; Sigma-Aldrich, St. Louis, MO, USA).
4. PBS.
5. RPMI 1640 medium supplemented with 10 % FSC.
6. 6-Well culture plates, BD Falcon* tissue culture plates.
7. Incubator with 5 % CO_2.
8. Human Interleukin-6 Quantikine ELISA Kit (R&D Systems).
9. Human Interleukin-8 Quantikine ELISA Kit (R&D Systems).
10. ELISA plate reader.

3 Methods

3.1 Monocyte Isolation (Fig. 1) (See Notes 2 and 3)

3.1.1 Blood Collection

1. Obtain IRB approval and informed consent for venipuncture (minimal risk human study) from healthy adult volunteers or cord blood from discarded placental immediately after birth.
2. Collect approximately 50 mL of venous blood (*see* **Note 4**).

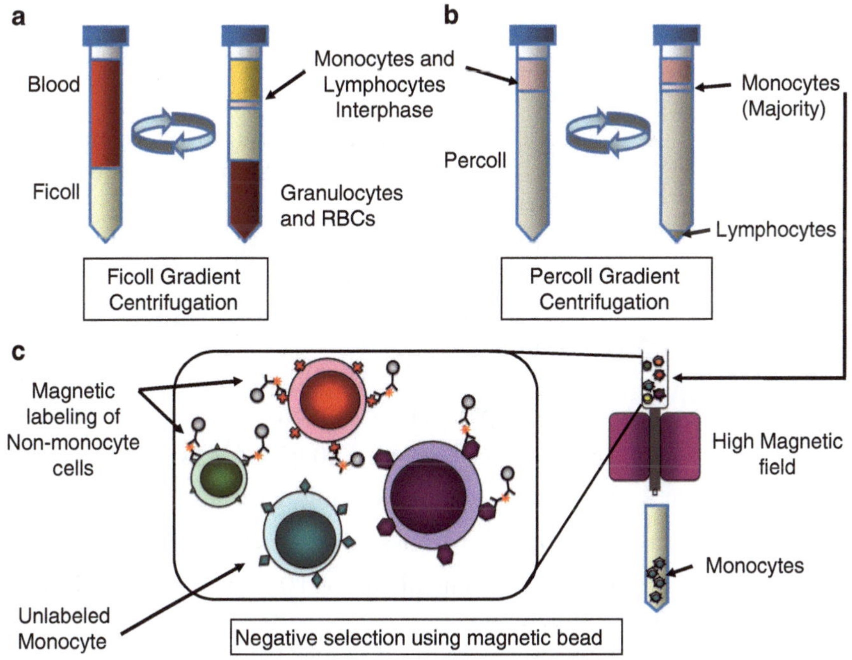

Fig. 1 Monocyte isolation. *Panel A* shows the separation of monocytes and lymphocytes (buffy coat) from polymorphonuclear leukocytes and RBCs during the Ficoll gradient centrifugation. *Panel B* shows separation of monocyte-enriched fraction from lymphocytes during Percoll gradient centrifugation. *Panel C* shows negative selection of monocytes using an antibody cocktail and magnetic bead cell separation of pure monocytes

3.1.2 Ficoll Gradient Centrifugation and Red Blood Cell Lysis

1. In a 15 mL tube, carefully layer 8 mL of heparinized blood over 5 mL of Ficoll-Paque Plus using a squeeze pipette.
2. Centrifuge at 400×*g* for 40 min.
3. Remove the interphase containing lymphocytes and monocytes into a new 15 mL tube (Fig. 1a).
4. Fill up the tube to 13 mL with PBS and centrifuge at 200×*g* for 10 min.
5. Discard supernatant, add 1× red blood cell lysing solution up to 13 mL mark, resuspend cell pellet, invert the tube 20 times for over 20 s, and centrifuge at 200×*g* for 10 min.
6. Discard supernatant, and add PBS up to 13 mL mark. Resuspend cell pellet gently, and invert five times to rinse cells. Centrifuge at 200×*g* for 10 min.
7. Discard supernatant, resuspend cell pellet in residual liquid, pool cells into one tube, and add PBS up to 13 mL mark. Remove 50 μL of cell suspension for counting, and centrifuge at 200×*g* for 10 min (*see* **Note 5**).

*3.1.3 Percoll Gradient Centrifugation (See **Note 6**)*

1. Discard supernatant. Resuspend cells at 50–60×10^6 cells/mL in medium.
2. Mix Percoll gradient solution very well before use (*see* **Note 7**). In a 15 mL tube, very carefully layer cells over Percoll gradient (*see* **Note 8**): 3 mL of cell suspension over 10 mL of Percoll or 1 mL of cell suspension over 3 mL of Percoll.
3. Spin at 580×*g* for 15 min.
4. Collect cells from the interphase into a 15 mL tube. Fill the tube up to 13 mL with PBS. Spin at 100–200×*g* for 10 min (Fig. 1b).
5. Discard supernatant. Add PBS up to 13 mL mark, resuspend cells, and invert five times to rinse cells. Spin as above. Repeat wash. Before the last wash, remove 50 μL of cell suspension for cell counting.

*3.1.4 Magnetic Microbead Monocyte Isolation (See **Note 9**)*

1. Aspirate supernatant completely.
2. Resuspend cell pellet in 30 μL of MACS buffer per 10^7 cells.
3. Add 10 μL of FcR blocking reagent per 10^7 cells.
4. Add 10 μL of biotin–antibody cocktail per 10^7 cells (*see* **Note 10**).
5. Mix well, and incubate for 10 min at 4–8 °C. Keep in the dark (refrigerator).
6. Add 10 μL of MACS buffer per 10^7 cells.
7. Add 20 μL of CD15 microbead per 10^7 cells.
8. Add 20 μL of anti-biotin microbead per 10^7 cells.
9. Mix well, and incubate as above for 15 min at 4–8 °C.

10. Wash cells by filling up the tube to 13 mL with MACS buffer. Invert few times to rinse cells, and then spin at 300×*g* for 10 min.
11. Aspirate the supernatant completely, and resuspend up to 10^8 cells in 500 μL of MACS buffer.
12. Proceed to magnetic separation with LS columns.
13. Place column in the magnetic field of a suitable MACS separator.
14. Prepare column by rinsing with 3 mL of MACS buffer. Use clean tube for collecting cells.
15. Apply cell suspension onto the column. Allow the cells to pass through, and collect effluent as fraction with unlabeled cells, representing the enriched monocyte fraction.
16. Wash column with 3 mL of MACS buffer. Perform washing steps by adding MACS buffer three times total, each time once the column reservoir is empty.
17. Collect the entire effluent in the same tube. This fraction represents the enriched monocyte cells.
18. Invert cells few times to mix well, and remove 50 μL for cell counting. Spin down the rest at 300×*g* for 10 min.
19. Count cells, and resuspend monocytes at 1×10^6 cells/mL (or desired concentration) in RPMI with 10 % FCS.
20. Monocyte viability and purity should be measured (*see* **Note 11**).

3.2 Antibody Neutralization (See Note 12)

1. Aliquot 1 mL of the monocyte suspension into wells of a 6-well culture plate (10^6 cells per well).
2. Incubate cells with IL-10 monoclonal antibody or IgG control antibody for 1 h (*see* **Note 13** to determine the concentration of antibodies).
3. Stimulate cells with LPS (10 ng/mL) and incubate for 18 h (*see* **Note 14**).
4. Collect cell culture supernatant after 18 h.
5. To measure cytokines by ELISA, usually 50–200 μL are aliquoted at specific time points and stored at −80 ° C.

4 Notes

1. Alcohol swab, tourniquet, 21 gauge butterfly needle, vacutainer 10 mL heparinized tubes (the same used clinically for complete blood counts), bandaid.
2. All reagents should be cell culture tested, endotoxin-free.
3. All steps should be performed at room temperature unless otherwise noted.

4. Monocytes represent anywhere between 0 and 9 % of circulating white blood cells. 50 mL of whole blood will yield on average 10^7 monocytes. Umbilical cord blood usually yields 50 % more monocytes than from adult whole blood. (There will be considerable variation in the number of monocytes collected based on the donor individual's white blood cell differential count. A lower number of monocytes per well can probably be used after the investigator's pilot experiments).
5. For cell counting: Incubate cell suspension with 0.4 % trypan blue dye (1:1) for 5 min. Load 10 μL of the stained cell suspension onto the hemocytometer, and count the cells under the light microscope. If there are too many cells (>100 cells/field), dilute the cell suspension accordingly before adding trypan blue.
6. Percoll gradient is performed when using cord blood only; this step should be skipped for adult blood.
7. Percoll gradient has to be at room temperature for use.
8. Cell suspension should be layered on the Percoll very slowly and carefully; failure to do so will result in lymphocyte contamination.
9. Remove all PBS before adding MACS buffer; keep MACS buffer on ice; keep cells during the labeling incubation at 4 °C in the dark (keep in refrigerator).
10. In order to yield good purity, make sure to use accurate ratio of cells and beads.
11. Flow cytometry is recommended for CD14+ cells to detect purity of monocytes which should be >90 %. Viability can be assessed by trypan blue exclusion and should be >95 %.
12. Experiments are done on the same day of isolation.
13. To determine the amount of IL-10 monoclonal antibody using a dose–response curve is necessary using five- to tenfold increments of antibody in LPS-stimulated MONOs. The endpoint will be a pro-inflammatory cytokine such as IL-8 (Fig. 2) and IL-6 (Fig. 3). The antibody is having an effect if IL-6 or IL-8 levels increase against a control sample due to the elimination of the negative feedback on pro-inflammatory cytokine release. Choose a submaximal dose of IL-10 monoclonal antibody to eliminate any nonspecific effects on the experimental assays. Use the same concentrations of IgG in parallel samples with the IL-10 antibody to also serve as a control for nonspecific effects. There should be no rise in pro-inflammatory cytokines with or without IgG as shown in Fig. 2.
14. If using LPS for monocyte stimulation, a suggested physiologic concentration would be 10 ng/mL of culture media [6].

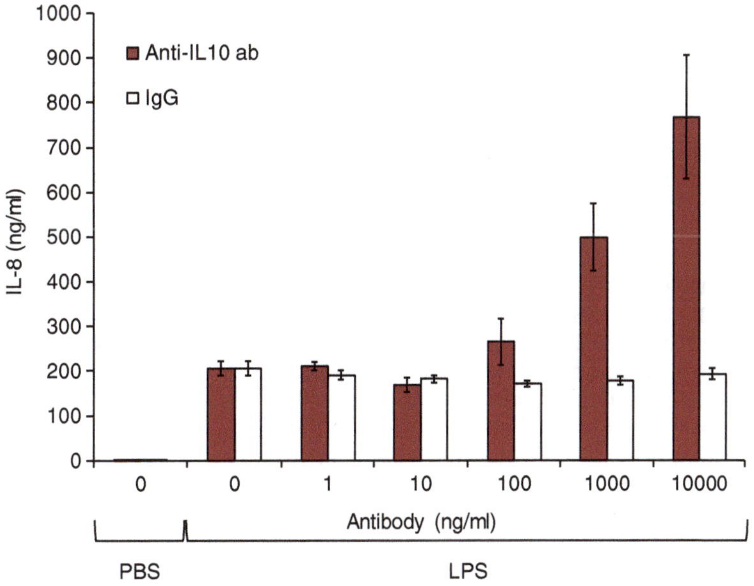

Fig. 2 LPS-induced IL-8 release with inhibition of endogenous IL-10 cytokine release using a monoclonal IL-10 antibody. Human monocytes (10^6/mL) were incubated with various doses of either IL10 monoclonal antibody (*solid bars*) or control IgG monoclonal antibody (*open bars*) for 1 h. Cells were than stimulated with either PBS control or LPS. IL-8 secretion was measured in cell culture supernatant after 18 h by ELISA

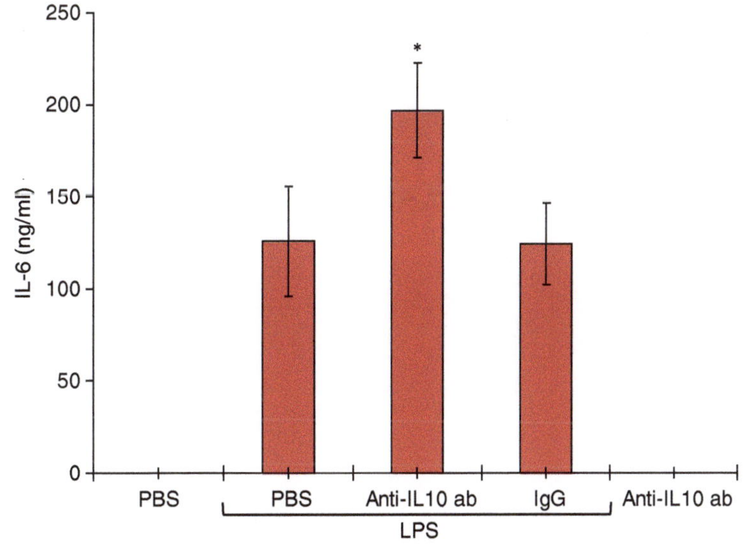

Fig. 3 LPS-induced IL-6 release with inhibition of endogenous IL-10 cytokine release using a monoclonal IL-10 antibody. Human monocytes (10^6/mL) were incubated with either IL-10 monoclonal antibody (1,000 ng/mL) or control IgG monoclonal antibody (1,000 ng/mL) for 1 h. Cells were than stimulated with either PBS control or LPS. IL-6 secretion was measured in cell culture supernatant after 18 h by ELISA. *Asterisk* indicates significant difference ($p < 0.01$) in IL-6 production between LPS-stimulated monocytes with or without IL-10 monoclonal antibody incubation

References

1. Gordon S, Taylor PR (2005) Monocyte and macrophage heterogeneity. Nat Rev Immunol 5:953–964
2. Davidson D, Patel H, Degoy AC et al (2013) Differential effect of exogenous interleukin-10 versus glucocorticoids on gene expression and pro-inflammatory cytokine release by polymorphonuclear leukocytes and monocytes of the newly born. Am J Transl Res 5:103–115
3. Chusid LA, Pereira-Argenziano L, Miskolci V et al (2010) Transcriptional control of cytokine release from monocytes of the newborn: effects of endogenous and exogenous interleukin-10 versus dexamethasone. Neonatology 97:108–116
4. Xing L, Remick DG (2003) Relative cytokine and cytokine inhibitor production by mononuclear cells and neutrophils. Shock 20:10–16
5. Akagawa KS (2002) Functional heterogeneity of colony-stimulating factor-induced human monocyte-derived macrophages. Int J Hematol 76:27–34
6. Romero R, Roslansky P, Oyarzun E et al (1988) Labor and infection. II. Bacterial endotoxin in amniotic fluid and its relationship to the onset of preterm labor. Am J Obstet Gynecol 158: 1044–1049
7. Davidson D, Miskolci V, Clark DC et al (2007) Interleukin-10 production after pro-inflammatory stimulation of neutrophils and monocytic cells of the newborn. Comparison to exogenous interleukin-10 and dexamethasone levels needed to inhibit chemokine release. Neonatology 92:127–133

Chapter 9

Cytokine-Induced Neutrophil Chemotaxis Assay

Dennis Davidson and Hardik Patel

Abstract

Chemotaxis is directed migration of a cell type to a distant chemoattractant. When this chemoattractant is a cytokine, the term chemokine is often used. Chemotaxis by neutrophils, specifically polymorphonuclear leukocytes (PMNs), plays a critical role in the innate immune response. On an equimolar basis, interleukin-8 (IL-8) is one of the most potent PMN chemokines known. This chapter describes an in vitro chemotaxis technique using PMNs and IL-8 which can serve the investigator as an established model from which new studies can be developed.

Key words Polymorphonuclear leukocyte, Neutrophil, Chemokine, Interleukin-8, Chemotaxis chamber, Cytokine, Chemotaxis assay

1 Introduction

The innate immune response, specifically the defense against a foreign agent (e.g., endotoxin, bacterial wall protein), usually begins with the activation of a tissue macrophage to produce a wide variety of mediators. The release of IL-8, one of the most potent chemokines produced by macrophages, then leads to the recruitment of neutrophils from the circulation to amplify the inflammatory response [1]. The inflammatory response can lead to host defense or tissue injury. Important clinical disorders involving this sequence of cellular events in the lung include adult respiratory distress syndrome (ARDS) and the initial phase of bronchopulmonary dysplasia (chronic lung disease) in premature infants [2, 3]. Using the chemotaxis assay described below, the investigator can assess the effect of various drugs and combination of endogenous mediators or the effect of development on chemotaxis. For example, virtually all studies comparing polymorphonuclear leukocyte (PMN) chemotaxis of the newborn to the PMN chemotaxis of the adult have demonstrated that chemotaxis is appreciably depressed for the newborn. This means that for the same concentration of chemokine (e.g., interleukin-8, IL-8) one may see more than 50 % less PMNs

Ivana Vancurova (ed.), *Cytokine Bioassays: Methods and Protocols*, Methods in Molecular Biology, vol. 1172,
DOI 10.1007/978-1-4939-0928-5_9, © Springer Science+Business Media New York 2014

of the newborn migrating under the assay conditions described below compared to PMNs of the adult [4]. These findings correlate with the clinical observation that the newborn is more vulnerable to infection than the adult [5]. The chemotaxis assay can also be used to test the effect of drugs on chemotaxis. For example the potent anti-inflammatory drug, dexamethasone (a glucocorticoid), does not appear to affect chemotaxis of PMNs directed to a chemokine such as IL-8. However, PMNs also produce IL-8 and its production is inhibited by dexamethasone. Using the chemotaxis assay system, it has been shown in vitro that the amplification of neutrophilic inflammation can be suppressed by dexamethasone for the adult and newborn [6, 7].

When quantitating chemotaxis in vitro a multiwell chemotaxis chamber is required. Chemotaxis is actually the total migration of cells from which the chemokinesis or random migration is subtracted. Migrating cells in culture media, in this case, a specific number of PMNs are placed in upper chambers. A membrane with a specific pore size is placed under this chamber through which the PMNs migrate, during an incubation period, and adhere to the lower surface where they can be counted. The chemokine in culture media (or biologic fluid), in this case IL-8, is placed in a lower chamber below the filter to give total migration. Chemokinesis is determined by placing the same concentration of IL-8 in the upper and lower chamber. Accordingly two sets of upper and lower wells are needed to quantitate chemotaxis. Since most studies will require dose (chemoattractant concentration) and chemotaxis responses, the possible comparison to other chemoattractants or pharmacologic interventions, and each condition to be done in triplicates, investigators will usually use a 48- or a 96-well micro-chemotaxis chamber.

2 Materials

2.1 Neutrophil Isolation (See Note 1)

1. Clinical supplies for venipuncture (*see* **Note 2**).
2. Ficoll-Paque PLUS (GE Healthcare Life Sciences).
3. 15 mL Conical test tubes.
4. Squeeze pipette (2 mL).
5. Benchtop centrifuge.
6. Phosphate buffer saline (PBS).
7. 5 % Dextran: 5 g Dextran in 100 mL of PBS.
8. RBC 10× lysing solution: 1.55 M NH_4Cl, 100 mM $NaHCO_3$, 1 mM EDTA, pH to 7.4.
9. Cell culture-tested water.
10. RPMI 1640 medium.
11. Fetal calf serum (FCS).

12. Trypan blue.
13. Hema 3 stain set.
14. Hemocytometer.
15. Light microscope with ocular grid.
16. Handheld cell counter.

2.2 Chemotaxis Assay and Accessories

1. Three tiered, 48 well, microchemotaxis chamber (Neuroprobe).
2. 2 and 5 μm pore polycarbonate filter membrane (Neuroprobe).
3. Centrifuge holders for lower chambers (for "drop-off" correction).
4. Large glass microscope slide for membrane placement.
5. Recombinant human IL-8 (*see* **Note 3**).
6. Tissue culture CO_2 incubator.

3 Methods

3.1 PMN Isolation (Fig. 1)

3.1.1 Blood Collection

1. Obtain IRB approval and informed consent for venipuncture (minimal risk human study) from healthy adult volunteers or cord blood from a discarded placenta immediately after birth.
2. Collect approximately 8 mL of venous blood (*see* **Note 4**).
3. In a sterile 15 mL tube, carefully layer 8 mL of heparinized blood (undiluted) over 5 mL of Ficoll-Paque PLUS using a pyrogen-free squeeze pipette.
4. Centrifuge at 400 × *g* for 40 min.
5. Discard layers covering the bottom red layer using squeeze pipette.
6. Pour PBS into red layer (RBC and PMN) to make 10 mL volume.
7. Mix thoroughly using a new squeeze pipette.

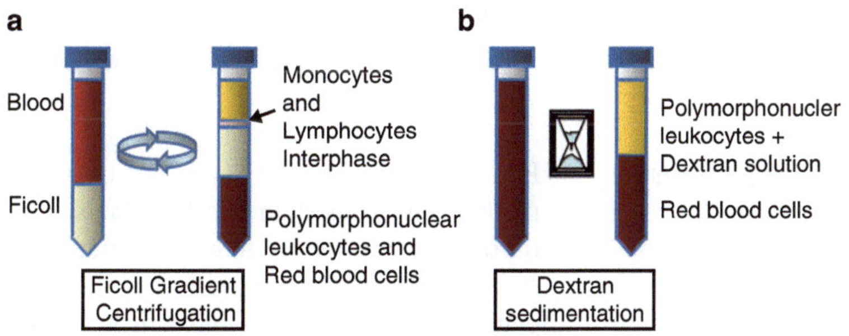

Fig. 1 Overview of PMN isolation technique. *Panel A* shows the separation of mononuclear cells from polymorphonuclear leukocytes using density gradient centrifugation. *Panel B* shows the separation of polymorphonuclear leukocytes from red blood cells by sedimentation

3.1.2 Dextran Sedimentation

1. In a sterile 50 mL tube, mix 5 mL of 5 % dextran and the above PMN suspension.
2. Mix thoroughly using same pipette.
3. Let PMNs settle by gravity for 60 min.
4. Transfer (yellow) supernatant to a 15 mL tube using new squeeze pipette (*see* **Note 5**).
5. Centrifuge at 300 × *g* for 5 min.
6. Pour off the supernatant to obtain a pellet.
7. Resuspend pellet in residual fluid by using pipette.

3.1.3 Red Cell Lysis

1. Prepare 1× lysing solution by mixing 10× lysing solution with sterile water, e.g., 3 mL of 10× solution + 27 mL of sterile water.
2. Add to the resuspended pellet 1× lysing solution up to the 13 mL mark.
3. Gently mix by inverting capped 15 mL tube 30×.
4. Centrifuge at 300 × *g* for 5 min.
5. Discard supernatant.

3.1.4 Washing

1. Add PBS to make 13 mL mark.
2. Centrifuge at 300 × *g* for 5 min.
3. Pour off supernatant.
4. Resuspend pellet in residual fluid using a new squeeze pipette.
5. Repeat washing twice more.

3.1.5 Resuspending Cells

1. Prepare 10 % FCS–RPMI 1640 medium (1 mL FCS + 9 mL RPMI) during the last wash.
2. After the last wash, discard supernatant and resuspend in 2 mL of 10 % FCS–RPMI.
3. Count cells, and dilute cells to volume so that concentration of cells is 3×10^6 cells/mL.
4. Continue to differential staining, and count (*see* **Note 6**).
5. Viability can be checked using trypan blue exclusion staining (*see* **Note 7**).

3.2 Chemotaxis Assay (Fig. 2)

Figure 2 is a schematic view of two adjacent wells in a three-tier microchemotaxis chamber. For PMNs, 50 μL (5×10^4 PMNs) can be used with a 5 μm pore filter membrane overlying the upper chamber. A 2 μm pore filter membrane is placed between the middle and lower plate to trap any neutrophils which "drop off" from the underside of the 5 μm pore filter membrane into the lower well. Once assembled, the chamber is placed in a 37 °C humidified incubator for 60 min.

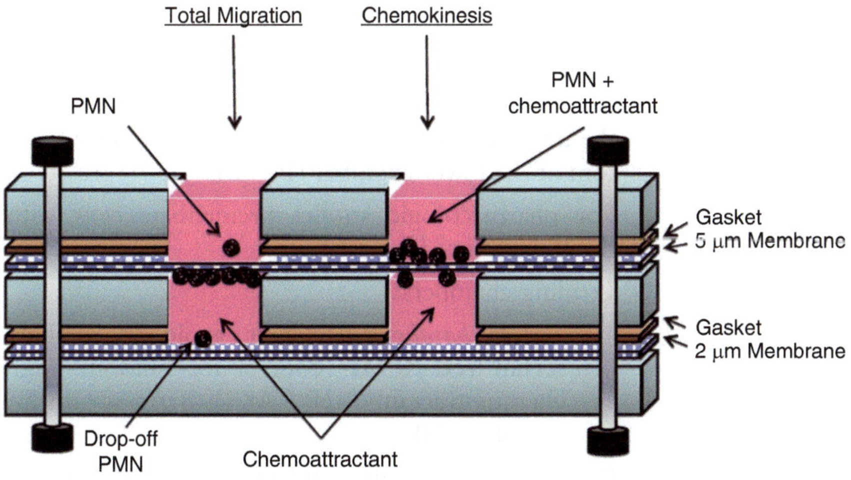

Fig. 2 Schematic 2-well section of a 48-well chemotaxis chamber. PMNs placed in the upper chamber will migrate down to the underside of the 5 μm pore filter membrane, which is then prepared for staining and cell counting. Usually less than 4 % of the migrating PMNs will drop off to the top of the 2 μm pore filter, through which they cannot pass. Chemotaxis (directed cell migration) is determined by subtracting total migration from chemokinesis (random cell migration). Note that for determining chemokinesis, the same concentration of chemoattractant must be in both the *upper* and *lower* wells

The underside of the 5 μm pore membrane contains the migrated PMNs. The upper side of the membrane will have PMNs that did not migrate. The upper side of the membrane is wiped off along a firm rubber straight edge supplied by the manufacturer before the membrane is stained. Blue-colored circles on the membrane can often be seen with the naked eye corresponding to the well when there is appreciable cell migration. The 5 μm pore membrane is then mounted on a glass slide for counting under the microscope. Cell counts are expressed as the number of cells per square millimeter or high-power field. Counting should be done from several areas within a well impression on the membrane, and the total number of cells can be extrapolated from several hundred cells counted.

The three-tiered microchemotaxis chamber is recommended to correct for possible drop-off PMNs from the underside of the membrane into the lower well. After the upper block of wells and 5 μm pore membrane is removed for staining, the combined lower and middle blocks are spun in specialized centrifuge holders as per the manufacturer's instructions. Drop-off is usually under 4 % of the cell counted on the 2 μm pore membrane for each well. PMNs cannot migrate through the 2 μm pore membrane. The number of cells that drop off can be added to the number counted on the undersurface of the 5 μm pore membrane (considered migrated cells). Results are usually displayed by showing total migration divided into its components of chemotaxis and chemokinesis [4] (*see* **Note 8**).

The protocol is described below:

1. Chemotaxis can be assessed in a 48-well chemotaxis chamber from Neuro Probe, with a 5 μm polycarbonate filter between the upper and lower chamber and a 2 μm filter lining the bottom of the lower chamber (Fig. 2).
2. For pharmacologic studies preincubate cells with selected drugs (*see* **Note 9**).
3. Loading chambers:
   (a) *Lower chamber* (load first, avoid air bubbles): 50 μL serum-free RPMI (control for random migration) or chemoattractants—10^{-8} M IL-8 (in serum-free RPMI) (*see* **Note 10**).
   (b) Cover lower wells with 5 μm membrane, gasket, and upper chamber assembly as per the manufacturer's instructions.
   (c) *Upper chamber*: 50 μL of preincubated cell suspension (3×10^6 cells/mL), containing the inhibitors, +/− chemoattractants (with chemoattractant for chemokinesis, without chemoattractant for total migration).
4. Incubate chamber for 1 h at 37 °C.
5. Discard the upper chamber contents by inverting the chamber on clean paper towels.
6. Disassemble upper chamber, gasket, and 5 μm pore membrane as per the manufacturer's instructions.
7. Wipe off non-migrated PMNs on the upper side of the 5 μm pore membrane as per the manufacturer's instructions.
8. Allow membrane to air-dry, and stain with Hema 3.
9. Centrifuge the bottom two-tier assembly with lower well content (400–500 × *g* for 10 min, RT) to collect drop-off cells on the 2 μm membrane.
10. Disassemble chamber, air-dry the 2 μm pore membrane, and stain with Hema 3.
11. Count cells on the filters under a light microscope. Neutrophil migration can be expressed as PMNs/high-power field.
12. Clean the chamber as per the manufacturer's instructions (*see* **Note 11**).

4 Notes

1. Neutrophils undergo apoptosis in 12–18 h; therefore, isolation and chemotaxis should be done on the same working day. All reagents should be cell culture tested, endotoxin-free.

2. Alcohol swab, tourniquet, 21 G butterfly needle, vacutainer 10 mL heparinized tubes (the same used clinically for complete blood counts), band-aid.
3. fMLP is a commonly used chemoattractant than can be used as a control in the same molarity as IL-8.
4. 8 mL of adult human blood will yield enough PMNs to do a 48-well experiment.
5. Be careful not to include any red layer; otherwise, it will result in too many contaminating red cells.
6. Hema 3 staining (standard method for white blood cell differential counting): Most of the contaminating white blood cells will be <5 % eosinophils, and the remainder should be almost all PMNs.
7. For cell counting: Incubate cell suspension with 0.4 % trypan blue dye (1:1) for 5 min. Load 10 μL of the stained cell suspension onto the hemocytometer, and count the cells under the light microscope. If there are too many cells (>100 cells/field), dilute the cell suspension accordingly before adding trypan blue.
8. For IL-8 the large majority of the PMN total migration is due to chemotaxis [4].
9. For pharmacological studies: Preincubate cells for 1 h +/− with selected drugs at appropriate concentration at 37 °C in pyrogen-free test tubes; control for no inhibitor was serum-free RPMI.
10. There should be an upward meniscus after adding 50 μL into the lower well which will help avoid air bubbles under the 5 μm pore membrane.
11. It is very important to soak the chambers overnight in *enzymatic* detergent to eliminate all contaminating proteins that might interfere with the next assay.

References

1. Kunkel SL, Lukacs N, Strieter RM (1995) Expression and biology of neutrophil and endothelial cell-derived chemokines. Semin Cell Biol 6:327–336
2. Pierrakos C, Karanikolas M, Scolletta S et al (2012) Acute respiratory distress syndrome: pathophysiology and therapeutic options. J Clin Med Res 4:7–16
3. Jobe AH (2011) The new bronchopulmonary dysplasia. Curr Opin Pediatr 23: 167–172
4. Tan ND, Davidson D (1995) Comparative differences and combined effects of interleukin-8, leukotriene B4, and platelet-activating factor on neutrophil chemotaxis of the newborn. Pediatr Res 38:11–16
5. Pahwa SG, Pahwa R, Grimes E et al (1977) Cellular and humoral components of monocyte and neutrophil chemotaxis in cord blood. Pediatr Res 11:677–680
6. Strieter RM, Kasahara K, Allen RM et al (1992) Cytokine-induced neutrophil-derived interleukin-8. Am J Pathol 141:397–407
7. Zentay Z, Sharaf M, Qadir M et al (1999) Mechanism for dexamethasone inhibition of neutrophil migration upon exposure to lipopolysaccharide in vitro: role of neutrophil interleukin-8 release. Pediatr Res 46:406–410

Chapter 10

An In Vitro One-Dimensional Assay to Study Growth Factor-Regulated Tumor Cell–Macrophage Interaction

Ved P. Sharma, Brian T. Beaty, Dianne Cox, John S. Condeelis, and Robert J. Eddy

Abstract

Growth factor-dependent pairing and motility between tumor cells and tumor-associated macrophages on extracellular matrix (ECM) fibers of the tumor microenvironment have been shown to enhance intravasation and metastatic spread of breast carcinomas. We describe an in vitro motility assay that combines time-lapse wide-field microscopy and micro-patterned linear adhesive substrates to reconstitute the in vivo behavior between macrophages, tumor cells, and ECM fibers in orthotopic rodent tumor models observed by intravital imaging. Commercially available linear stripes of 650 nm dye-labeled fibronectin microlithographed onto glass cover slips are sequentially plated with fluorescently labeled MTLn3 tumor cells and bone marrow-derived macrophages and time-lapse imaged for up to 8 h. Incubation with pharmacological inhibitors during the assay can identify important paracrine or autocrine signaling pathways involved in the macrophage–tumor cell interaction. This high-resolution motility assay will lead to a more detailed description of immune cell–tumor cell behavior as well as interrogating additional cell types within the tumor microenvironment which use cytokine/growth factor paracrine signaling interactions to facilitate intravasation and metastasis.

Key words EGF, CSF-1, Breast carcinoma cells, Macrophage, Live-cell imaging, Time-lapse microscopy

1 Introduction

The establishment of distant secondary tumors or metastases from the primary tumor is correlated with poor prognosis in breast cancer patients. During metastasis, cells escape the primary tumor, undergo invasion by migrating through the surrounding tissue stroma, and eventually intravasate through the vascular endothelium. Once in the circulation, the tumor cells can extravasate and seed the formation of secondary tumors in various target organs [1]. Although tumor cells can undergo random and directional motility, the process of invasion and dissemination are more efficient when the tumor cells undergo directed cell migration [2].

Ivana Vancurova (ed.), *Cytokine Bioassays: Methods and Protocols*, Methods in Molecular Biology, vol. 1172, DOI 10.1007/978-1-4939-0928-5_10,

The phenomenon of chemotaxis, defined as directional migration towards an increasing gradient of a chemokine or a growth factor, is a common form of directed cell migration occurring in tumors. Chemotaxis not only influences the migratory behavior of tumor cells but also serves to define the tumor microenvironment [1].

The migratory behavior of individual cells and cell–cell interactions within the tumor microenvironment can be observed in vivo with high spatial and temporal resolution by intravital multiphoton microscopy. Intravital imaging of mouse mammary tumors has revealed that tumor cell migration is coordinated by the surrounding tumor microenvironment composed of a variety of cell types including fibroblasts, adipocytes, vascular endothelial cells, and infiltrating immune cells [1]. Recently, a role has been uncovered for macrophages during both in vitro and in vivo mammary tumor cell invasion and metastasis that is dependent on a paracrine interaction between epidermal growth factor (EGF) secreted by macrophages and colony-stimulating factor (CSF-1) secreted by tumor cells. Inhibitors of EGF or CSF-1 signaling disrupt this interaction and decrease tumor cell velocity and protrusion, validating the requirement for an intact paracrine loop [3–5].

In addition to the cellular component, the tumor microenvironment comprises the extracellular matrix secreted by various tumor-associated cell types including fibroblasts and endothelial cells. Intravital imaging using second harmonic generation has revealed that extracellular matrix molecules, such as type I collagen, are arranged into a network of linear fibers approximately 2–5 μm in width. MTLn3 tumor cells expressing GFP not only increase their velocity on collagen fibers, but in addition they exhibit a coordinated streaming motility in close association with these fibers [6, 7]. Intriguingly, these steams were found to contain co-migrating macrophages organized into arrays of alternating tumor cells and macrophages [1] while the migration of tumor cells within streams is more efficient than random motility with accompanying increases in various cell motility parameters including velocity, net path length, and directionality [2, 4, 7]. The EGF/CSF-1 paracrine loop is critical for the co-migration of tumor cells and macrophages in streams and is thought to play a significant role in organizing and promoting directed migration to the vasculature and subsequent intravasation during metastasis [2, 8].

In order to further investigate the ability of the tumor microenvironment to organize macrophage–tumor cell streaming behavior observed by intravital imaging [8], we developed an in vitro motility assay that employs a linear micro-patterned adhesive stripe to mimic the fibrillar nature of the tumor extracellular matrix. We found that MTLn3 cells plated onto 2.5 or 5 μm wide stripes of fibronectin or type I collagen micro-patterned onto a glass cover slip supported high-velocity cell motility at rates comparable to in vivo values measured on collagen fibers of similar width.

We observed that MTLn3 cells plated on fibronectin or type I collagen stripes migrated with higher velocity than on planar matrix-coated substrates and displayed enhanced lamellipodial protrusion and increased motility upon local interaction and pairing with bone marrow-derived macrophages (BMMs) [9].

In this chapter, we describe technical details on the use and application of commercially available and customizable cover slips micro-patterned with stripes of extracellular matrix (ECM) of varying widths as an in vitro motility assay. This simplified approach uses conventional time-lapsed, wide-field fluorescence microscopy and is designed to analyze multiple adjacent fields simultaneously to maximize data collection from a single micro-patterned cover slip. Furthermore, this assay can be used to quantify individual cell motility parameters such as speed, persistence, lamellipodial protrusion, and retraction at high resolution as well as explore the effect of various ECM molecules on these parameters. Our findings validate the use of linear micro-patterned substrates to reconstitute in vivo tumor cell–macrophage pairing and stream formation and can be applied to autocrine and paracrine chemokine/growth factor signaling in a variety of tumor-associated cells.

2 Materials

2.1 Preparation of CYTOO Motility Chips

1. CYTOOchip Motility FN650 (CYTOO Inc., cat# 10-031-10-06).
2. CYTOOchip Motility A (CYTOO Inc., cat# 10-031-00-06).
3. CYTOOchamber 4 wells (CYTOO Inc., cat# 30-011).
4. Silicone glue (DAP Aquarium Sealant, 100 % silicone).
5. Glass-bottom 35 mm petri dish (MatTek, cat# P35G-1.5-14-C).
6. Razor blade.
7. Phosphate buffer saline (PBS).
8. FITC-conjugated bovine collagen I (Sigma) solution: Dilute FITC-conjugated bovine collagen in PBS to 40 μg/ml.

2.2 MTLn3 Preparation

1. MTLn3 [10]: TagRFP-cortactin-expressing MTLn3 [11].
2. Culture medium: MEM-alpha (Life Technologies) containing 5 % fetal bovine serum (FBS) and penicillin/streptomycin solution.
3. 0.05 % Trypsin–EDTA.
4. Imaging medium: L-15 (Life Technologies, cat# 21083) containing 5 % FBS.

2.3 BMM Preparation

1. Murine BMMs [12].
2. BMM culture medium: MEM-alpha (Life Technologies) containing 15 % FBS, 1 % penicillin/streptomycin, and 36 ng/ml (2.5 nM) recombinant human CSF-1.

3. Bacterial petri dishes (Falcon).
4. CellTracker™ Green CMFDA (Life Technologies, cat# C7025): Each vial of CellTracker™ Green contains 50 μg dye. Make 5 mM stock solution by adding 21.5 μl of DMSO and mixing by pipette 3–4 times. The stock solution can be kept at −20 °C for up to 3 months.

2.4 Live Imaging of Tumor Cell–Macrophage Interaction

1. Imaging medium: L-15 medium (Life Technologies) containing 5 % FBS.
2. EGFR inhibitor, Iressa (AstraZeneca).
3. EGFR inhibitor, Tyrphostin AG 1478 (Cell Signaling Tech., cat#9842).
4. MCSF receptor antibody, AFS98 (Novus Biologicals, cat# NBP1-43363).
5. Epi-fluorescence microscope (DeltaVision), equipped with a 20x objective, a heat enclosure to maintain temperature at 37 °C, a CoolSNAP HQ^2 CCD camera and a high precision X and Y scanning stage for capturing time-lapsed images from multipoint locations.

3 Methods

3.1 CYTOOchip-MatTek Dish Assembly

A CYTOO cover slip, with dimensions of 20×20 mm^2, attached to the bottom of a 35 mm glass-bottom petri dish having a 14 mm well in the center (*see* **Notes 1–3**; Fig. 1), as follows:

1. Carefully remove the glass cover slip from a 35 mm glass-bottom MatTek dish using a razor blade.
2. With the bottom side of the MatTek dish facing up, apply small amount of silicone glue all around the hole.
3. Remove the CYTOOchip Motility out of the blister pack and place, ECM side down, centered on the hole with a pair of sterile forceps. Gently press with the forceps to ensure a good seal with the applied silicone glue. Leave the dish at room temperature for 6 h to let the glue dry. Store dishes at 4 °C, and use them within 1–2 days.

3.2 Collagen I Coating of CYTOOchip Motility A

1. Remove the activated CYTOOchip out of the blister pack, and glue it to the bottom of a 35 mm glass-bottom dish, as described in Subheading 3.1.
2. Put 200 μl of the FITC-conjugated collagen solution in the center well to cover the activated CYTOOchip. Incubate for 2 h at room temperature.
3. Wash three times with PBS, store at 4 °C, and use within 1–2 days.

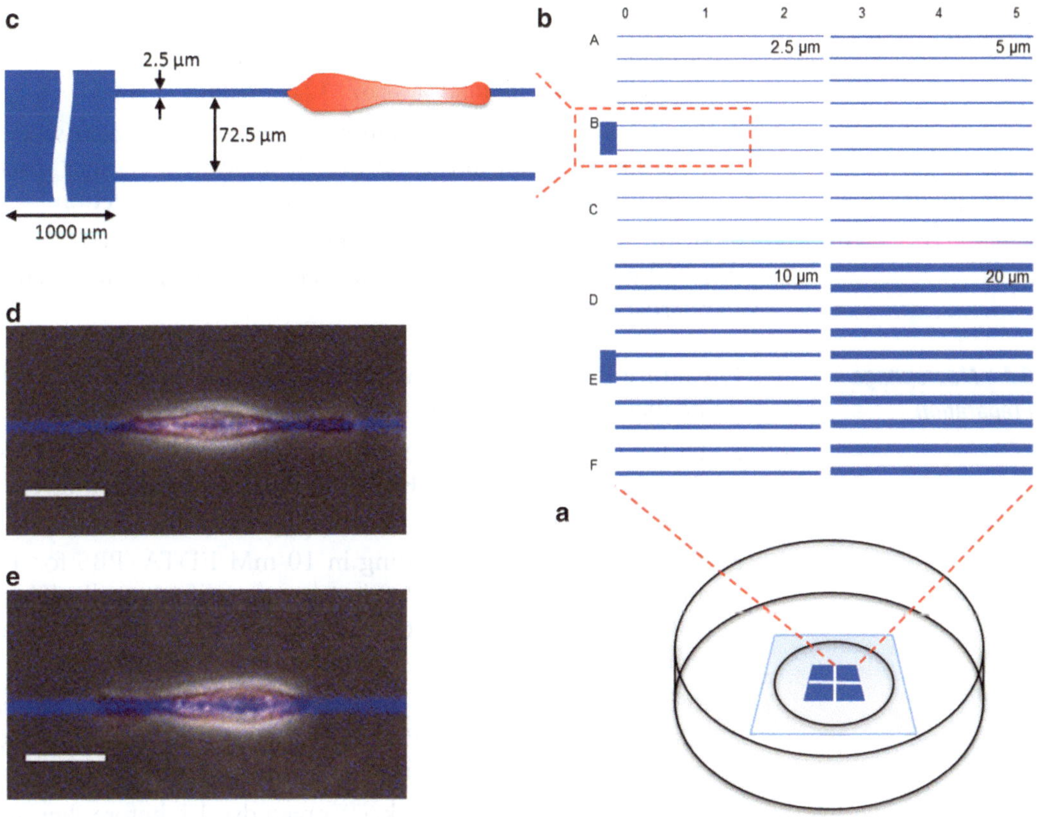

Fig. 1 An overview of the in vitro motility assay utilizing linear adhesive micro-patterned substrates. (**a**) A cartoon showing a single 20 × 20 mm², 175 μm (#1.5) thick CYTOOchip™ Motility FN650 (CYTOO, Inc.) affixed to a 35 mm diameter, 14 mm microwell dish (MatTek Corporation). Each chip is micro-patterned with four quadrants of parallel linear stripes of fibronectin labeled with 650 nm dye in widths ranging from 2.5, 5, and 10 to 20 μm with 192 stripes for each width. (**b**) A zoomed area of one quadrant of the CYTOOchip showing fibronectin-650 stripes of different widths. The chip includes an area of planar 2D area of fibronectin-650 that is continuous with the 2.5 μm linear stripes so that a direct comparison of 1D and 2D motility can be made. (**c**) A zoomed area of CYTOOchip with 2.5 μm fibronectin-650 stripes continuous with the 2D area. A cartoon of an elongated cell attached to the fibronectin-650 stripe is also shown. (**d**, **e**) TagRFP-cortactin-expressing MTLn3 cells showing elongated morphology on (**d**) 2.5 μm and (**e**) 5 μm fibronectin-650 linear stripes. Merge images of phase, RFP, and far-red channels are shown. Scale bars, 25 μm

3.3 Preparation of Carcinoma Cells

1. Put 2 ml of culture media on ECM (FN650 or FITC-collagen I)-coated CYTOOchip-MatTek dish, and incubate in the cell culture incubator for 30 min.
2. Trypsinize a 10 cm culture dish containing either parental or TagRFP-cortactin-expressing MTLn3 cells with 1 ml of 0.05 % trypsin–EDTA solution for 5 min in a cell culture incubator at 37 °C containing 5 % CO_2. Add 9 ml of culture media to stop the reaction.

3. Aspirate media from the CYTOOchip-MatTek dish, put 2 ml of MTLn3 cells (5×10^4 cells/ml) into the dish, and transfer it to a cell culture incubator at 37 °C containing 5 % CO_2.
4. After 3–4 h, check the cells using a tissue culture microscope. Cells should nicely line up in the 1D areas with enough space in between for the cells to freely move in either direction.
5. Wash floating cells once with imaging media, aspirate, then add 2 ml of imaging media, and transfer to an epi-fluorescence microscope.

3.4 Macrophage Preparation

1. Culture BMMs in bacterial petri dishes at 37 °C in a 5 % CO_2 incubator [13] (*see* **Note 4**).
2. To render BMMs quiescent [12], deprive cells of CSF-1 by changing to CSF-1-free BMM medium and incubating cells for 16 h prior to experiment.
3. Harvest BMMs by incubating in 10 mM EDTA/PBS for 10 min at 37 °C in a 5 % CO_2 incubator to lift the cells. Then pellet cells by centrifugation at $400 \times g$ for 5 min at room temperature.
4. Resuspend BMMs in CSF-1-free culture medium, and transfer cells to a 15 ml centrifuge tube. BMMs can be kept suspended in the tube inside cell culture incubator for up to 6 h.
5. Label BMMs with CellTracker™ green dye 1 h before they are ready to be added to carcinoma cells on the CYTOOchip. Add the dye at 1:1,000 (final concentration, 5 μM) to 5×10^6 BMMs in 1 ml of media, and incubate for 30 min in a cell culture incubator.
6. Wash two times, and resuspend BMMs in 2 ml of imaging media (5×10^4 cells/ml).

3.5 Live Imaging of Tumor Cell–Macrophage Interaction

1. To maintain good focus during live imaging, turn the microscope heat ON at least 2–3 h before starting to image.
2. Place the CYTOO chamber with attached TagRFP-cortactin MTLn3 cells on the microscope. Adjust the focus on the cells, and set up for multiple-position imaging (*see* **Note 5**). For longer duration time lapses, imaging of multiple adjacent fields is recommended because cells frequently move in and out of the field (*see* **Note 6**).
3. After 2–3 h of time-lapse imaging of TagRFP-cortactin MTLn3 cells alone (Figs. 1d, e and 2) to establish baseline carcinoma cell motility, add 1×10^5 BMMs and continue imaging.
4. BMMs start attaching to the dish (mostly on the ECM but also on nonadhesive areas) within 5 min of addition. Continue imaging for 6 h. Instances of tumor cell–macrophage pairing (Fig. 3a, b) and streaming, consisting of an alternating pattern

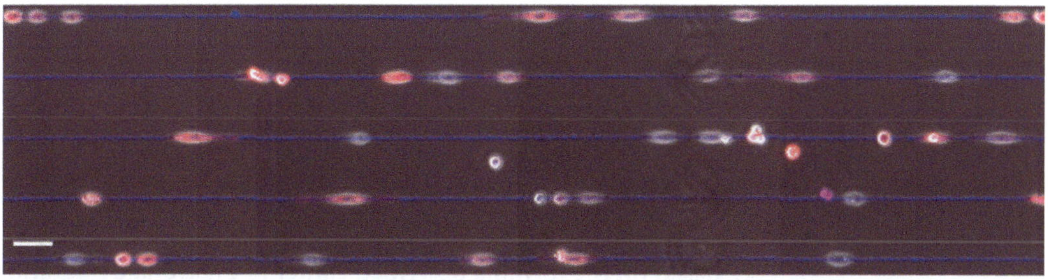

Fig. 2 Demonstration of combining multiple adjacent fields of view in ImageJ. TagRFP-cortactin-expressing MTLn3 cells were plated on CYTOOchip Motility FN650 for 3–4 h. Adjacent fields of view with MTLn3 cells on linear stripes were imaged in phase, RFP, and far-red channels on a DeltaVision microscope with a 20× objective (field-of-view width = 329 μm). Merged images (phase, RFP, and far-red channels) of four adjacent fields, totaling a length of 1,316 μm, were combined as described in Subheading 3.6. Scale bar, 50 μm

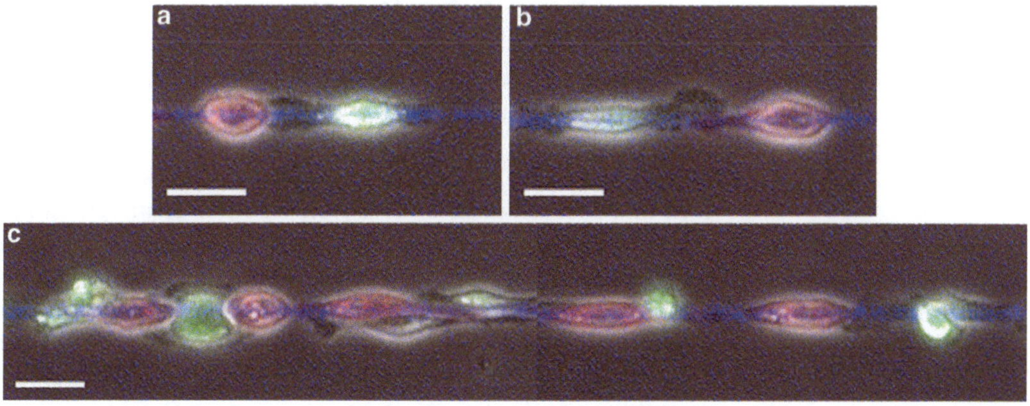

Fig. 3 Examples of tumor cell–macrophage pairing and streaming on 1D adhesive substrates. TagRFP-cortactin-expressing MTLn3 cells were plated for 3–4 h on CYTOOchip Motility FN650. After time-lapse imaging for 1.5–2 h, CellTracker green-labeled BMMs were added and imaged for an additional 6 h. (**a**) Single-tumor cell (*red*) and macrophage (*green*) pairing on 2.5 μm fibronectin-650 (*blue*) stripe. (**b**) Single-tumor cell (*red*) and macrophage (*green*) pairing on 5 μm fibronectin-650 (*blue*) stripe. (**c**) Multiple-tumor cells (*red*) and macrophages (*green*) were capable of forming streams of alternating tumor cells and macrophages on fibronectin-650 stripe (*blue*) like those observed in vivo [8, 9]. Scale bars, 25 μm

of tumor cell and macrophages (Fig. 3c), can be seen starting from 1 h after the addition of BMMs.

5. To evaluate the role of EGF–CSF-1 paracrine signaling pathway, EGFR inhibitors (1 μM Iressa or 5 μM AG-1478) or CSF-1 receptor function blocking antibody, AFS98 (50 μg/ml), is added at the indicated concentrations and imaging is continued for 6 h.

3.6 Analysis of Time-Lapse Movies for Cell Motility Parameter Calculations

1. Open fluorescent time-lapse movies in ImageJ. Multiple time-lapse movies of adjacent areas can be stitched together in ImageJ by combining two fields at a time with the command Image > Stacks > Tools > Combine. A montage of 6–7 adjacent fields totaling approximately 2 mm of continuous stripe was regularly analyzed (Fig. 2).

2. For measuring carcinoma cell motility parameters, threshold the RFP-cortactin channel using the command Image > Adjust > Threshold.
3. A custom ImageJ macro (available upon request), based on outlining a cell with the Wand tool and determining the cell centroid in each frame, was used to calculate the average and maximum tumor cell speed.
4. Based on the formula described earlier [14], average migration persistence was calculated as speed/[1 + (100/360) × angle], where angle is the directional change in degrees. On 1D stripes, the angle value is either 0 or 180, depending on whether the cell persists in the same direction or changes direction.
5. For displaying the images of tumor cell–macrophage interaction, fluorescent and phase channels were merged using the command Image > Color > Merge Channels.

4 Notes

1. As an alternate to attaching the CYTOO cover slip to the bottom of a MatTek dish, 1- or 4-well CYTOO chambers can be used, where CYTOO cover slip is sandwiched between the bottom frame and the main body of the CYTOO chamber and held firmly by the magnetic force. Make sure that the silicone gasket is attached to the main body of the chamber before attaching it to the bottom frame; this ensures a leakproof assembly.
2. When the main body of the CYTOO chamber is brought closer to the bottom frame, the parts have a tendency to snap together due to magnetic force, and this can cause coverslip breakage. A new user is advised to assemble the chamber without the cover slip to get used to the strength of the magnetic force.
3. A 4-well CYTOO chamber combined with multiple-position imaging is very useful for testing up to four different conditions on a single CYTOOchip in one imaging session. The only drawback is that all the 2D areas on the chip are covered by the chamber walls and therefore not available for imaging.
4. Do not use regular tissue culture-treated dish for BMMs or the cells will adhere too tightly and will be difficult to replate. BMMs should be cultured in bacterial petri dishes.
5. On a DeltaVision epi-fluorescence microscope, equipped with a 20× objective and Photometrics CoolSnap HQ^2 CCD camera, the width of a field of view is about 329 μm. Up to 6–8 fields were imaged by successively moving each field by this distance in *x*–*y* direction.

6. One of the challenges in analyzing thresholded time-lapse movies is when adjacent cells touch each other. To overcome this, cells were manually separated by drawing a black line (value 0, same as the background) between the cell boundaries with a pencil tool in every frame where cells were touching each other.

Acknowledgment

We thank the people from the Condeelis, Segall, and Cox laboratories for helpful discussions. This work was supported by CA150344 (R.J.E. and V.P.S.), CA100324 (J.S.C. and D.C.) and a postdoctoral fellowship from Susan G. Komen for the Cure© KG111405 to V.P.S.

References

1. Roussos ET, Condeelis JS, Patsialou A (2011) Chemotaxis in cancer. Nat Rev Cancer 11: 573–587
2. Roussos ET, Balsamo M, Alford SK et al (2011) Mena invasive (MenaINV) promotes multicellular streaming motility and transendothelial migration in a mouse model of breast cancer. J Cell Sci 124:2120–2131
3. Goswami S, Sahai E, Wyckoff JB et al (2005) Macrophages promote the invasion of breast carcinoma cells via a colony-stimulating factor-1/epidermal growth factor paracrine loop. Cancer Res 65:5278–5283
4. Wyckoff J, Wang W, Lin EY et al (2004) A paracrine loop between tumor cells and macrophages is required for tumor cell migration in mammary tumors. Cancer Res 64:7022–7029
5. Wyckoff JB, Wang Y, Lin EY et al (2007) Direct visualization of macrophage-assisted tumor cell intravasation in mammary tumors. Cancer Res 67:2649–2656
6. Sidani M, Wyckoff J, Xue C et al (2006) Probing the microenvironment of mammary tumors using multiphoton microscopy. J Mammary Gland Biol Neoplasia 11:151–163
7. Wang W, Wyckoff JB, Frohlich VC et al (2002) Single cell behavior in metastatic primary mammary tumors correlated with gene expression patterns revealed by molecular profiling. Cancer Res 62:6278–6288
8. Patsialou A, Bravo-Cordero JJ, Wang Y et al (2013) Intravital multiphoton imaging reveals multicellular streaming as a crucial component of in vivo cell migration in human breast tumors. Intravital 2:e25294, http://dx.doi.org/10.4161/intv.25294
9. Sharma VP, Beaty BT, Patsialou A et al (2012) Reconstitution of in vivo macrophage-tumor cell pairing and streaming motility on one-dimensional micro-patterned substrates. Intravital 1:77–85
10. Segall JE, Tyerech S, Boselli L et al (1996) EGF stimulates lamellipod extension in metastatic mammary adenocarcinoma cells by an actin-dependent mechanism. Clin Exp Metastasis 14:61–72
11. Sharma VP, Eddy R, Entenberg D et al (2013) Tks5 and SHIP2 regulate invadopodium maturation, but not initiation, in breast carcinoma cells. Curr Biol 23:2079–2089
12. Stanley ER (1997) Murine bone marrow-derived macrophages. Methods Mol Biol 75:301–304
13. Ishihara D, Dovas A, Hernandez L et al (2013) Wiskott-Aldrich syndrome protein regulates leukocyte-dependent breast cancer metastasis. Cell Rep 4:429–436
14. Farina KL, Wyckoff JB, Rivera J et al (1998) Cell motility of tumor cells visualized in living intact primary tumors using green fluorescent protein. Cancer Res 58:2528–2532

Chapter 11

Chemotactic Responses by Macrophages to a Directional Source of a Cytokine Delivered by a Micropipette

Michael Cammer and Dianne Cox

Abstract

Macrophages, which are organized throughout every tissue, represent a key component of the immune system and the recruitment of macrophages to specific sites is important in normal host defense. However, when inappropriately recruited macrophages may damage or destroy healthy tissue; this is seen in several autoimmune diseases such as arthritis. Many cytokines, including CSF-1 and chemokines, are often upregulated in inflamed tissues and can induce the directional migration of macrophages towards the highest concentration of the cytokine in a process called chemotaxis. Chemokines were first described as chemoattractant cytokines synthesized at sites of inflammation that stimulate the directional migration of leukocytes and mediate inflammation. Whereas specific receptors for chemoattractants reside over the entire cell surface, macrophages can detect very shallow chemotactic gradients leading to spatially defined responses to the chemoattractant such as the extension of directed protrusions leading to cell migration. In this chapter we describe a method for the localized delivery of chemoattractants via a micropipette needle to macrophages in culture followed by methods for imaging and an outline of quantifying macrophage responses.

Key words Chemokines, CSF-1, Chemotaxis, Protrusions, Migration, Macrophage

1 Introduction

Sites of infection or injury secrete inflammatory cytokines that act as chemoattractants orchestrating the recruitment of monocytes/macrophages. Macrophages sense and migrate along external gradients of soluble factors. This is directed migration or chemotaxis. During migration cells cycle through a set of actin-dependent events that include the extension of an F-actin rich protrusion towards the source of the chemoattractant, followed by stabilization of the protrusion via integrin-mediated attachment to the underlying extracellular substrate, and the actomyosin-based contraction of the body and detachment at the rear.

Cell migration has been observed since the development of the light microscope beginning with Leeuwenhoek in the seventeenth

Ivana Vancurova (ed.), *Cytokine Bioassays: Methods and Protocols*, Methods in Molecular Biology, vol. 1172,
DOI 10.1007/978-1-4939-0928-5_11, © Springer Science+Business Media New York 2014

century. Since then cellular migration has been observed in many different organisms and cell types including the founding studies by Elie Metchnikoff using immune cells. It has only been since the last half of the twentieth century that the tools have been developed to quantify this process.

The Boyden chamber has become one of the most used tools to assess cell motility and invasion. The classical Boyden chamber consists of two compartments where the cells contained in the upper compartment are separated by a membrane containing pores from the lower compartment containing a chemoattractant. This membrane represents a physical barrier that cells can only overcome by active migration. Since its initial introduction in 1962 to study neutrophil chemotaxis [1], a number of different Boyden chamber devices have been developed and used to study chemotaxis by different cell types including tumor cells [2]. While this is a quantitative method to measure chemotaxis and can provide information on the importance of specific molecules involved in chemotaxis, it is unfortunately an end point assay with low resolution.

Since then several different types of assays have been developed that allow for cellular imaging during chemotaxis, including the Zigmond and Dunn chambers that contain two wells separated by a bridge where the cells can be images during migration [3, 4]. More recently more sophisticated microfluidic devices have been developed (reviewed in [5]). However, many researchers have used a micropipette to rapidly generate a chemotactic gradient to measure the chemotactic properties of a wide variety of cell types ranging from yeast to *Dictyostelium* amoebae to mammalian cells including neuronal, cancer, and immune cells [6–9].

There are essentially two types of assays using a micropipette to stimulate cells. One is to look at the behavior of a population of cells. The second is to study the behavior of individual cells. The former may be done in many cell culture systems. The second requires high numerical aperture optics and mastery of the first method (*see* **Note 1**) [10]. In this chapter we describe a method for the first application, using a micropipette to deliver a local gradient of chemoattractant to induce the migration of bone marrow-derived macrophages (BMMs) (Fig. 1). We describe looking at a field approximately 400 × 300 μm using a 10× phase contrast objective.

2 Materials

2.1 Cell Culture and Reagents

1. 10 cm bacterial dishes (do *not* use regular tissue culture-treated dish for BMMs or the cells will adhere too tightly and will be difficult to replate). Smaller dishes can be used for the actual imaging experiment (35 mm). A 10× objective should provide acceptable images even through plastic.

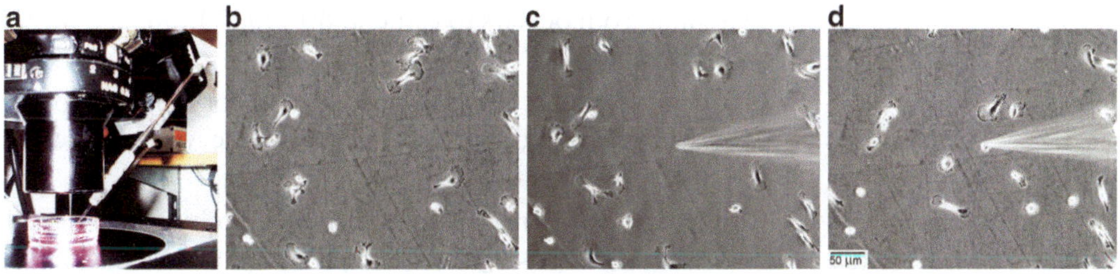

Fig. 1 (**a**) Overview of instrumentation. Eppendorf micromanipulator with Femtotips II micropipette in 35 mm dish on a heated stage plate. Condenser allows sufficient space for micromanipulator and micropipette. Example data: cells imaged by phase contrast. (**b**) Before. (**c**) Micropipette added. (**d**) End of time lapse

2. Insulin syringes: 3 ml syringe with a 23 g needle.
3. BMM medium: Minimal Essential Medium Alpha (α-MEM) with L-glutamine, 15 % Fetal Bovine Serum (FBS), 1 % Penicillin/Streptomycin.
4. Recombinant mouse CSF-1 for BMM growth.
5. CX3CL1 (Fractalkine) (R&D Systems).
6. Sterile 10 mM EDTA/PBS.
7. If there is no CO_2 control for the microscope, 10 mM HEPES can be added to the medium or Leibovitz's L-15 medium can be used.
8. Cell culture grade light mineral oil optional, e.g., Sigma M5310, to cover media in dish to prevent evaporation while being gas permeable.
9. Chemokine to put in micropipette, for example we use a final concentration of 50 ng/ml CX3CL1 in medium for BMMs.
10. Fluorescent tracer optional. We use 70 kDa fluoroscein conjugated dextran at a final concentration of 100 µg/ml. Examples are fluoroscein, Alexa 488, Oregon green, rhodamine, Alexa 568.
11. 0.22 µm filters (Millipore).

2.2 Instrumentation

Any inverted microscope with a camera may be used (*see* **Notes 2** and **3**). The parts listed here are one example and alternatives are discussed in Subheading 4. **Note 2** explains the benefits of phase contrast specifically and **Note 3** describes why fluorescent labeling is avoided.

2.2.1 Microcopy

1. Olympus IX71 inverted microscope with planapo phase contrast 10× NA objective. A plan corrected objective is important for imaging a large field with consistent focus and spatial response across the field. This technique is easier with a microscope that does not have a 100/0 beam splitter when

switching from oculars to camera. 80/20, 70/30, and 50/50 are preferable.

2. Optional: green interference filter for light source. Monochromatic light in the middle of the visible spectrum may improve resolution.
3. Condenser with manually selectable positions preferable because a slight misalignment may be required to compensate for a change in the surface of the media due to surface tension curvature where the micropipette enters.
4. Optional: epifluorescence for monitoring flow of chemokine mixed with fluorescent dye.
5. Camera. We use Cooke Sensicam and Photometrics Coolsnap HQ. Any camera with a large chip and software for time-lapse imaging may be used (*see* **Note 4**). If cells are light sensitive, shutter recommended to close light between exposures.
6. Anti-vibration platform. Needle will shake or even crash into bottom of dish if there are vibrations.
7. Software: MicroManager is free, operates on multiple computer platforms, drives a wide range of microscope hardware made by many manufacturers, and has exemplary online support. Any microscope and camera control software that can perform time-lapse imaging may be substituted.
8. Warm inserted stage on microscope (Harvard Apparatus).

2.2.2 Microinjection

1. Eppendorf PatchMan NP2 microinjector with electronic joystick XYZ control and continuous pressure unit. This is an integrated one-stop-shopping solution. However, the system may be broken into separate components from different vendors to save on cost.
2. Eppendorf microloader tips (5242 956.003 or 930001007).
3. Femtotip II microinjection capillaries or your own pulled capillaries.
4. Eppendorf Reference fixed-volume micropipettes or equivalent for the volume to load in needle (1 μl, 2 μl or 5 μl).

3 Methods

3.1 BMM Isolation, Cell Culturing and Plating

1. Start with two intact femurs and tibiae from one mouse kept in cold α-MEM (*see* **Note 5**).
2. Using a 3 ml syringe with a 23 g needle, flush femurs and tibiae with 2 ml of BMM medium (8 ml for the whole animal) in a clean sterile 10 cm dish. You should be able to see cell clumps falling down from the bone. Bones now should appear quite white and not red/pinkish as they were before.

3. Add CSF-1 to a final concentration of 2.5 nM for growth and incubate at 37° in a 5 % CO_2 incubator. In general, BMMs are said to be mature once they attach and spread on the bacterial plastic and are frequently used between day 7 and 14 after isolation.
4. Harvest BMMs by incubating in 10 mM EDTA/PBS for 10 min at 37° in a 5 % CO_2 incubator to lift the cells. Then pellet the cells by centrifugation at 400 ×*g* for 5 min at 4 °C.
5. BMMs must be replated into a 35 mm dish at 30–50 % confluency at least 1–2 days prior to the experiment since they need time to attach (*see* **Notes 6** and 7). A minimum of 2 ml in a 35 mm dish is recommended. Cells can be made more quiescent by removing CSF-1 from the media 16 h prior the experiment.
6. Prior to the experiment replace BMM medium with phenol red-free medium with 10 mM HEPES if not using a CO_2 chamber (*see* **Note 8**).

3.2 Equipment Set-Up

1. Turn environmental control on 1 h in advance to stabilize microscope stage and objectives (*see* **Note 9**).
2. Turn on microinjection apparatus.
3. Turn on microscope, camera, and software.

3.3 Load Needle

1. Prepare chemokine at required concentration with optional fluorescent dye for tracking the flow from the needle tip (*see* **Note 10**).
2. We recommend filtering the chemokine to remove particulates using 0.22 μm pores but Eppendorf reports spinning at 10,000 ×*g* for 15 min may be sufficient.
3. Load the chemokine into a microloader tip. Look at the tip to make sure fluid has been taken up since sometimes these tips do not load. If tip cannot be loaded, try tightening it on the pipette. If it still won't load, discard it and try another.
4. Insert the microloader tip down the large opening of the micropipette until it stops due to tapering at the tip. Be sure to keep the microloader straight. Some people prefer holding the micropipette in one hand and the microloader in the other. Others like to have the micropipette held in an upright position, rest their elbows on the bench, and lower the microloader into the capillary.
5. Slowly eject the contents of the microloader while pulling the microloader out of the micropipette.
6. Make sure fluid extends all the way to the tapered tip. If at any time you think you may have touched the tip of the micropipette or brushed it against something, most likely it is broken.

3.4 Perform Experiment

1. Have microscope environmental system stabilized and micro-injector ready.
2. Put dish of cells on stage.
3. Remove top and align for Kohler illumination.
4. Mount the chemokine loaded micropipette in the micromanipulator such that the tip is in the light beam from the condenser to the dish and only a mm or so from the top of the media. The micromanipulator controls may only have a small range such as 1–2 cm so it is important to allow plenty of play remaining using the micromanipulator controls (*see* **Note 11**).
5. Start pressure on the needle (*see* **Note 12**). We use 18–40 hPa.
6. Drop needle below surface of media. This should be done quickly to prevent clogging at the micropipette tip due to evaporation.
7. Optional: Using a transfer pipette slowly lay light mineral oil over top of media by moving the pipette tip to encourage even coverage.
8. When micropipette breaks the surface turn condenser ring slightly out of position to get phase rings to align and to restore Kohler illumination.
9. Use the oculars to look for and position the micropipette.
10. Drop needle slowly until you begin to see a slightly darker band. A trick is to move the needle back and forth in the Y axis, or apparently up and down in the eyepieces, to see it better.
11. Raise the focus above the cells so that when you drop the needle into the focus plane it will still be far above the bottom of the dish.
12. Bring tip into focus.
13. Drop focal plane so cells are in focus.
14. Move to an extreme position, for instance near the edge of the dish which may be moved by hand, where you will not be imaging. This is because initiating flow from the capillary tip is sometimes tricky and you do not want to contaminate cells with chemokine before imaging. Beware that plastic dishes may not be truly flat so you have to be extra careful to raise the micropipette before moving the dish. If you raise it straight up, even if you cannot see it, it should be simple to lower back into the field.
15. Check needle flow. Using phase contrast you should be able to see a slight change in density a few microns from the needle tip. If the chemokine solution includes a fluorophore, a short exposure with binning at 4×4 or 8×8 should show the flow with minimal exposure of light (Fig. 2).

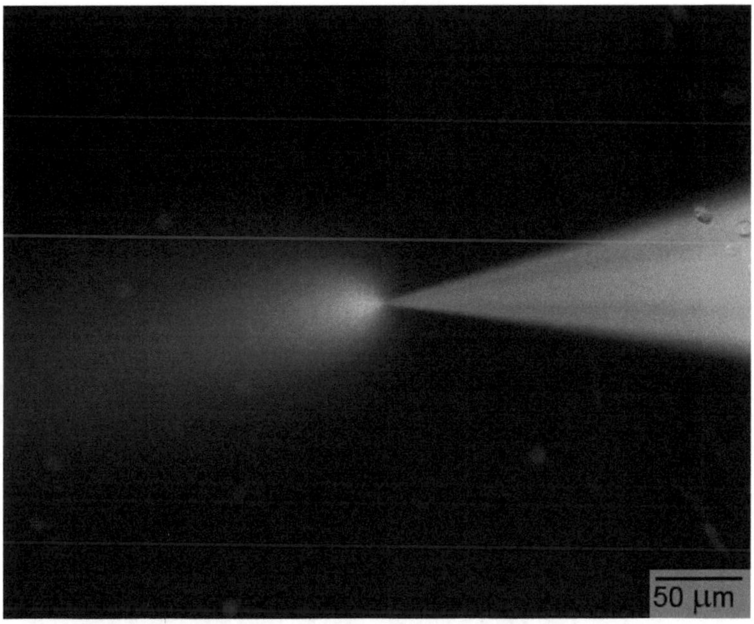

Fig. 2 Fluorescence image shows good flow from micropipette

16. If there is no observable flow from the needle, try the clean function if using an Eppendorf microinjector. This boosts the pressure to initially begin the flow. Another trick is to very lightly touch the tip of the capillary to the bottom of this dish and drag it. This may break the hole larger, but you have to be careful not to break too much off because the flow will be too great.
17. Move needle up and to the edge of visual field.
18. Find field of cells for the experiment with clear spot on side where needle enters.
19. Start recording for at least a few minutes to get baseline of cell movement. Of course you also need to perform separate control recordings where the conditions are exactly the same except the chemokine is missing from the micropipette cocktail.
20. Move needle into field and down near substrate, 20–40 μm above the bottom works well. If the light is shuttered, throw manual shutter to on position so you can see through oculars. Alternative, if this cannot be done, stop time-lapse collection and move needle into position, start a new time-lapse collection, and at the end concatenate the sequences. The timing will not be precise, but with imaging at 5–15 min intervals over 1–3 h, this may be acceptable.
21. Image for required time. Periodic checks of focus and flow from micropipette are recommended.

3.5 Analysis

Whereas there are many parameters of cell motility that may be measured, in this assay we focus on whether the cells move towards the chemokine, randomly, or away. This may be established by locating the initial and ending position of each cell and measuring these distances in reference to the micropipette tip (Fig. 3a, d). While this may be partially automated using ImageJ [11] or other software, the measurements may be performed with minimal computer experience by recording the beginning and end of each cell's X, Y location and entering into a spreadsheet to calculate the distances to the micropipette tip. Excel, Open Office, and Google Spreadsheet all handle this Pythagorean calculation fine. If the calculations are done using pixel coordinates, at the final step you may multiply by a constant to convert units to microns. We have also written a routine to color code these direct paths as temperature maps. This provides a quick view of the population's behavior with regard to the micropipette tip (Fig. 3b). The aggregate

a

b

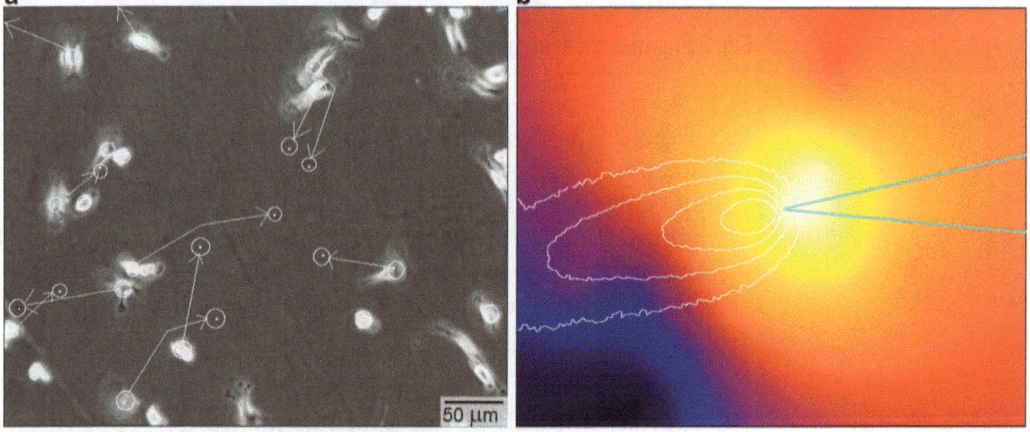

c

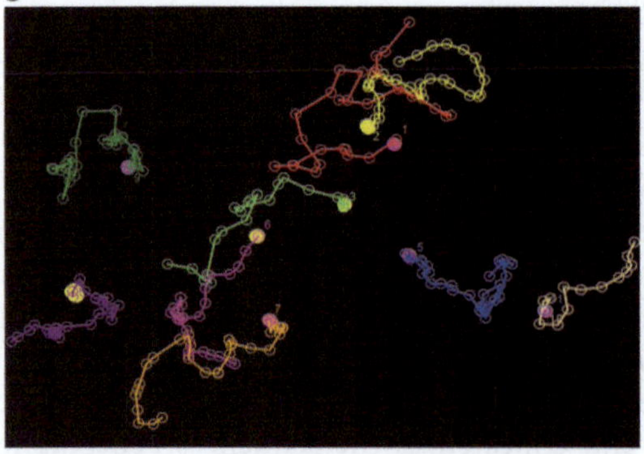

d

cell starting		cell ending	
X	Y	X	Y
409	55	397	189
399	101	369	170
149	178	143	178
103	240	101	241
71	227	124	212
152	348	10	361
3	401	73	347
52	452	50	432
160	484	270	377
164	315	344	252
234	427	256	287
455	378	447	380
491	328	408	309

needle tip location	
X	Y
338	247

distance from needle tip			
start	ending	difference	movement
205	83	122	122
158	83	75	75
201	207	-6	
235	237	-2	
268	217	51	51
212	347	-136	-136
369	283	85	85
352	342	10	
296	147	150	150
187	8	179	179
208	91	117	117
176	172	4	
173	94	80	80

=SQRT(POWER((xc-B4),2) + POWER((yc-C4),2))

Fig. 3 (**a**) Starting and ending positions of cells marked manually. Coordinates used to create the temperate map (**b**) which shows the aggregate paths of the cells as flow where *white* is convergence and *black* is divergence and (**d**) the calculations of netpaths using a spreadsheet. The *lines* in B show flow from the micropipette as pictured in Fig. 2. (**c**) Paths generated by MTrackJ [12] at 10 min intervals as one example of analysis by tracking software

information of each cell's location change from start to end (net-path) defines a trend but does not provide information regarding the nature of the motility, for instance whether it is fast and meandering or slow and direct (*see* **Note 13**).

There are plugins in ImageJ (Fig. 3c) and features in software packages such as Volocity and Imaris that provide more detailed analysis of paths. Users of these software report that they are successful at automatic detection of moving cells to make motility measurements simple. Certainly all these software are effective when cells are isolated from one another. However, where cells touch, which occurs in most cases, these types of software fail to accurately follow individual paths. In our experience the accuracy of properly identified cells outweighs time savings claimed by automated feature recognition without manual confirmation.

4 Notes

1. This method may be used for imaging at high resolution in single cells. 40× N.A. 1.3 and 60× N.A. 1.4 have been used successfully but require practice and finesse to find the micropipette.
2. Phase contrast is recommended because it provides an excellent overall view of both the cells and the micropipette. It is also very forgiving if the microscope focus slips. Brightfield is not high contrast enough to image cells and difficult to judge the tip of the needle. Nomarski may be acceptable but may impede automated image segmentation. Reflectance imaging provides excellent contrast for computer segmentation when the focus is held precisely at the coverslip but does not work well at low numerical aperture, renders the micropipette invisible, and where cells may have discrete contacts rather than adhering flat, they are mismeasured as multiple cells.
3. Imaging fluorescently labeled cells may simplify the task of automated segmentation. We describe methods here for unlabeled cells because either cells have to express genetically encoded fluorescent proteins or be stained with fluorophores such as CSFE, DiI, or Celltracker Violet. Most monocytes are not amenable to fluorescent protein transfection. Labeling cells with small molecules may alter their behaviors. In both cases, exposure to excitation for fluorescence is more likely to damage the cells than exposure to low levels of light for transmitted imaging.
4. A relatively inexpensive camera that also works well for fluorescence microscopy is the QiClick. sCMOS cameras may allow collection of larger fields at the same magnification.

5. Cleaned bones can be shipped on ice overnight and used to generate BMMs.
6. Culture dish must be compatible with the microscope objective. Microscopy purists will insist this means a #1.5 coverslip bottom, which is standard for most microscope objectives. This method uses lower magnifications, such as 20× and below with low numerical aperture lenses, and plastic dishes are acceptable. Also, special long working distance objectives with collars to adjust for varied materials and refractive indices are available.
7. Each cell type needs its own optimization including whether the substrate needs to be pretreated or coated.
8. Any standard cell culture media acceptable. If modifying protocol to watch fluorescence, media without phenol red is highly recommended.
9. Environmental control may be required for correct timing of biological functions. We have used two different types of temperature and pH control. (a) Most commonly heated stage with cells in L15 or HEPES buffered media as described above. (b) Complete enclosure around microscope with forced heated air or premixed 5 % CO_2 in air bubbled through water to maintain humidity. We find this requires 3 h to stabilize or it may be kept on all the time.
10. The appropriate concentration that needs to be loaded into the needle must be determined empirically for each chemokine. We have found that 10 ng/ml CSF-1 works well in the needle but it should be noted that expression of the CSF-1R must be upregulated by incubation in CSF-1 free media 16 h prior to the experiment.
11. Micropipette positioning may be provided by a few different methods ranging from complete electronic to manual control. Examples are Harvard Apparatus, World Precision Instruments, or modifying optics parts such as from Thorlabs or Newport Instruments. Single joystick control makes the method very simple, but a manual micromanipulator may be sufficient.
12. An inexpensive alternative for applying constant pressure to the micropipette is a syringe pump. This operates by a motor slowly turning a screw that pushes the plunger on a syringe. An alternative may be a gravity based system, but we have not tried this.
13. If you see a trend of cells streaming in a given direction in parallel, likely there is an issue of flow in the culture system. This may be due to uneven heating. Environmental control needs to be uniform to prevent artifacts due to flow pushing cells or dispersing the chemokine.

Acknowledgments

This work was supported by NIH grant GM071828.

References

1. Boyden S (1962) The chemotactic effect of mixtures of antibody and antigen on polymorphonuclear leucocytes. J Exp Med 115: 453–466
2. Wass JA, Varani J, Piontek GE et al (1981) Characteristics of the chemotactic factor-mediated cell swelling response of tumor cells. J Natl Cancer Inst 66:927–933
3. Zigmond SH (1977) Ability of polymorphonuclear leukocytes to orient in gradients of chemotactic factors. J Cell Biol 75:606–616
4. Zicha D, Dunn GA, Brown AF (1991) A new direct-viewing chemotaxis chamber. J Cell Sci 99:769–775
5. Paguirigan AL, Beebe DJ (2008) Microfluidics meet cell biology: bridging the gap by validation and application of microscale techniques for cell biological assays. Bioessays 30: 811–821
6. Futrelle RP, Traut J, McKee WG (1982) Cell behavior in Dictyostelium discoideum: preaggregation response to localized cyclic AMP pulses. J Cell Biol 92:807–821
7. Bailly M, Condeelis JS, Segall JE (1998) Chemoattractant-induced lamellipod extension. Microsc Res Tech 43:433–443
8. Gerisch G, Keller HU (1981) Chemotactic reorientation of granulocytes stimulated with micropipettes containing fMet-Leu-Phe. J Cell Sci 52:1–10
9. Vallier LG, Segall JE, Snyder M (2002) The alpha-factor receptor C-terminus is important for mating projection formation and orientation in Saccharomyces cerevisiae. Cell Motil Cytoskeleton 53:251–266
10. Ishihara D, Dovas A, Park H et al (2012) The chemotactic defect in wiskott-Aldrich syndrome macrophages is due to the reduced persistence of directional protrusions. PLoS One 7(1):e30033
11. Rasband WS (1997/2013) ImageJ. U. S. National Institutes of Health, Bethesda, MD, USA, http://imagej.nih.gov/ij/
12. Meijering E, Dzyubachyk O, Smal I (2012) Methods for cell and particle tracking. Methods Enzymol 504:183–200

Chapter 12

Assessment of Phagocytic Activity of Cultured Macrophages Using Fluorescence Microscopy and Flow Cytometry

Lokesh Sharma, Wenjun Wu, Sanjay L. Dholakiya, Samir Gorasiya, Jiao Wu, Ravikumar Sitapara, Vivek Patel, Mao Wang, Michelle Zur, Shloka Reddy, Nathan Siegelaub, Katrina Bamba, Frank A. Barile, and Lin L. Mantell

Abstract

Phagocytosis is the process by which phagocytes, including macrophages, neutrophils and monocytes, engulf and kill invading pathogens, remove foreign particles, and clear cell debris. Phagocytes and their ability to phagocytose are an important part of the innate immune system and are critical for homeostasis of the host. Impairment in phagocytosis has been associated with numerous diseases and disorders. Different cytokines have been shown to affect the phagocytic process. Cytokines including TNFα, IL-1β, GM-CSF, and TGF-β1 were found to promote phagocytosis, whereas high mobility group box-1 (HMGB1) inhibited the phagocytic function of macrophages. Here, we describe two commonly used methods to assess the phagocytic function of cultured macrophages, which can easily be applied to other phagocytes. Each method is based on the extent of engulfment of FITC-labeled latex minibeads by macrophages under different conditions. Phagocytic activity can be assessed either by counting individual cells using a fluorescence microscope or measuring fluorescence intensity using a flow cytometer.

Key words Phagocytosis, Macrophages, Hyperoxia, Flow cytometry, HMGB1

1 Introduction

The process of phagocytosis was first described by Elie Metchnikoff more than 100 years ago [1]. Phagocytosis (from Ancient Greek *phago*, meaning "eating," *kytos*, meaning "cell" and *-osis*, meaning "process") is a process by which phagocytes engulf and kill invading microorganisms [2]. It is a major mechanism of the immune system used to eliminate pathogens, foreign particles, and cell debris within the body [3]. Phagocytosis is a multistep phenomenon that starts with the recognition of the target using surface receptors [3]. Recognition is followed by membrane protrusion to form a

Ivana Vancurova (ed.), *Cytokine Bioassays: Methods and Protocols*, Methods in Molecular Biology, vol. 1172, DOI 10.1007/978-1-4939-0928-5_12,

phagocytic cup around the target and engulf it into the cell to make a phagosome [4]. Then, phagosomes fuse with lysosomes, forming phagolysosomes. Phagolysosomes contain lytic enzymes and chemicals such as peroxidase, lysozyme, hydrolytic enzyme, and hydrogen peroxide which causes oxidative burst for digestion of its contents [5]. The last step of phagocytosis is elimination of the phagolysosome contents by exocytosis [3].

Phagocytosis provides the first line of host defense against harmful pathogens and serves as a link between the innate immune response and the initiation of adaptive immune responses [6]. In mammals, phagocytosis is mainly carried out by macrophages, neutrophils, monocytes, dendritic cells, and mast cells, which are often referred as professional phagocytes [7]. Macrophages are first-responder cells in the innate immune response, although during some infections recruited neutrophils play a major role in the clearance process [8, 9].

In addition, phagocytosis plays an important role in the clearance of dead cells (efferocytosis) and, thereby, in maintaining normal tissue homeostasis and remodeling [3, 10]. Failure of this process causes increased accumulation of necrotic materials, which ultimately damage the surrounding tissues and contribute to disease development [11]. Therefore, phagocytosis remains an important physiological process and any interruption in the process will disturb the immune balance.

Previous studies suggest that defects in the innate immune response against invading pathogens are associated with macrophage dysfunction [12, 13]. The impairment in phagocytic capacity of alveolar macrophages, neutrophils, and monocytes has been associated with the pathogenesis of a number of airway diseases such as COPD, cystic fibrosis (CF) and asthma as well as multiple sclerosis, Alzheimer's disease and chronic granulomatous disease [14–17]. Phagocytic function is also impaired upon exposure to cigarette smoke and supraphysiological concentrations of oxygen (hyperoxia). Studies have shown that hyperoxia can impair macrophage phagocytic function by oxidizing actin cytoskeleton and thereby inhibiting the formation of phagocytic cup [18, 19]. In addition, prolonged exposure to hyperoxia can impair the respiratory burst, which is essential in bacterial killing by macrophages [20]. This hyperoxia-induced dysfunction of phagocytes can reduce the host ability to clear invading pathogens, leading to increased susceptibility in the host to infections by opportunistic pathogens [21]. Hyperoxia-induced phagocytic dysfunction is implicated in the development of ventilator-associated pneumonia (VAP), a leading cause of morbidity and mortality in patients receiving mechanical ventilation [19, 21–23]. Thus, it is important to assess the phagocytic function of phagocytes under different pathophysiological conditions in order to understand the pathogenesis of these diseases.

Cytokines are elevated in patients with infections, such as CF, and may contribute to the regulation of macrophage functions [24]. Many cytokines such as TNFα, IL-1β, GM-CSF, and TGF-β1 have been shown to promote the phagocytic function of macrophages [25]. Markedly elevated levels of extracellular high mobility group box 1 (HMGB1), a recently discovered inflammatory cytokine [26], have also been shown in patients with CF and VAP, as well as in the animal models of these diseases [21, 27, 28]. High levels of extracellular HMGB1 can directly impair the phagocytic activity of alveolar macrophages, leading to impaired bacterial clearance in animal models of CF and VAP [21, 27]. Interestingly, high levels of extracellular HMGB1 can also impair the clearance of apoptotic neutrophils by macrophages, which may contribute to the persistent presence of excessive inflammation in some diseases [29]. These studies demonstrate the important role of extracellular HMGB1 in controlling phagocytosis of invading pathogens and apoptotic cells. Therefore, blocking extracellular HMGB1 or inhibiting its accumulation in the extracellular milieu using either antibodies or pharmacological agents may be clinically useful in reducing susceptibility to infections and persistent inflammation in patients [30].

Currently, there are a number of assays available to assess phagocytic activity in vitro. Most of these assays include using fluorescence labeled latex beads, zymosan or *E. coli* particles, and microscopes or flow cytometers to measure the phagocytic activity of particular phagocytes. This chapter describes the materials, methods, and notes for both assays using fluorescence microscopy and flow cytometry. This protocol should be readily applicable to analyze the phagocytic activity of different types of phagocytes.

2 Materials

2.1 Cells and Cell Culture

1. Mouse macrophages like RAW 264.7 cells (ATCC, Manassas, VA), cultured according to standard cell culture conditions [31].
2. DMEM cell culture medium supplemented with 10 % fetal bovine serum (FBS), 1 % glutamine, 100 U/ml penicillin–streptomycin.
3. Polystyrene 6 or 12-well Plates.
4. Micro cover glass, 18 cmm circle.

2.2 Phagocytic Particles, Opsonins, and Solutions (See Note 1)

1. FITC-labeled latex minibeads (Polysciences), stored at 4 °C.
2. Fetal bovine serum (FBS), stored at ≤−20 °C.
3. IgG from human serum, stored at ≤−20 °C.
4. Phosphate Buffered Saline (PBS).
5. Trypan Blue Solution 0.4 % (w/v) in PBS.

6. 4 % paraformaldehyde (freshly prepared, or stored at −20 °C for longer storage).
7. 4-6-Diamidino-2-phenylindole, dihydrochloride (DAPI), protect from light, store at ≤20 °C.
8. Rhodamine phalloidin (Molecular Probes), MW ~1,250, dissolved in methanol. Protect from light, store at ≤20 °C.

2.3 Detection Methods

1. Fluorescence microscope.
2. Flow cytometer.

3 Methods

3.1 Procedure Using Fluorescence Microscopy

This section describes the protocol for detecting fluorescence labeled particles engulfed by cells using a fluorescence microscope. This protocol has been used in published studies by the authors and other investigators [19, 21, 27, 32]. In this protocol, RAW 264.7 cells (a macrophage-like cell line) and FITC-labeled latex beads were used. Assays for phagocytosing other particles should be similar, except that the step where addition of FITC-labeled beads is replaced by the addition of other particles (*see* **Notes 2** and **3**).

1. Before seeding the cells, place glass coverslip into wells of 12-well plate after flame- sterilizing the coverslip by dipping in 100 % alcohol and passing through flame. Alternatively, cells can be seeded directly into the wells if plates can be placed on the stage of the microscope for further analysis (*see* **Note 4**).
2. Seed RAW 264.7 cells at 10^4 cells/cm^2. The seeding cell density depends upon the treatment period. This seeding density is optimal for treatment for 24 h. For longer treatments, less number of cells should be seeded (*see* **Note 5**).
3. Before starting the phagocytosis assay, FITC labeled beads should be opsonized by mixing with appropriate opsonizing material for 1 h. Opsonization can be done by either using FBS or IgG, depending upon the nature of the study. The volume of FBS or IgG can vary, but should be enough to cover the beads. Number of beads can range from 10 to 100 times of the cell number.
4. Following opsonization, suspend the beads into normal culture media and add to the plates with RAW cells. Allow phagocytosis to proceed for 1–2 h.
5. At the end of the phagocytosis, remove culture plate from incubator and keep on ice for 20–30 s to give thermal shock to stop the phagocytosis.
6. Remove the medium and wash the cells twice with cold PBS to remove all the non-phagocytosed beads.

7. Add 0.04 % Trypan Blue solution for 30 min to quench the fluorescence of beads remained outside of the cells (*see* **Note 6**).
8. Wash the cells twice with PBS.
9. Fix the cells using 4 % paraformaldehyde for 10 min.
10. Wash again twice with PBS.
11. To visualize the cell nuclei, incubate cells with DAPI solution (typically, 1:1,000 dilution of 14.3 mM stock solution in PBS) for 10 min (*see* **Note 7**).
12. Wash the cells twice with PBS.
13. Add rhodamine phalloidin (5 U/ml) to stain the actin cytoskeleton for at least 2 h. Best results are obtained when incubated overnight at 4 °C.
14. Wash twice with PBS.
15. Remove the coverslips from cell culture plates and mount them carefully on glass slides. If it is done in wells, you can directly take the pictures in the wells.
16. For quantification and data representation, usually 5–10 pictures of random fields should be taken. Approximately 200 cells should be counted for number of phagocytosed beads. Phagocytic activity can be represented as beads/cell or expressed as % of cells having a minimum number of beads (Fig. 1).

3.2 Procedure Using Flow Cytometry

Use of flow cytometry to assess the phagocytic ability of cells can overcome many of the limitations in methods using fluorescence microscopy. Application of flow cytometry significantly increases the number of cells analyzed and reduces the effort of taking pictures and manual counting of phagocytosed particles, which makes the assay robust and convenient. It has been described in many published studies using a wide range of cell types and multiple phagocytic particles [33, 34].

3.2.1 Sample Preparation

1. Seed the RAW 264.7 cells into 6-well plate with density 10^4 cells/cm^2.
2. Perform the similar phagocytosis assay as mentioned in the above protocol using FITC-labeled latex beads.
3. Stop phagocytosis by adding ice cold PBS or putting the plate on ice for 30 s.
4. Wash off the extracellular beads by multiple washings (*see* **Note 8**).
5. Remove the cells from the well by trypsinization or gentle scrapping using a cell scrapper.
6. Additional wells containing no phagocytic particles (only cells) and only phagocytic particles (FITC-labeled beads here) should be used as controls.
7. Centrifuge the cells and fix with 4 % paraformaldihyde.

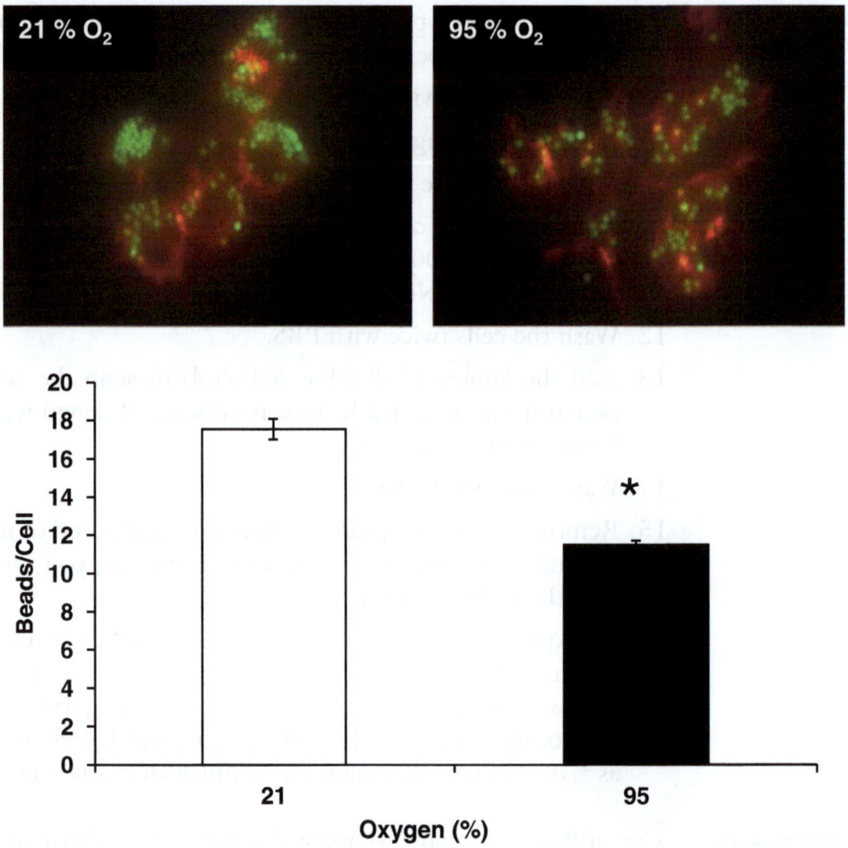

Fig. 1 Phagocytosis assay showing a reduction in the phagocytic activity of macrophages upon exposure to 95 % oxygen for 24 h. Data are derived from independent experiments ($n=4$). $*P<0.05$ compared to 21 % oxygen

8. Wash off the paraformaldehyde and resuspend the cells in 1 ml PBS.
9. The cells should be ready for flow cytometry analysis.

3.2.2 Data Analysis

1. Analyze the phagocytosis by flow cytometry using appropriate filters depending upon the labeling of the phagocytic particles; in this case FL-1 for FITC-labeled latex beads (Accuri c6 flow cytometry, Becton Dickinson).
2. Fluorescence intensities of cells and phagocytic particles should be assessed as controls.
3. In case of mixture of cells, phagocytosis of individual cell type can be assessed using fluorescence labels against cell specific markers.
4. Count at least 30,000 events. Data analysis can be performed using any flow cytometry software. Here, we used Cflow software (Becton Dickinson).
5. Phagocytosis can be assessed by measuring increase in fluorescence intensity compared with controls (Fig. 2).

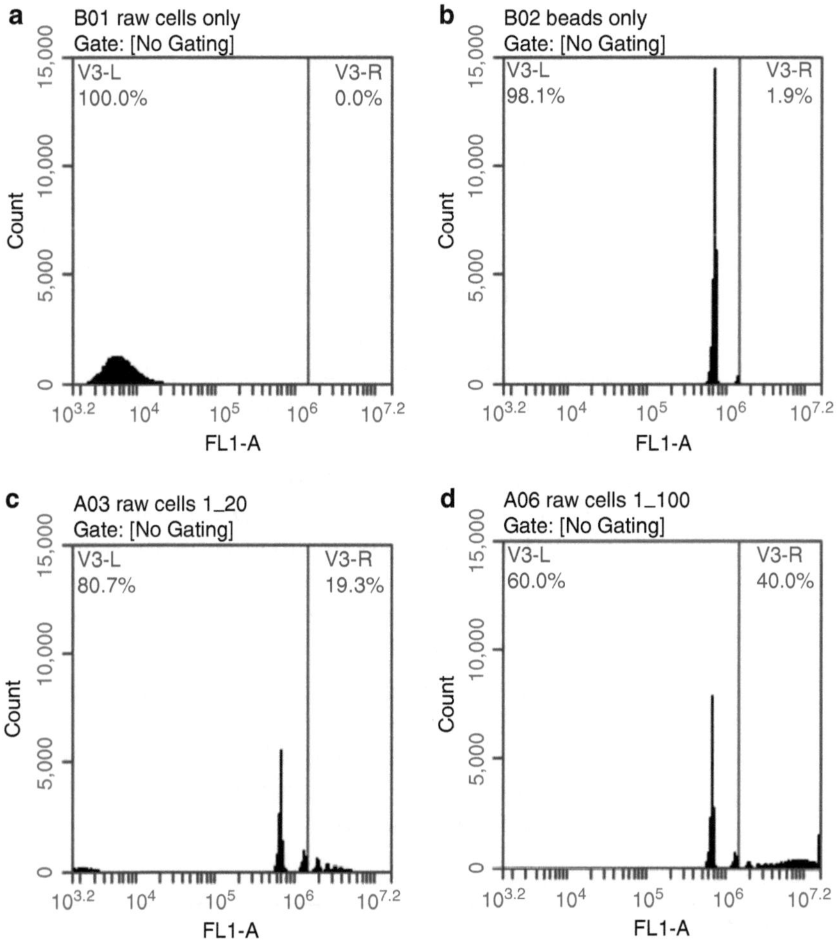

Fig. 2 Flow-cytometric analysis of phagocytosis was assessed using two different ratios of cells to beads. Only cells (**a**) and only beads (**b**) were used as controls. Note that there is a shift in fluorescence from 1.9 % (**b**) to (19.3 %, **c**) upon addition of beads to cultured macrophages at 20 beads/cell; to (40.0 %, **d**) at 100 beads/cells

4 Notes

1. All the solutions used in the experiment, such as DAPI and phalloidin, should be freshly prepared.
2. Protocol mentioned above was done using FITC-labeled latex beads. Beads can be replaced by other fluorescence labeled particles (such as fluorescence labeled *E. coli* bioparticles or zymosan particles). Such particles are readily available.
3. The entire procedure, which includes fluorescence (phagocytic particles, DAPI, rhodamine, etc.), should be done in the dark or where the exposure to light is minimal.
4. Extreme care is required to place the coverslip into wells and equal care is needed when taking it out. It might break easily.

Alternatively, chamber slides (Cat # 354102, BD Falcon) can be used to avoid handling coverslips.

5. The number of cells should be decided according to the cells used and the duration of the experiment. Cells that grow fast will soon form colonies and cells in the center of colonies will not be able to access the particles to phagocytose. On the other hand, terminally differentiated cells, such as alveolar macrophages do not grow like RAW cells, so that the final number of cells will be the same as seeded.
6. For quenching, ethidium bromide can be used as an alternative to the trypan blue. In case of some particles, such as *E. coli* bioparticles, which show fluorescence only upon engulfment, quenching is not needed.
7. For nuclear and cytoskeleton staining, antibodies can also be used. However, phalloidin and DAPI staining are very promising and are usually better than antibody staining. DAPI has not been shown in the pictures presented in Fig. 1. However, it is important to assess the number of cells, especially when there are clusters of cells present.
8. For flow-cytometric analysis, the extracellular beads should be washed off properly. Otherwise, beads will be a major proportion of the events counted, affecting quality of the results. This problem can be circumvented by using particles having only intracellular fluorescence.

Acknowledgments

This work was supported by grants (LLM) from National Heart and Blood Institute (HL093708) and St. John's University.

References

1. Metchnikoff I (1884) A disease of Daphnia caused by a yeast. A contribution to the theory of phagocytes as agents for attack on disease-causing organisms. Archiv Pathol Anat Physiol Klin Med 96:177–195
2. Flannagan RS, Jaumouillé V, Grinstein S (2012) The cell biology of phagocytosis. Annu Rev Pathol 7:61–98
3. Underhill DM, Goodridge HS (2012) Information processing during phagocytosis. Nat Rev Immunol 12:492–502
4. Swanson JA (2008) Shaping cups into phagosomes and macropinosomes. Nat Rev Mol Cell Biol 9:639–649
5. Garin J, Diez R, Kieffer S et al (2001) The phagosome proteome insight into phagosome functions. J Cell Biol 152:165–180
6. Greenberg S, Grinstein S (2002) Phagocytosis and innate immunity. Curr Opin Immunol 14: 136–145
7. Rabinovitch M (1995) Professional and non-professional phagocytes: an introduction. Trends Cell Biol 5:85–87
8. Martin TR, Frevert CW (2005) Innate immunity in the lungs. Proc Am Thorac Soc 2:403
9. Zhang P, Summer WR, Bagby GJ et al (2000) Innate immunity and pulmonary host defense. Immunol Rev 173:39–51

10. Savill J, Dransfield I, Gregory C et al (2002) A blast from the past: clearance of apoptotic cells regulates immune responses. Nat Rev Immunol 2:965–975
11. Hodge S, Hodge G, Scicchitano R et al (2003) Alveolar macrophages from subjects with chronic obstructive pulmonary disease are deficient in their ability to phagocytose apoptotic airway epithelial cells. Immunol Cell Biol 81:289–296
12. Berenson CS, Garlipp MA, Grove LJ et al (2006) Impaired phagocytosis of nontypeable Haemophilus influenzae by human alveolar macrophages in chronic obstructive pulmonary disease. J Infect Dis 194:1375–1384
13. McClure CD, Schiller NL (1996) Inhibition of macrophage phagocytosis by Pseudomonas aeruginosa rhamnolipids in vitro and in vivo. Curr Microbiol 33:109–117
14. Smith ME (2001) Phagocytic properties of microglia in vitro: implications for a role in multiple sclerosis and EAE. Microsc Res Tech 54:81–94
15. Andrews T, Sullivan KE (2003) Infections in patients with inherited defects in phagocytic function. Clin Microbiol Rev 16:597–621
16. Taylor A, Finney-Hayward T, Quint J et al (2010) Defective macrophage phagocytosis of bacteria in COPD. Eur Respir J 35:1039–1047
17. Donnelly LE, Barnes PJ (2012) Defective phagocytosis in airways disease. Chest J 141:1055–1062
18. O'Reilly PJ, Hickman-Davis JM, Davis IC et al (2003) Hyperoxia impairs antibacterial function of macrophages through effects on actin. Am J Respir Cell Mol Biol 28:443–450
19. Morrow DMP, Entezari-Zaher T, Romashko J III et al (2007) Antioxidants preserve macrophage phagocytosis of Pseudomonas aeruginosa during hyperoxia. Free Radic Biol Med 42:1338–1349
20. Suttorp N, Simon LM (1983) Decreased bactericidal function and impaired respiratory burst in lung macrophages after sustained in vitro hyperoxia. Am Rev Respir Dis 128: 486–490
21. Patel VS, Sitapara RA, Gore A et al (2013) High Mobility Group Box-1 mediates hyperoxia-induced impairment of Pseudomonas aeruginosa clearance and inflammatory lung injury in mice. Am J Respir Cell Mol Biol 48:280–287
22. Phipps JC, Aronoff DM, Curtis JL et al (2010) Cigarette smoke exposure impairs pulmonary bacterial clearance and alveolar macrophage complement-mediated phagocytosis of Streptococcus pneumoniae. Infect Immun 78:1214–1220
23. Davis KA (2006) Ventilator-associated pneumonia: a review. J Intensive Care Med 21:211–226
24. Hartl D, Latzin P, Hordijk P et al (2007) Cleavage of CXCR1 on neutrophils disables bacterial killing in cystic fibrosis lung disease. Nat Med 13:1423–1430
25. Ren Y, Savill J (1995) Proinflammatory cytokines potentiate thrombospondin-mediated phagocytosis of neutrophils undergoing apoptosis. J Immunol 154:2366–2374
26. Lotze MT, Tracey KJ (2005) High-mobility group box 1 protein (HMGB1): nuclear weapon in the immune arsenal. Nat Rev Immunol 5:331–342
27. Entezari M, Weiss DJ, Sitapara R et al (2012) Inhibition of HMGB1 enhances bacterial clearance and protects against P. aeruginosa pneumonia in cystic fibrosis. Mol Med 18:477–485
28. van Zoelen MA, Ishizaka A, Wolthuis EK et al (2008) Pulmonary levels of high-mobility group box 1 during mechanical ventilation and ventilator-associated pneumonia. Shock 29:441–445
29. Liu G, Wang J, Park YJ et al (2008) High mobility group protein-1 inhibits phagocytosis of apoptotic neutrophils through binding to phosphatidylserine. J Immunol 181:4240
30. Kennedy TP, Nelson S (2013) Hyperoxia, HMGB1, and ventilator-associated pneumonia: reducing risk by practicing what we teach. Am J Respir Cell Mol Biol 48:269–270
31. Barile FA (2013) Cell culture methodology. In: Principles of toxicology testing, vol 2, 2nd edn, CRC press, Boca Raton, pp 163–186
32. Drevets DA, Campbell PA (1991) Macrophage phagocytosis: use of fluorescence microscopy to distinguish between extracellular and intracellular bacteria. J Immunol Methods 142:31–38
33. Dunn P, Tyrer H (1981) Quantitation of neutrophil phagocytosis, using fluorescent latex beads. Correlation of microscopy and flow cytometry. J Lab Clin Med 98:374–381
34. Lehmann AK, Sørnes S, Halstensen A (2000) Phagocytosis: measurement by flow cytometry. J Immunol Methods 243:229–242

Chapter 13

Assessment of Cytokine-Modulated Proteasome Activity

Christopher J. Kirk, Saul R. Powell, and Edmund J. Miller

Abstract

This chapter presents two methods for assessment of proteasome function. The first is a modification of the standard fluorogenic peptide cleavage assay which takes into account the effect of ATP on proteasome activity. This method is described in both its macro and high throughput micro-assay forms. The second is the *Pro*teasome *C*onstitutive *I*mmuno-*S*ubunit (active site) *E*LISA or ProCISE method. ProCISE is a modification of active site directed probe analysis and allows for convenient differentiation between active constitutive and immuno-subunits. While the utility of measuring proteasome activity and its relationship to cytokine action and inflammation are clear, the assessment and interpretation is not always straightforward. Therefore, we also discuss the pitfalls of the standard fluorogenic assay, particularly in the interpretation of results obtained, and the advantages of the newer, ProCISE assay.

Key words Proteasome, Immunoproteasome, ProCISE, Chymotryptic activity, ATP-dependent activity

1 Introduction

Mammalian cells have a number of major pathways for the degradation of proteins including lysosomal proteases [1], calcium-dependent calpains [2], and proteasomes [3]. The ubiquitin-proteasome system (UPS) is the major non-lysosomal, ATP dependent route for the degradation of proteins, removing denatured, misfolded, damaged, or mistranslated proteins from cells. However it is now evident that the proteasome is more than a cellular "waste disposal" system. The UPS serves as a highly complex, tightly regulated system that selectively degrades abnormal and short-lived proteins [4], and plays a critical role in a variety of highly interconnected cellular pathways, making proteasomal activity critical to normal cell homeostasis, and the regulation of inflammatory and stress responses (Fig. 1). Of particular importance is the role of the UPS in cytokine signaling, where it is a key regulator of inflammatory cytokine production [5, 6]. Altered, or inappropriate UPS activity has been linked to the pathogenesis of a variety of disease states including cardiovascular disease [7, 8],

Ivana Vancurova (ed.), *Cytokine Bioassays: Methods and Protocols*, Methods in Molecular Biology, vol. 1172, DOI 10.1007/978-1-4939-0928-5_13, © Springer Science+Business Media New York 2014

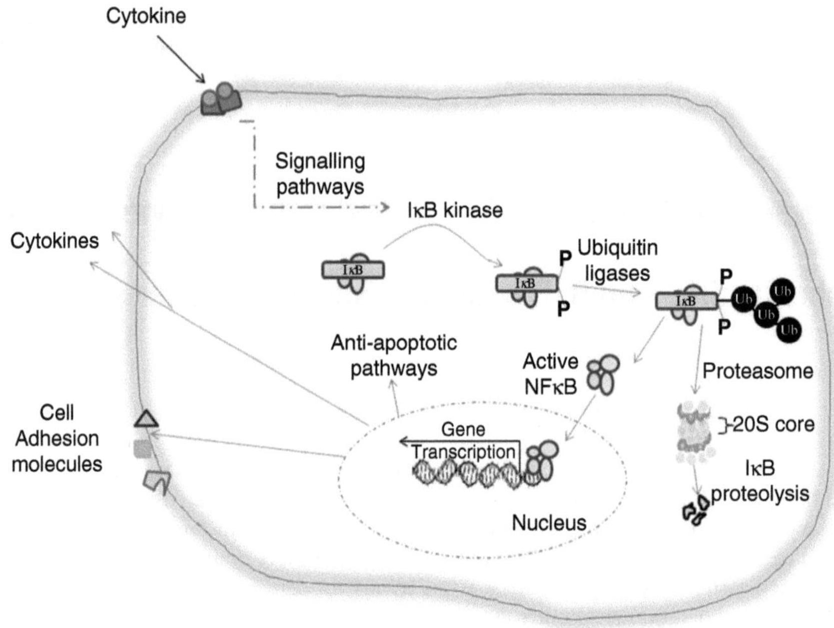

Fig. 1 Schematic of proteasome involvement with cytokine signaling. The UPS is a complex, highly regulated system that selectively degrades abnormal and short-lived proteins and plays a critical role in cytokine signaling

inflammatory bowel disease [9, 10], ischemia-reperfusion injury [11, 12], type-1 diabetes [13, 14], pulmonary fibrosis [15], Alzheimer's disease [16], graft rejection [17], and other immunopathologies [18–22], as well as changes noted in aging [23].

Proteasomes are large macromolecular complexes derived from at least 14 discrete subunits (α1–7 and β1–7) that assemble in ring structures [24, 25]. The 26S proteasome consists of two subcomplexes, the 20S proteolytic core and 2 regulatory particles. The 19S regulatory particles cap either end of the 20S and confer specificity for ubiquinated proteins. Alternatively, the 20S can be capped by the 11S activator ring consisting of PA28α and PA28β subunits [26, 27]. Docking of an 11S to a 20S core increases its proteolytic capacity without affecting catalytic subunit content [28, 29]. There are two major classes of proteasomes: the constitutive proteasome (c20S), expressed ubiquitously throughout the body, and the immunoproteasome (i20S), expressed primarily in cells of hematopoietic origin or in nonhematopoietic cells exposed to inflammatory cytokines [30–32]. In the c20S, proteolytic activities are encoded in the β5, β1 and β2 subunits and are characterized based on their substrate specificities as chymotrypsin-like (CT-L), caspase-like, and trypsin-like, respectively. In the i20S, the LMP7, LMP2, and MECL1 subunits replace β5, β1 and β2, respectively. Furthermore, while newly assembled proteasomes may contain a mixture of inducible β1*i* and constitutive subunits [24], all proteasome active-site subunits contain an N-terminal threonine as their catalytic nucleophile, which distinguishes them from other classes of proteases [33]. Thus, there are many variants of the proteasome, some of which may be tissue specific [34–37]. However, the precise roles of individual proteasome isoforms remain largely undefined.

Since almost all basic cellular functions, including cell surface receptor induced signal transduction, control of the cell cycle, and cellular stress responses involve proteasome activity, it is perhaps not surprising that altered proteasome function has been linked to a variety of disease states. Thus inhibitors of proteasome function have been developed, to study and rectify some of these changes. Both synthetic inhibitors, including peptide-aldehydes and -vinyl sulfones, and natural inhibitors such as lactacystin and epoxyketones have been studied. Each of these inhibitors binds to, and directly inhibits, active sites within the 20S core particle of the proteasome structure, with various degrees of specificity for particular catalytic subunits [38–40]. Some of these inhibitors have been effective clinically particularly in multiple myeloma, Waldenström macroglobulinemia, low-grade non-Hodgkin lymphomas, primary amyloidosis and other malignancies [41, 42]. In addition, protease inhibitors can have direct effects on inflammatory pathways, controlling the degradation of IκB, and the activity of NFκB, which in turn controls the expression of inflammatory mediators such as tumor necrosis factor, interleukin-1, and intracellular and vascular adhesion molecules [43].

While the utility of measuring proteasome activity and its relationship to cellular pathways is clear, the assessment and interpretation is not always straightforward [44]. In this chapter, two methods for assessment of proteasome function are presented. The first is a modification of the standard fluorogenic peptide cleavage assay described by Reinheckel et al. [45] that we have modified [46] to take into account the effect of ATP on proteasome activity and will be described both as a macro-assay and a high throughput micro-assay. The second and newest method is the *Pro*teasome *C*onstitutive *I*mmuno-*S*ubunit (active site) *E*LISA or ProCISE method first introduced in 2009 [47]. ProCISE is a modification of active site directed probe analysis and allows for convenient differentiation between active constitutive and immuno-subunits.

2 Materials

2.1 Fluorogenic Small Peptide Cleavage Assay

Prepare all solutions in distilled deionized water (d_2H_2O). Prepare and store all solutions at room temperature unless otherwise indicated.

1. HEPES buffer: 50 mM HEPES, pH 7.5. Add about 200 mL of d_2H_2O to a 500 mL glass beaker or graduated cylinder. Weigh out 6.50 g of HEPES sodium salt (>99.5 %) and transfer to the beaker. 20 mM KCl. Weigh out 0.745 g of KCl

($>$99 %) and transfer to the beaker. 5 mM $MgCl_2 \cdot 6H_2O$. Weigh out 0.508 g of $MgCl_2 \cdot 6H_2O$ ($>$99 %) and transfer to the beaker. 1 mM Dithiothreitol (DTT). Weigh out 0.077 g of DTT and transfer to the beaker. Bring volume to about 450 mL with d_2H_2O and correct pH to 7.5 with dilute HCl. Bring to 500 mL final volume (*see* **Note 1**).

2. 50 mM ATP stock: Weigh out and add 0.02755 g of adenosine triphosphate, disodium salt ($>$99 %) per mL of HEPES buffer (*see* **Note 2**).
3. Macroassay ATP dilutions: Dilute 60 μL of 50 mM ATP stock to 2 mL with HEPES buffer to yield 1.5 mM ATP solution. Do serial 1:2 dilutions with HEPES buffer to yield 0.75, 0.375, and 0.188 mM final concentrations.
4. Microassay ATP dilutions: Dilute 35 μL of 50 mM ATP stock to 2 mL with HEPES buffer to yield 0.875 mM ATP solution. Do serial 1:2 dilutions with HEPES buffer to yield 0.438, 0.219, and 0.110 mM final concentrations.
5. 0.48 mM Lactacystin: Lactacystin (ENZO Life Sciences) comes in a 200 μg vial. Add 1.1 mL of HEPES buffer to yield 0.48 mM stock solution. Keep frozen at −20 °C until used.

   For Macroassay, dilute 0.48 mM lactacystin 1:2 with HEPES buffer to yield 0.24 mM solution. For Microassay, dilute 0.48 mM lactacystin 1:3.4 with HEPES buffer to yield 0.140 mM solution.
6. 4.3 mM Suc-LLVY-AMC (substrate for chymotryptic activity): Weigh out 5 mg of Suc-LLVY-AMC (ENZO Life Sciences) in 1.511 mL of DMSO in glass-stoppered tube. Store refrigerated until used.

   For Macroassay, dilute 4.3 mM substrate 1:10 with HEPES Buffer to yield 0.43 mM final concentration.

   For Microassay, dilute 4.3 mM substrate 1:17.2 with HEPES buffer to yield 0.25 mM final concentration.
7. 2.16 mM Z-LLE-AMC (substrate for caspase-like activity): Add 2.65 mg of Z-LLE-AMC (ENZO Life Sciences) to 1.85 mL of DMSO in glass stoppered tube. Store at −20 °C for up to 3 months.

   For Macroassay, dilute 2.16 mM substrate 1:2 with HEPES buffer to yield 1.08 mM final concentration.

   For Microassay, dilute 2.16 mM substrate 1:3.4 with 50 mM HEPES to yield 0.63 mM final concentration.

2.2 ProCISE Assay

2.2.1 Buffers, Plates Capture Beads, and Detection Reagents

1. Lysis buffer: 20 mM Tris–HCl, pH 8.0, 5 mM EDTA.
2. Assay buffer: 20 mM Tris–HCl, pH 8.0, 0.5 mM EDTA.
3. ELISA buffer: Phosphate buffered saline (PBS) containing 1 % bovine serum albumin (w/v), 0.1 % Tween-2 (v/v).

4. SDS buffer: 2 % SDS (w/v) in ddH_2O.
5. 8 M guanidine hydrochloride denaturant: 38.2 g of guanidine hydrochloride in 21 mL of Assay Buffer; this will yield 50 mL of 8 M solution. Complete dissolution may require ≥1 h.
6. 96-Well or 384-well Multiscreen DV opaque, non-sterile filter plates (Millipore).
7. Streptavidin-sepharose beads.
8. ELISA pico substrate (Pierce/Thermo Fisher).

2.2.2 Proteasome Subunit Antibodies and Secondary Antibodies

1. Monoclonal antibodies recognizing β1, β2, LMP7, and LMP2 (Biomol or Enzo Lifesciences) and MECL-1 (Santa Cruz Biotechnology).
2. Rabbit polyclonal antibodies against human β5 generated in-house but commercial sources can be used.
3. Horseradish peroxidase (HRP)-conjugated goat anti-rabbit, and goat anti-mouse (Jackson ImmunoResearch) and HRP-conjugated rabbit anti-goat (Invitrogen) antibodies.

2.2.3 Proteasome Active Site Probe and Equipment

1. PR-584 was manufactured as described elsewhere [48]. Stored as a 10 mM stock solution in DMSO.
2. Vacuum manifold with plate fixture (Millipore).
3. Plate shaker with variable speed.
4. GENios-Basic plate reader (Tecan Austria GmBH).

3 Methods

3.1 Fluorogenic Small Peptide Cleavage Assay

This is the most commonly employed assay and follows cleavage of proteasome-specific peptides linked to a fluorophore, usually amido-4-methylcoumarin (AMC) or β-napthylamide (βNA). This assay forms the basis for many of the commercially available kits for determination of proteasome activity including the Chemicon and Enzo 20S Proteasome Assay kits. Use of a peptide linked to the luminescent probe, aminoluciferin, in combination with a buffer optimized for cell permeabilization has been used to monitor proteasome activity in cell preparations [49] and is the basis for the commercially available Proteasome-Glo™ (Promega) assay system. Typically, these assays are applied to *ex vivo* systems containing either a purified proteasome preparation or a tissue isolate. Most of these assay kits do not take into account the requirement by the proteasome for ATP in order to observe maximal activity [50, 51]. On the other hand some investigators have suggested adding 2 mmol/L ATP or higher [45, 52] to purified or semi-purified preparations. We have observed that at least in rat tissue isolates, 2 mM ATP is too high and actually inhibits proteasome [46].

To observe maximal proteasome activity which is representative of total proteasome activity requires much lower assay mixture ATP concentrations and is dependent on the amount of tissue lysate protein. Purchasing commercial kits is not necessary as the procedure for determining proteasome activity in tissue homogenates and lysates is quite straightforward. Carry out all procedures and methods at room temperature unless otherwise indicated.

3.1.1 Preparation of Tissue Lysate

1. Homogenize frozen tissue in polytron-like tissue homogenizer at low power settings (*see* **Note 3**) in 10 volumes of ice-cold HEPES buffer taking care to prevent heating of sample (*see* **Note 4**).
2. Centrifuge the tissue homogenate at 10,000 × *g* for 30 min at 4 °C.
3. Carefully pipette the supernatant and discard the pellet.
4. Keep the supernatant at 4 °C and immediately adjust for protein content. The supernatant should not be frozen and should be used as rapidly as possible for determination of peptidase activities (*see* **Note 5**).

3.1.2 Macro-assay Procedure

The suggested protocols are scaled for a macroassay with a final reaction mixture volume of 240 μL. All reactions should be run in duplicate minimum.

1. Add 30 μg protein in 10 μL of the supernatant to 200 μL of HEPES buffer to a series of 9–12 × 75 mm glass tubes labeled 1–9 (*see* **Note 6**). Allow protein to equilibrate at room temperature for a few minutes.
2. Add ATP dilutions to tubes (*see* **Note 7**). Add 10 μL of HEPES buffer to tubes 1 and 6 (final, 0 μM ATP). Add 10 μL of 1.5 mM ATP to tube 2 (final, 62.5 μM ATP). Add 10 μL of 0.75 mM ATP to tubes 3 and 7 (final, 31.8 μM ATP). Add 10 μL of 0.375 mM ATP to tubes 4 and 8 (final, 15.9 μM ATP). Add 10 μL of 0.188 mM ATP to tubes 5 and 9 (final, 8 μM ATP).
3. Add 10 μL of 0.24 mM lactacystin to tubes 6–9 (*see* **Note 8**). Add 10 μL of HEPES buffer to tubes 1–5. Allow tubes to preincubate at room temperature for 10 min.
4. Add 10 μL of 0.25 mM Suc-LLVY-AMC (18 μM final) *or* 10 μL of 1.08 mM Z-LLE-AMC (45 μM final) to bring the final volume of the reaction mixture to 240 μL (*see* **Note 9**).
5. Incubate the reaction mixture at 37 °C for 30 min.
6. Add 300 μL of ice-cold ethanol to quench the reaction and let sit at room temperature for 10 min.
7. Add an additional 1 mL of d_2H_2O in 10 min (*see* **Note 10**).
8. Read using fluorescence spectroscopy at excitation wavelength, 380 nm, and emission wavelength, 440 nm.

3.1.3 Microassay Procedure

The suggested protocols are scaled for a microassay designed for a 96 well plate with a final reaction mixture volume of 140 μL. All reactions should be run in duplicate minimum (*see* **Note 11**).

1. Add 30 μg of protein in 10 μL of the supernatant to 100 μL of HEPES buffer to each well in a single row labeled A–H of a 96 well plate. Allow protein to equilibrate at room temperature for a few minutes.
2. Add ATP dilutions to wells. Add 10 μL of HEPES buffer to wells A and F (final, 0 μM ATP). Add 10 μL of 0.875 mM ATP to well B (final, 62.5 μM ATP). Add 10 μL of 0.438 mM ATP to wells C and G (final, 31.8 μM ATP). Add 10 μL of 0.219 mM ATP to wells D and H (final, 15.9 μM ATP). Add 10 μL of 0.110 mM ATP to well E (final, 8 μM ATP).
3. Add 10 μL of 0.14 mM lactacystin to wells F–H. Add 10 μL of HEPES buffer to wells A–E. Allow reaction to pre-incubate at room temperature for 10 min.
4. Add 10 μL of 0.43 mM Suc-LLVY-AMC (18 μM final) *or* 10 μL of 0.63 mM Z-LLE-AMC (45 μM final) to bring the final volume of the reaction mixture to 140 μL.
5. Incubate the reaction mixture at 37 °C for 30 min.
6. Add 100 μL of ice-cold ethanol to quench the reaction and let sit at room temperature for 10 min.
7. Read using fluorescent microplate reader at excitation wavelength, 380 nm, and emission wavelength, 440 nm.

3.1.4 Calculations

Three different proteasome activities can be calculated (*see* Fig. 2).

(a) *Non-ATP dependent*: the so-called basal activity that is observed in the absence of added ATP.

(b) *Total activity*: defined as the maximal activity in the presence of the optimal ATP concentration.

(c) *ATP-dependent activity*: defined as the total activity minus the non-ATP activity.

Specific activity can be determined by running the reaction against an AMC standard. Differentiation of these different activities is important as changes may reflect alterations not only in proteasome function but also possible damage to, or shifts in the regulatory particles that cap the 20S proteasome [53, 54].

3.1.5 Pitfalls

Interpretation of results obtained with the use of these peptides can be misleading. Many investigators make the assumption that simply assaying proteasome activity with a fluorogenic peptide substrate accurately reflects 26S-proteasome activity. This is incorrect because 26S-proteasome activity is dependent on the presence of ATP and should be assessed over a range of ATP concentrations as described. Many factors can lead to spurious, inconsistent

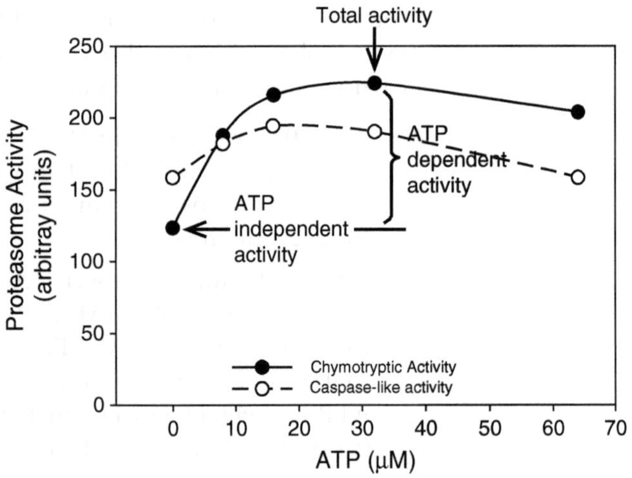

Fig. 2 Representative graph illustrating relationship between ATP concentration and proteasome activity. Depicted here is the chymotryptic activity determined using Suc-LLVY-AMC as a substrate and the caspase-like activity using Z-LLE-AMC as a substrate. In this example, the ATP independent chymotryptic activity is 125 AU. The total chymotryptic activity is 225 AU. The ATP dependent activity is 100 AU (225–125). Caspase-like activity would be calculated in the same way

determination of proteasome activity. These include the presence of detergents, repetitive freeze–thaw cycles, and even the fluorogenic peptide itself which may trigger channel opening [51]. Kissalev has suggested determining multiple peptidase activities of the proteasome using multiple fluorogenic peptides that target different activities [51]. Finally, estimation of the individual proteolytic activities (chymotryptic vs. tryptic vs. caspase-like), does not differentiate between the relative contributions of the constitutive or immunoforms of the catalytic subunits. Making this distinction is important and can alter interpretation of results as these two proteasome isoforms have different functions. The constitutive proteasome is primarily involved with intracellular protein quality control [7, 55], while the immunoproteasome is involved with immune function and regulation of inflammatory pathways [5, 6, 56]. At present the only way to distinguish between these is ProCISE which is described in the next section.

3.2 ProCISE Assay

The standard method for determining proteasome activity uses biochemical substrates, such as Leu-Leu-Val-Tyr-7-methyl-4-aminocoumarin (LLVY-AMC) for determining CT-L activity [57]; however, these substrates cannot differentiate between activity derived from active sites of the c20S (β5) and i20S (LMP7). In addition, the role of the non-CT-L active sites and their inhibition by proteasome inhibitors cannot be easily determined because of the lack of appropriate experimental tools. With these issues in mind, the ProCISE (proteasome constitutive *immunoproteasome*

Table 1
Sample concentration ranges for ProCISE

Tissue	Sample concentration (mg/mL)
Adrenal gland	1–3
Brain	5–7
Heart	4
Spleen	4
Kidney	5
Liver	7
Blood	10
Skin	3
Ear	2–3
Tumor	1–3

subunit ELISA) assay was developed based on enzyme-linked immunosorbent assay (ELISA) and designed to measure the amounts and inhibition of proteasome active sites in cells and tissue samples [47, 58]. ProCISE quantitatively measures the levels of individual proteasome subunit active sites within cell lysates via the addition of a biotinylated active-site probe that covalently binds to all active sites, capture of probe-bound active sites on streptavidin beads. The ProCISE assay involves the following steps: (1) incubation of the probe with activated 20S (in tissue lysates), (2) denaturation with 6 M guanidine hydrochloride, (3) addition of streptavidin-coated beads, (4) extensive bead washing, (5) addition of proteasome subunit-specific primary antibody followed by a secondary HRP-conjugated antibody, and (6) luminescence-based detection.

3.2.1 Preparation of Tissue Lysates and Probe Addition

1. Tissues should be washed with PBS and stored frozen at −80 °C prior to analysis.
2. Homogenize tissue in Lysis Buffer in 1.5 or 2 mL microcentrifuge tubes.
3. Spin homogenates at maximum speed in a microcentrifuge for 15 min at 4 °C.
4. Dilute supernatant samples with Lysis Buffer in microcentrifuge tubes or a polypropylene 96-well plate and determine protein concentration. Final concentrations ranges vary by tissue type (Table 1). For triplicate samples across all 6 subunits, 230 μL of lysate is required.

5. Mix 390 μL of Lysis Buffer and 10 μL of 10 mM probe in a microcentrifuge tube
6. Add 4.7 μL of probe dilution to each 230 μL of lysate supernatant. Mix well.

Incubate at 25 °C for 2 h. Use a water bath for incubations in microcentrifuge tubes and a plate incubator for 96-well plates.

3.2.2 Denaturant, Bead, and Plate Preparation

1. Perform during 2-h probe incubation.
2. Make 8 M guanidine hydrochloride denaturant.
3. Wash beads. Wash 3.5 mL of resuspended streptavidin beads with 15 mL of ELISA Buffer in a 50 mL conical tube. Pipette up and down to mix and rinse pipette. Q.S. to 50 mL with ELISA Buffer and centrifuge at ~2,000 × g for 5 min. Decant supernatant and repeat wash two times more. Resuspend beads with 14 mL of ELISA Buffer and place on tube rocker at room temperature.
4. Equilibrate filter plates. Label 6 filter plates, one for each antibody to be used: Beta 5, Beta 1, Beta 2, LMP7, LMP2, or MECL1. Add 200 μL of ELISA Buffer to each well using the Matrix WellMate liquid dispenser. Allow plates to sit ≥30 min before use to allow membranes to hydrate and to block wells with BSA.

3.2.3 Filter Plate Loading

1. Remove the 200 μL of ELISA Buffer from each filter plate using the vacuum manifold. Firmly blot the bottom of each plate on a paper towel after vacuum filtration, to ensure that no liquid remains at the bottom of the filters that could allow fluid to wick through from the well.
2. Add 20 μL of washed beads to each well of the 6 filter plates using an electronic repeating 12-channel pipettor (e.g., fill starting at row H and ending with row A). Pour the beads into a reagent reservoir and agitate the bead suspension in between filling each plate.
3. Add 70 μL of 8 M guanidine hydrochloride to each well of the Beta 5, Beta 1, and Beta 2 plates using an electronic repeating 12-channel pipettor (starting at row H and ending with row A). Constitutive subunit plates are prepared first, then the immuno subunit plates, to minimize the time that beads are exposed to denaturant in the absence of sample. Proceed to next step immediately.
4. Using an electronic repeating multichannel pipettor, transfer 10 μL of each mouse testing sample to each well of the constitutive subunit filter plates containing beads and denaturant. Perform triplicate wells for each sample on each of the 3 constitutive plates (e.g., dispense 10 μL into wells A1-H1, then A2-H2, then A3-H3 of plate "Beta 5," then plate "Beta 1," then plate "Beta 2").

The same pipette tips can be used for all 9 volumes of sample added to the plates (3 replicates × 3 plates).

5. Repeat the addition of guanidine hydrochloride (*see* **step 3** of Subheading 3.2.3) to each well of the LMP7, LMP2, and MECL1 plates.
6. Immediately proceed to sample loading of the immuno subunit plates (*see* **step 4** of Subheading 3.2.3).
7. Place all plates on a titer plate shaker for 1 h at a high speed, to ensure that all samples mix well with denaturant (e.g., setting ~6–7 on a shaker with settings from 1 (minimum speed) to 10 (maximum speed)).

3.2.4 Plate Washing and Antibody Incubations

1. Remove unbound material by filtration on the vacuum manifold. Blot bottom of plates on a paper towel. Fill wells with 200 μL/well ELISA Buffer.
2. Wash a total of five times with 200 μL/well ELISA Buffer. After each fill, place the plates on a plate shaker for 1–2 min, using a moderate setting (~3–4). Blot bottom of plates on a paper towel after each vacuum filtration. For pauses, plates can be incubated in ELISA Buffer on the plate shaker.
3. In 15 mL conical tubes, dilute each *primary* antibody in ELISA Buffer (*see* **Note 12**). . Ensure that all antibody stocks are well-mixed prior to use. Following dilution, invert tube several times to mix thoroughly. Add 100 μL of antibody solution to each well of the appropriate subunit plate, using an electronic repeating multichannel pipettor (starting at row H and ending with row A).
   (a) Beta 5 (1:2,000) = 6 μL antibody into 12 mL ELISA Buffer.
   (b) Beta 1 (1:1,000) = 12 μL antibody into 12 mL ELISA Buffer.
   (c) Beta 2 (1:1,000) = 12 μL antibody into 12 mL ELISA Buffer.
   (d) LMP7 (1:2,000) = 6 μL antibody into 12 mL ELISA Buffer.
   (e) LMP2 (1:1,000) = 12 μL antibody into 12 mL ELISA Buffer.
   (f) MECL1 (1:5,000) = 2.4 μL antibody into 12 mL ELISA Buffer.
4. Cover the plates with lids and incubate overnight at 4 °C on a plate shaker, using a moderate speed (~3–4).
5. The next day, remove the primary antibody solution by vacuum filtration and wash a total of five times with 200 μL/well ELISA Buffer. After each fill, place the plates on a plate shaker

for 1–2 min, using a moderate setting (~3–4). Blot bottom of plates on a paper towel after each vacuum filtration.

6. In 15 mL conical tubes, dilute each *secondary* antibody in ELISA Buffer. Ensure that all antibody stocks are well-mixed prior to use. Following dilution, invert tube several times to mix thoroughly. Add 100 μL of antibody solution to each well of the appropriate subunit plate, using an electronic repeating multichannel pipettor (starting at row H and ending with row A).
   (a) Beta 5 (1:5,000, HRP-goat anti-rabbit) = 2.4 μL antibody into 12 mL ELISA Buffer.
   (b) Beta 1 (1:1,000, HRP-goat anti-mouse) = 12 μL antibody into 12 mL ELISA Buffer.
   (c) Beta 2 (1:1,000, HRP-goat anti-mouse) = 12 μL antibody into 12 mL ELISA Buffer.
   (d) LMP7 (1:5,000, HRP-goat anti-mouse) = 2.4 μL antibody into 12 mL ELISA Buffer.
   (e) LMP2 (1:5,000, HRP-goat anti-mouse) = 2.4 μL antibody into 12 mL ELISA Buffer.
   (f) MECL1 (1:5,000, HRP-rabbit anti-goat) = 2.4 μL antibody into 12 mL ELISA Buffer.
7. Cover the plates with lids and incubate for 2 h at room temperature on a plate shaker, using a moderate speed (~3–4).
8. Remove the secondary antibody solution by vacuum filtration and wash a total of five times with 200 μL/well ELISA Buffer. After each fill, place the plates on a plate shaker for 1–2 min, using a moderate setting (~3–4). Blot bottom of plates on a paper towel after each vacuum filtration.

3.2.5 Detection and Data Analysis

1. Set up a luminescence reading method on the plate reader (*see* **Note 13**).
2. Make 64 mL total of a 1:1 mixture of the two reagents in the SuperSignal ELISA Pico Chemiluminescent Substrate kit (e.g., 16 mL luminol/enhancer solution + 16 mL stable peroxide solution in each of *two* 50 mL conical tubes). Starting with one plate at a time, add 100 μL of substrate to each well using an electronic repeater multichannel pipettor. Incubate at RT on a plate shaker at a moderate setting (~3–4) for an optimal amount of time for signal generation (*see* **Note 14**). Time the addition of substrate and plate reading carefully to ensure that no more than one plate requires reading at a time.
3. Normalize RLU to protein concentrations. Analyze luminescence values (RLU) for samples (e.g., relative to controls) utilizing a graphing/mathematical software.

4 Notes

1. Some investigators have suggested the use of 10–20 % glycerol or 0.25 mmol sucrose to stabilize the complex [45, 59]. We find that if the lysate is used immediately, addition of these makes little difference.
2. We recommend that assay solutions and dilutions are made fresh.
3. We use a Polytron® tissue homogenizer. Since heart tissue is quite fibrous and can be difficult to disrupt we pulverize pieces of heart muscle that have been frozen at −70 °C and then suspend in 10 volumes of HEPES isolation and assay buffer. The power setting for heart muscle is 2–3 for 5 s which is repeated 4 times with intervening 30 s hiatuses. Depending on the fibrosity of the tissue, investigators will have to adjust the power setting and length of homogenization accordingly.
4. Polytron®-like homogenizers will generate a lot of heat while disrupting a tissue. Keep the homogenizing tube immersed in ice water.
5. Supernatant should not be frozen and then used as this leads to loss of ATP-dependent activity. Also supernatant should not be allowed to stand for long periods of time as this will also lead to loss of activity.
6. We have detected activity with as little as 10 μg of heart tissue isolate protein, however a minimum of 30 μg tissue isolate protein is suggested. Investigators using tissues from other organs will have to determine the optimal tissue concentration for their sample.
7. Peptidase activities should be determined over a series of four sequential dilutions of ATP so that the maximal activation is obtained. Maximal activation will depend on the particular sample and protein content. We have observed that for 30 μg rat heart protein, adding sufficient concentrated ATP in 10 μL to yield final concentrations of 0, 8, 16, 32, and 64 μM generally will yield maximal activation of the proteasome. There are species and tissue variability so investigators will need to determine maximal activating concentrations of ATP for their particular samples.
8. To confirm specificity of the assay, a separate series of samples that includes the 0 μM and 2–3 other ATP concentrations around the expected optimal should be preincubated with a proteasome peptidase inhibitor. For the 30 μg example above, chymotrypsin-like activity is inhibited with 10 μM lactacystin (final). Caspase-like activity can be inhibited with 10 μM Z-Pro-Nle-Asp-CHO (ENZO Life Sciences catalogue # BML-ZW9490-0100) (final). Little benefit is derived from the addition of higher concentrations of proteasome inhibitors.

9. We do not routinely determine trypsin-like activity because of high background nonproteasome-mediated cleavage of substrates.
10. In heart tissue lysates we have observed that cleavage of substrates was linear over this time period. For other tissues investigators will have to determine the linear time period.
11. We routinely limit the number of wells per sample to no more than 8. This allows for 5 concentrations of ATP and an additional 3 wells for inhibitor-treatment. This way the samples can be arranged in a single 8 well row allowing for the simultaneous determination of 6 different samples if analyzed in duplicate.
12. Primary and secondary antibody dilutions should be optimized for each new vendor or lot of antibody.
13. Note that some instruments may require equilibration time following changes in gain settings. For each plate reader utilized, prior empirical optimization should be done to establish gain settings, etc. to ensure that the full dynamic range of the instrument is utilized, without saturating signal.
14. For each plate reader utilized, time course or kinetic studies should also be performed to determine the optimal luminescence substrate incubation time for each subunit (luminescence generation will peak, then recede over time).

References

1. Dean RT (1979) Lysosomes and protein degradation. Ciba Found Symp 75:139–149
2. Murachi T, Tanaka K, Hatanaka M et al (1980) Intracellular Ca2+-dependent protease (calpain) and its high-molecular-weight endogenous inhibitor (calpastatin). Adv Enzyme Regul 19: 407–24
3. Arrigo AP, Tanaka K, Goldberg AL et al (1988) Identity of the 19S 'prosome' particle with the large multifunctional protease complex of mammalian cells (the proteasome). Nature 331:192–4
4. Wolf DH, Hilt W (2004) The proteasome: a proteolytic nanomachine of cell regulation and waste disposal. Biochim Biophys Acta 1695: 19–31
5. Angeles A, Fung G, Luo H (2012) Immune and non-immune functions of the immunoproteasome. Front Biosci 17:1904–16
6. Ebstein F, Kloetzel PM, Kruger E et al (2012) Emerging roles of immunoproteasomes beyond MHC class I antigen processing. Cell Mol Life Sci 69:2543–58
7. Wang X, Li J, Zheng H et al (2011) Proteasome functional insufficiency in cardiac pathogenesis. Am J Physiol Heart Circ Physiol 301: H2207–19
8. Powell SR, Herrmann J, Lerman A et al (2012) The ubiquitin-proteasome system and cardiovascular disease. Prog Mol Biol Transl Sci 109:295–346
9. Schmidt N, Gonzalez E, Visekruna A et al (2010) Targeting the proteasome: partial inhibition of the proteasome by bortezomib or deletion of the immunosubunit LMP7 attenuates experimental colitis. Gut 59:896–906
10. Conner EM, Brand S, Davis JM et al (1997) Proteasome inhibition attenuates nitric oxide synthase expression, VCAM-1 transcription and the development of chronic colitis. J Pharmacol Exp Ther 282:1615–22
11. Takaoka M, Itoh M, Hayashi S et al (1999) Proteasome participates in the pathogenesis of ischemic acute renal failure in rats. Eur J Pharmacol 384:43–6
12. Zaouali MA, Bardag-Gorce F, Carbonell T et al (2013) Proteasome inhibitors protect the steatotic and non-steatotic liver graft against cold ischemia reperfusion injury. Exp Mol Pathol 94:352–9

13. Zu L, Bedja D, Fox-Talbot K et al (2010) Evidence for a role of immunoproteasomes in regulating cardiac muscle mass in diabetic mice. J Mol Cell Cardiol 49:5–15
14. Otoda T, Takamura T, Misu H et al (2013) Proteasome dysfunction mediates obesity-induced endoplasmic reticulum stress and insulin resistance in the liver. Diabetes 62:811–24
15. Weiss CH, Budinger GR, Mutlu GM et al (2010) Proteasomal regulation of pulmonary fibrosis. Proc Am Thorac Soc 7:77–83
16. Opattova A, Filipcik P, Cente M et al (2013) Intracellular degradation of misfolded tau protein induced by geldanamycin is associated with activation of proteasome. J Alzheimers Dis 33:339–48
17. Palmowski MJ, Gileadi U, Salio M et al (2006) Role of immunoproteasomes in cross-presentation. J Immunol 177:983–90
18. Wang J, Maldonado MA (2006) The ubiquitin-proteasome system and its role in inflammatory and autoimmune diseases. Cell Mol Immunol 3:255–61
19. Christmann RB, Mathes A, Affandi AJ et al (2013) Thymic stromal lymphopoietin is up-regulated in the skin of patients with systemic sclerosis and induces profibrotic genes and intracellular signaling that overlap with those induced by interleukin-13 and transforming growth factor beta. Arthritis Rheum 65:1335–46
20. Connor AM, Mahomed N, Gandhi R et al (2012) TNFalpha modulates protein degradation pathways in rheumatoid arthritis synovial fibroblasts. Arthritis Res Ther 14:R62
21. Vereecke L, Beyaert R, van Loo G (2009) The ubiquitin-editing enzyme A20 (TNFAIP3) is a central regulator of immunopathology. Trends Immunol 30:383–91
22. Kanazawa N (2012) Rare hereditary autoinflammatory disorders: towards an understanding of critical in vivo inflammatory pathways. J Dermatol Sci 66:183–9
23. Grune T, Jung T, Merker K et al (2004) Decreased proteolysis caused by protein aggregates, inclusion bodies, plaques, lipofuscin, ceroid, and 'aggresomes' during oxidative stress, aging, and disease. Int J Biochem Cell Biol 36:2519–30
24. Jung T, Grune T (2012) Structure of the proteasome. Prog Mol Biol Transl Sci 109:1–39
25. Qureshi N, Morrison DC, Reis J (2012) Proteasome protease mediated regulation of cytokine induction and inflammation. Biochim Biophys Acta 1823:2087–93
26. Song X, von Kampen J, Slaughter CA et al (1997) Relative functions of the alpha and beta subunits of the proteasome activator, PA28. J Biol Chem 272:27994–8000
27. Zhang Z, Krutchinsky A, Endicott S et al (1999) Proteasome activator 11S REG or PA28: recombinant REG alpha/REG beta hetero-oligomers are heptamers. Biochemistry 38:5651–8
28. Stohwasser R, Salzmann U, Giesebrecht J et al (2000) Kinetic evidences for facilitation of peptide channelling by the proteasome activator PA28. Eur J Biochem 267:6221–30
29. Whitby FG, Masters EI, Kramer L et al (2000) Structural basis for the activation of 20S proteasomes by 11S regulators. Nature 408:115–20
30. Glynne R, Powis SH, Beck S et al (1991) A proteasome-related gene between the two ABC transporter loci in the class II region of the human MHC. Nature 353:357–60
31. Martinez CK, Monaco JJ (1991) Homology of proteasome subunits to a major histocompatibility complex-linked LMP gene. Nature 353: 664–7
32. Nandi D, Jiang H, Monaco JJ (1996) Identification of MECL-1 (LMP-10) as the third IFN-gamma-inducible proteasome subunit. J Immunol 156:2361–4
33. Groll M, Ditzel L, Lowe J et al (1997) Structure of 20S proteasome from yeast at 2.4A resolution. Nature 386:463–71
34. Tanahashi N, Murakami Y, Minami Y et al (2000) Hybrid proteasomes. Induction by interferon-gamma and contribution to ATP-dependent proteolysis. J Biol Chem 275: 14336–45
35. Murata S, Sasaki K, Kishimoto T et al (2007) Regulation of CD8+ T cell development by thymus-specific proteasomes. Science 316: 1349–53
36. Calise J, Powell SR (2013) The ubiquitin proteasome system and myocardial ischemia. Am J Physiol Heart Circ Physiol 304:H337–49
37. Drews O, Wildgruber R, Zong C et al (2007) Mammalian proteasome subpopulations with distinct molecular compositions and proteolytic activities. Mol Cell Proteomics 6:2021–31
38. Zhou HJ, Aujay MA, Bennett MK et al (2009) Design and synthesis of an orally bioavailable and selective peptide epoxyketone proteasome inhibitor (PR-047). J Med Chem 52:3028–38
39. Screen M, Britton M, Downey SL et al (2010) Nature of pharmacophore influences active site specificity of proteasome inhibitors. J Biol Chem 285:40125–34
40. Kirk CJ (2012) Discovery and development of second-generation proteasome inhibitors. Semin Hematol 49:207–14

41. Rajkumar SV, Richardson PG, Hideshima T et al (2005) Proteasome inhibition as a novel therapeutic target in human cancer. J Clin Oncol 23:630–9
42. Chen D, Frezza M, Schmitt S et al (2011) Bortezomib as the first proteasome inhibitor anticancer drug: current status and future perspectives. Curr Cancer Drug Targets 11: 239–253
43. Pahl HL, Baeuerle PA (1996) Control of gene expression by proteolysis. Curr Opin Cell Biol 8:340–7
44. Gotze S, Bose A, Abele D et al (2013) Pitfalls in invertebrate proteasome assays. J Exp Biol 216: 1351–1354
45. Reinheckel T, Sitte N, Ullrich O et al (1998) Comparative resistance of the 20S and 26S proteasome to oxidative stress. Biochem J 335 (Pt 3):637–42
46. Powell SR, Davies KJ, Divald A (2007) Optimal determination of heart tissue 26S-proteasome activity requires maximal stimulating ATP concentrations. J Mol Cell Cardiol 42:265–9
47. Muchamuel T, Basler M, Aujay MA et al (2009) A selective inhibitor of the immunoproteasome subunit LMP7 blocks cytokine production and attenuates progression of experimental arthritis. Nat Med 15:781–7
48. Bennett MK, Buchholz TJ, Demo SD et al (2006) Compounds for enzyme inhibition. US Patent No: 2006/0088471 A1, USA
49. Moravec RA, O'Brien MA, Daily WJ et al (2009) Cell-based bioluminescent assays for all three proteasome activities in a homogeneous format. Anal Biochem 387:294–302
50. Kisselev AF, Goldberg AL (2001) Proteasome inhibitors: from research tools to drug candidates. Chem Biol 8:739–58
51. Kisselev AF, Goldberg AL (2005) Monitoring activity and inhibition of 26S proteasomes with fluorogenic peptide substrates. Methods Enzymol 398:364–78
52. French SW, Mayer RJ, Bardag-Gorce F et al (2001) The ubiquitin-proteasome 26s pathway in liver cell protein turnover: effect of ethanol and drugs. Alcohol Clin Exp Res 25:225S–229S
53. Divald A, Kivity S, Wang P et al (2010) Myocardial ischemic preconditioning preserves postischemic function of the 26S proteasome through diminished oxidative damage to 19S regulatory particle subunits. Circ Res 106: 1829–38
54. Powell SR, Samuel SM, Wang P et al (2008) Upregulation of myocardial 11S-activated proteasome in experimental hyperglycemia. J Mol Cell Cardiol 44:618–21
55. Wang X, Robbins J (2006) Heart failure and protein quality control. Circ Res 99: 1315–28
56. Ferrington DA, Gregerson DS (2012) Immunoproteasomes: structure, function, and antigen presentation. Prog Mol Biol Transl Sci 109: 75–112
57. Lightcap ES, McCormack TA, Pien CS et al (2000) Proteasome inhibition measurements: clinical application. Clin Chem 46:673–83
58. Parlati F, Lee SJ, Aujay M et al (2009) Carfilzomib can induce tumor cell death through selective inhibition of the chymotrypsin-like activity of the proteasome. Blood 114: 3439–47
59. Rodgers KJ, Dean RT (2003) Assessment of proteasome activity in cell lysates and tissue homogenates using peptide substrates. Int J Biochem Cell Biol 35:716–27

Chapter 14

Evaluation of the Adverse Effect of Low Concentration of Cadmium on Interleukin-4 Induced Class Switch Recombination in Burkett's Lymphoma Raji Cell Line

Vladimir Poltoratsky

Abstract

Affinity maturation of B lymphocytes, a process that includes somatic hypermutation and class switch recombination, initiates global DNA rearrangements. The interruption of this process has an adverse effect on human health and results in immunodeficiency and autoimmune disease. Class switch recombination is a fundamental factor of the human adaptive immunity. Evaluation of the class switch recombination efficiency is an important component of laboratory diagnostic of immunotoxic components. Here, we describe a method for testing the efficiency of the class switch recombination. Cultivation of Raji Burkett's lymphoma cell line with anti-CD40 antibodies and recombinant interleukin-4 (IL-4) triggers a cascade of signal transduction network events that lead to switching the immunoglobulin isotopes from IgM to IgE. This chapter describes the methodology of class switch recombination assay for assessment of the effect of the environmental pollutants in toxicological laboratory diagnostics.

Key words Interleukin-4, Class switch recombination, Cadmium toxicity, Flow cytometry, ELISA spot assay, NF-κB signaling, STAT6 signaling

1 Introduction

Adaptive immune response is the third line of defense of the immune system. After being activated by antigen and T helper cells, B cells modify the DNA segments encoding the variable and constant parts of immunoglobulins to improve recognition and effective functions of antibodies. In humoral immune response, the isotype switching modulates ability of antibody to interact with complement complex and isotype-specific surface receptors. Switching of the IgM class to IgG is required for neonatal immunity, complement activation, Fc receptor phagocyte response, switching to IgA for mucosal immunity, and switching to IgE for immunity against helminthes and type I hypersensitivity [1–6].

In vivo T cell-dependent class switch recombination (CSR) in B cells is induced by activation of the CD40 B cell receptor with T cell

Ivana Vancurova (ed.), *Cytokine Bioassays: Methods and Protocols*, Methods in Molecular Biology, vol. 1172,
DOI 10.1007/978-1-4939-0928-5_14, © Springer Science+Business Media New York 2014

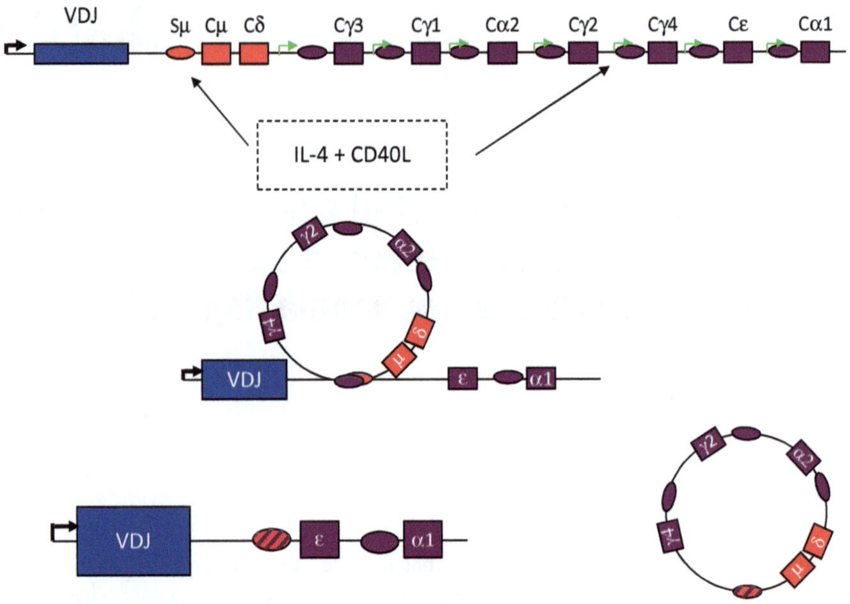

Fig. 1 Schematic illustration of class switch recombination

CD195 (CD40L) and activation of the B cell surface receptors by cytokines secreted by CD4 T cells. In vitro IgM to IgE CSR in B cells can be triggered by activation of CD40 with anti-CD40 antibodies and IL-4 receptor (IL-4R) with recombinant IL-4 [7–10].

Activation of IL-4R triggers signal transduction pathway that leads to the phosphorylation, dimerization, and translocation of transcriptional factor STAT6 to the nucleus. In nucleus, STAT6 is activated by proteolysis, and in consort with other activation nuclear factors, participates in the expression of IgE silent transcript promoter and activation induced cytosine deaminase (AID) [11–13]. It is believed that initiation of the silent transcription remodels the chromatin and exposes the ssDNA sites for AID- mediated deamination of the deoxycytosines [14, 15].

An interaction of CD40 with CD195 triggers the signal transduction pathway that leads, through TRAF2 to TRAF6-mediated signaling, to activation and translocation of NFκB to the nucleus [16, 17]. Activation NFκB is required for AID expression [11–13]. Initiation of the sterile RNA transcript of GC-rich switched regions upstream the IgH chain C regions Cμ, Cγ, Cε, and Cα is believed to provide ssDNA regions acceptable to AID. AID deaminates deoxycytidine (dC) to form DNA lesion, deoxyuracil (dU) [14, 15, 18, 19]. Proteins of the base excision repair cascade excise dU from DNA and produce single strand DNA breaks. Mismatch repair (MMR) proteins produce double-strand DNA breaks that initiate recombination between switch regions upstream of Cμ and other activated C regions (Fig. 1).

Down-regulation of the MMR pathway diminishes the CSR efficiency [20–22]. On the other hand, it has been shown that MMR is inhibited by cadmium, an environmental pollutant [23, 24]. Traces amounts of cadmium inhibit the ATPase activity of the Msh2-Msh6 complex [23, 25]. Environmental studies demonstrated the immunotoxicological effect of cadmium. Examination of factory workers exposed to cadmium demonstrated the avert effect of this pollutant on the innate and adaptive immunity [26–28]. Here, we provide the protocol for testing the effect of cadmium on the efficiency of the IgM to IgE class switched recombination utilizing a Burkett's lymphoma cell line Raji as a model system.

2 Materials

2.1 Cell Culture

1. Human Burkett's lymphoma cells (American Type Cell Collection) (*see* **Note 1**).
2. Complete RPMI medium: RPMI supplemented with 10 mM Hepes, pH 7.5, 10 % fetal bovine serum (FBS, heat inactivated), 2 mM L-glutamine, 50 μM β-mercaptoethanol, 50 μg/mL penicillin, and 50 μg/mL streptomycin (*see* **Note 2**).
3. Mouse anti-human CD40 antibodies.
4. Recombinant human IL-4 protein (hIL-4).
5. 10 mM cadmium solution: Dissolve 18.34 mg of $CdCl_2$ in 10 mL of deionized water. Sterilize solution by filtration through 0.45 μM filter and store at 4 °C.
6. Hausser™ Bright-Line™ Phase hemacytometer (Hausser Scientific).
7. Trypan Blue dye.
8. 75 cm^2 culture flasks and 6-well Plates.

2.2 Flow Cytometry Analysis

1. Goat anti-human IgM-Texas Red conjugate (Southern Biotech).
2. Mouse anti-human IgE (Fc)-biotin conjugate (Southern Biotech).
3. Streptavidin-Quantum Red conjugate (Fisher Scientific).
4. Hank's Balanced Salt Solution (HBSS).
5. Phosphate Buffered Saline (PBS), pH 7.4: Dissolve 8.0 g NaCl, 1.44 g Na_2HPO_4, 0.24 g KH_2PO_4, and 0.2 g KCl in 1 L of deionized water, adjust pH to 7.4, and sterilize buffer through 0.2 μm filter.
6. 4 % paraformaldehyde in PBS.
7. Flow cytometer.

2.3 Enzyme Linked Immunosorbent Assay (ELISA) Spot Assay

1. Capture antibody: mouse anti-human IgE (Southern Biotech) (*see* **Note 3**).
2. Detection antibody: mouse anti-human IgE-biotin conjugate (Southern Biotech) (*see* **Note 3**).
3. Streptavidin-alkaline phosphatase conjugate.
4. NBT/BCIP Substrate (Thermo Scientific).
5. 96-well tissue culture plates.
6. Coating buffer: Dissolve 3.03 g Na_2CO_3 and 6.0 g $NaHCO_3$ in 1 L of deionized water; adjust pH to 9.6 with sodium hydroxide. Sterilize buffer through 0.2 μm filter.
7. Washing buffer: PBS buffer, pH 7.4, supplemented with 0.05 % (V/V) Tween 20 (PBST).
8. Blocking buffer: Dissolve 2 mg bovine serum albumin (BSA) in 1 mL of PBS.
9. Dilution buffer: Dissolve 1 mg BSA in 1 mL of PBST.

3 Methods

In this section, we describe the method for quantification of the class switch recombination in the Burkett's lymphoma B cells. The method is based on activation of the CD40 and IL-4R surface receptors on B cells. Activation of CD40, the member of the tumor necrosis factor superfamily, is caused by crosslinking with antibodies against this receptor. This method is schematically depicted in Fig. 2. Figure 3 illustrates the effect of cadmium on the IgE class switch recombination. We describe two methods of detection of the CSR: flow cytometry and ELISA-spot assay. While the flow cytometry analysis is more accurate and robotic, the ELISA spot assay does not require expensive equipment, such as flow cytometer, and could be performed in any laboratory.

3.1 Cell Culture

1. Grow human Burkett's lymphoma Raji cells in RPMI complete medium to a concentration 2×10^5 cells per mL.

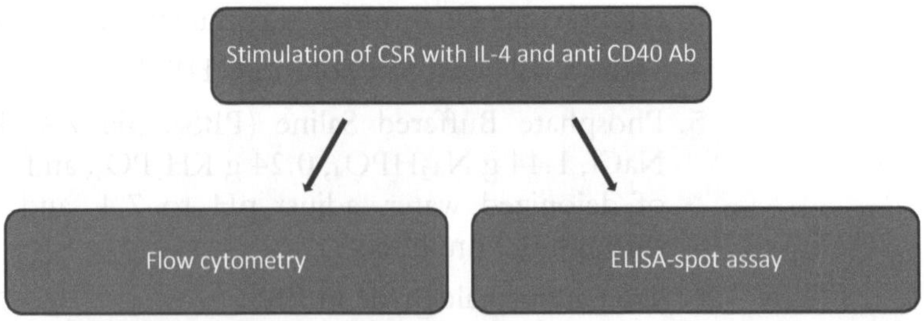

Fig. 2 Schematic illustration of the protocol

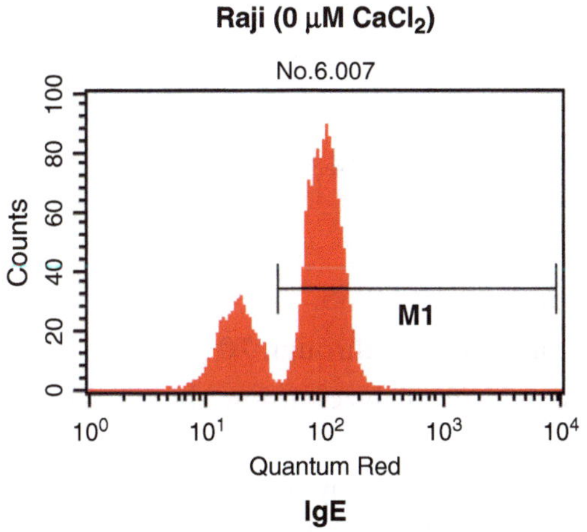

Fig. 3 Flow cytometry analysis of cadmium effect on class switch recombination

2. Count the cells stained with Trypan Blue with hemacytometer and split them (1×10^4 cells/well) into 4 wells of a 6 well-plate (*see* **Note 4**).
3. Mark wells as 1, 2, 3, and 4. Add cadmium to a final concentration of 10 μg/mL to wells number 2 and 4.
4. Add anti-CD40 antibody and human recombinant IL-4 at final concentrations of 5 μg/mL and 5 ng/mL, respectively, to wells number 3 and 4 (*see* **Note 5**).
5. Incubate the plate for 4 days in a humidified CO_2 tissue culture incubator.
6. After 4 days, wash the cells with fresh media and analyze the efficiency of the CSR by flow cytometry or ELISA-spot assays.

3.2 Staining Cells for Flow Cytometry Analysis

1. Wash the cells with HBSS twice and resuspend the cell pellets in 1 mL of HBSS solution. Put aside 50 μl of the cell suspension to use as an unstained cell control.
2. Add 5 μg of anti-human IgE-biotin conjugate
3. Tap tubes to mix and incubate for 10 min on ice.
4. Wash the cells with 8 mL of HBSS and spin at 600 × *g* 4 °C for 5 min. Pour off supernatant.
5. Repeat washing twice. Pour off supernatant and resuspend pellets in 1 mL of HBSS.
6. Add 5 μL of Streptavidin-Quantum Red conjugate.
7. Tap tubes to mix and incubate for 10 min on ice, covered from light.
8. Wash the cells with 8 mL of HBSS and spin at 600 × *g* 4 °C for 5 min. Pour off supernatant.
9. Repeat washing.

10. Pour off supernatant and resuspend pellets in 1 mL of 4 % paraformaldehyde in PBS.
11. Incubate for 5 min.
12. Wash cells with 8 mL of HBSS and spin at 600 × *g* 4 °C for 5 min. Pour off supernatant.
13. Repeat washing.
14. Resuspend the cells in 1 mL of HBSS.
15. Just before sorting, filter samples through 70 μm mesh.
16. Analyze at flow cytometry analyzer using the appropriate control to set gates.
17. The efficiency of CSR is calculated by dividing the number of IgE-positive cells in well number 4 to the number of IgE-positive cells in well number 3, and multiplying the obtained number by 100 (Fig. 3).

3.3 ELISA Spot Assay

1. Prior to experiment, prepare the plates coated with IgE. Dilute mouse anti-human IgE capture antibody 1:500 in Coating Buffer. For one plate, dilute 25 μg of capture antibody in 12.5 mL of Coating Buffer. Add 100 μL of diluted anti-IgE per well in 96-well plate under sterile conditions, seal it, and keep at 4 °C overnight.
2. Wash wells two times with 200 μL of sterile PBS solution per well.
3. To block nonspecific binding and reduce background, add 200 μL of blocking solution per well.
4. Seal the plate and incubate overnight at 4 °C, or at room temperature for 1 h on a plate shaker.
5. Wash the wells two times with 200 μL of sterile PBS solution per well. Add 100 μL of sterile PBS solution per well, seal and keep plates at 4 °C.
6. At the beginning of ELISA-spot assay, collect cells and wash them four times with sterile PBS. Resuspend cells in RPMI complete medium and count cell concentration in hemacytometer with Trypan Blue dye (*see* **Note 4**).
7. Dilute 5.5×10^3 cells in 11 mL of RPMI complete medium and plate 100 μL per well. Incubate overnight at humidified CO_2 chamber at 37 °C.
8. Wash plates five times with 400 μL of PBST per well and get rid of remaining solution by dumping the inverted plates on a paper towel (*see* **Note 6**).
9. Add 200 μL of 2 % BSA in PBS and incubate for 2 h on a plate shaker (200 rpm) at room temperature.
10. Wash the plates five times with 400 μL of PBST per well, and get rid of remaining solution by dumping the inverted plates on a paper towel.

11. Dilute anti IgE-biotin conjugated detecting antibody 1:500 in Dilution Buffer. For one plate, dilute 12 μL of the detecting antibody in 6 mL of Dilution Buffer. Add 50 μL of diluted anti-IgE per well in a 96-well plate, seal it and keep at 4 °C overnight.
12. Wash the plates five times with 400 μL of PBST per well, and get rid of remaining solution by dumping the inverted plates on paper towel.
13. Dilute anti streptavidin-alkaline phosphatase conjugate detecting antibody 1:5,000 in Dilution Buffer. For one plate, dilute 2 μg of the detecting antibody in 10 mL of Dilution Buffer. Add 50 μL of diluted streptavidin-alkaline phosphatase conjugate per well in 96-well plate, seal it, and incubate at room temperature for 1 h on a plate shaker with 200 rpm.
14. Wash the plates seven times with 400 μL of PBST per well, and get rid of remaining solution by dumping inverted plates on paper towel.
15. Wash the plates two times with 400 μL of PBS per well, and get rid of remaining solution by dumping inverted plates on paper towel.
16. Add 50 mL of NBT/BCIP alkaline phosphatase substrate per well in 96-well plate, seal it, and keep at room temperature, covered from light, from 10 min to 3 h (*see* **Note 7**).
17. Wash the plates two times with 400 μL of PBST per well, and get rid of remaining solution by dumping inverted plates on paper towel. Wash the plates two times with 400 μL of distilled water per well and get rid of remaining solution by dumping inverted plates on paper towel (*see* **Note 8**).
18. Dry the wells and count spots under inverted microscope.
19. The efficiency of CSR is calculated by dividing the number of IgE-positive cells in well number 4 to the number of IgE-positive cells in well number 3, and multiplying the obtained number by 100.

4 Notes

1. Burkett's Lymphoma Raji is a fast growing suspension cell line that must be passaged at a 1:10 ratio twice weekly. Cells start losing viability at high density ($>1 \times 10^6$ cells/mL).
2. Cells must be grown in a 5 % CO_2 humidified atmosphere at 37 °C. Medium must be supplemented with Hepes to increase the maximum buffering range of the medium, and β- mercaptoethanol to minimize B lymphocytes clamping. The complement system in FBS must be inactivated by incubating serum

at 56 °C for 30 min. You can also use commercial heat inactivated serum.

3. Capture and detection antibodies must recognize different epitopes of IgE antibodies.
4. Mix cell suspension with Trypan Blue. Count the unstained cells in the chamber; dead cells will be stained blue. The ratio of the uncolored to blue cells should be more than 95 %. Cultures that contain significant amount of dead cells will not perform well in this assay.
5. It is a good practice to add additional control treatments, such as treating the cells with only anti-CD40 or hIL-4. These cells should demonstrate minimal CSR.
6. After washings, check the plates under inverted microscope for presence of cells. It is important to remove all cell debris from wells. Remaining cell fragments form false positive signals.
7. Developing time for alkaline phosphatase is determined empirically. Placing the plates in 37 °C incubator will shorten incubation time.
8. It is important to wash the wells in the last step with distilled water. If the wells are washed with PBST, then salt will crystallize on the wells.

References

1. Carroll MC, Isenman DE (2012) Regulation of humoral immunity by complement. Immunity 37:199–207
2. Kawakami T, Galli SJ (2002) Regulation of mast-cell and basophil function and survival by IgE. Nat Rev Immunol 2:773–786
3. Macpherson AJ, Geuking MB, McCoy KD (2012) Homeland security: IgA immunity at the frontiers of the body. Trends Immunol 33:160–167
4. Nimmerjahn F, Ravetch JV (2011) FcgammaRs in health and disease. Curr Top Microbiol Immunol 350:105–125
5. Pabst O (2012) New concepts in the generation and functions of IgA. Nat Rev Immunol 12:821–832
6. Roopenian DC, Akilesh S (2007) FcRn: the neonatal Fc receptor comes of age. Nat Rev Immunol 7:715–725
7. Pene J, Rousset F, Briere F et al (1988) IgE production by normal human B cells induced by alloreactive T cell clones is mediated by IL-4 and suppressed by IFN-gamma. J Immunol 141:1218–1224
8. Shapira SK, Vercelli D, Jabara HH et al (1992) Molecular analysis of the induction of immunoglobulin E synthesis in human B cells by interleukin 4 and engagement of CD40 antigen. J Exp Med 175:289–292
9. Vercelli D, Jabara HH, Arai K et al (1989) Endogenous IL-6 plays an obligatory role in IL-4-dependent human IgE synthesis. Eur J Immunol 19:1419–1424
10. Zhang K, Clark EA, Saxon A (1991) CD40 stimulation provides an IFN-gamma- independent and IL-4-dependent differentiation signal directly to human B cells for IgE production. J Immunol 146:1836–1842
11. Crouch EE, Li Z, Takizawa M et al (2007) Regulation of AID expression in the immune response. J Exp Med 204:1145–1156
12. Dedeoglu F, Horwitz B, Chaudhuri J et al (2004) Induction of activation-induced cytidine deaminase gene expression by IL-4 and CD40 ligation is dependent on STAT6 and NFκB. Int Immunol 16:395–404
13. le Huong T, Kobayashi M, Nakata M et al (2013) In vivo analysis of Aicda gene regulation: a critical balance between upstream enhancers and intronic silencers governs appropriate expression. PLoS ONE 8:e61433

14. Poltoratsky V, Goodman MF, Scharff MD (2000) Error-prone candidates vie for somatic mutation. J Exp Med 192:F27–F30
15. Poltoratsky VP, Wilson SH, Kunkel TA et al (2004) Recombinogenic phenotype of human activation-induced cytosine deaminase. J Immunol 172:4308–4313
16. Chang TP, Vancurova I (2013) NFκB function and regulation in cutaneous T-cell lymphoma. Am J Cancer Res 3:433–445
17. DiDonato JA, Mercurio F, Karin M (2012) NFκB and the link between inflammation and cancer. Immunol Rev 246:379–400
18. Muramatsu M, Kinoshita K, Fagarasan S et al (2000) Class switch recombination and hypermutation require activation-induced cytidine deaminase (AID), a potential RNA editing enzyme. Cell 102:553–563
19. Petersen-Mahrt SK, Harris RS, Neuberger MS (2002) AID mutates *E. coli* suggesting a DNA deamination mechanism for antibody diversification. Nature 418:99–103
20. Wiesendanger M, Kneitz B, Edelmann W et al (2000) Somatic hypermutation in MutS homologue (MSH)3-, MSH6-, and MSH3/MSH6-deficient mice reveals a role for the MSH2-MSH6 heterodimer in modulating the base substitution pattern. J Exp Med 191:579–584
21. Schrader CE, Edelmann W, Kucherlapati R et al (1999) Reduced isotype switching in splenic B cells from mice deficient in mismatch repair enzymes. J Exp Med 190:323–330
22. Vora KA, Tumas-Brundage KM, Lentz VM et al (1999) Severe attenuation of the B cell immune response in Msh2-deficient mice. J Exp Med 189:471–482
23. Clark AB, Kunkel TA (2004) Cadmium inhibits the functions of eukaryotic MutS complexes. J Biol Chem 279:53903–53906
24. Jin YH, Clark AB, Slebos RJ et al (2003) Cadmium is a mutagen that acts by inhibiting mismatch repair. Nat Genet 34:326–329
25. Banerjee S, Flores-Rozas H (2005) Cadmium inhibits mismatch repair by blocking the ATPase activity of the MSH2-MSH6 complex. Nucleic Acids Res 33:1410–1419
26. El-Sayed LH, Ghoneim HM, El-Sayed MH et al (2003) Immunological abnormalities in workers exposed to pollutants at an Egyptian copper company. Immunopharmacol Immunotoxicol 25:473–490
27. Frenkel K, Karkoszka J, Cohen B et al (1994) Occupational exposures to Cd, Ni, and Cr modulate titers of antioxidized DNA base autoantibodies. Environ Health Perspect 102(Suppl 3):221–225
28. Palus J, Rydzynski K, Dziubaltowska E et al (2003) Genotoxic effects of occupational exposure to lead and cadmium. Mutat Res 540:19–28

Chapter 15

A Mix-and-Measure Assay for Determining the Activation Status of Endogenous Cdc42 in Cytokine-stimulated Macrophage Cell Lysates

Veronika Miskolci, Désirée Spiering, Dianne Cox, and Louis Hodgson

Abstract

Cytokine stimulations of leukocytes many times result in transient activation of the p21 Rho family of small GTPases. The role of these molecules during cell migration and chemotaxis is well established. The traditional approach to study the activation dynamics of these proteins involves affinity pull-downs that are often cumbersome and prone to errors. Here, we describe a reagent and a method of simple "mix-and-measure" approach useful for determining the activation status of endogenous Cdc42 GTPase from cell lysates.

Key words Cdc42 activity, Biosensor, MeroCBD, Fluorometry, Mix-and-measure

1 Introduction

Activation status of the Rho family GTPases is often cumbersome to determine using traditional approaches including affinity-reagent based pull-down experiments [1]. These are often technically involved and challenging approaches, and thus could lead to possible artifacts due to the procedural variability and difficulty. Here, we describe a facile, "mix and measure" assay for determining the activation status of Cdc42 GTPase from whole-cell lysates. The approach we discuss here is based on the system described previously [2], in which a fluorescent biosensor that could detect the activation status of endogenous Cdc42 in living cells was used also in a lysate-based assay for neutrophils [2]. This biosensor (meroCBD) is based on a GTPase binding domain derived from Wiskott Aldrich Syndrome protein (WASp) with an organic dye capable of fluorescence modulation dependent on the local solvent polarity changes. The small binding domain is derivatized with a merocyanine dye, and upon binding of this dye-labeled domain to the target (endogenous, active Cdc42), the local solvent polarity changes from aqueous to more hydrophobic increasing the

Ivana Vancurova (ed.), *Cytokine Bioassays: Methods and Protocols*, Methods in Molecular Biology, vol. 1172,
DOI 10.1007/978-1-4939-0928-5_15, © Springer Science+Business Media New York 2014

fluorescence emission [3, 4]. This sensor system is useful for in vitro characterization of the GTP-loading state of Cdc42 such as for the guanine nucleotide exchange assays, as well as for live-cell imaging experiments where the activation dynamics of Cdc42 can be monitored in real time [2]. Here, we discuss step-by-step, how to produce this meroCBD biosensor, characterize it in vitro, and then to use it to measure the endogenous Cdc42 activity changes in macrophage lysates during the time course of a cytokine stimulation experiment.

2 Materials

2.1 MeroCBD Preparation

1. LB medium.
2. Carbenicillin 100 mg/mL stock.
3. IPTG 1 M stock.
4. BL21(DE3) competent bacteria (*see* **Note 1**).
5. Talon Buffer: 50 mM NaH_2PO_4 (monobasic), 5 mM $MgCl_2$, 10 % glycerol, pH 7.4; must be at 4 °C.
6. Dialysis/labeling Buffer: 50 mM NaH_2PO_4 (monobasic), pH 7.4.
7. Storage Buffer: 50 mM Tris–HCl, 50 mM NaH_2PO_4 (monobasic), 5 mM $MgCl_2$, pH 7.4.
8. Amicon Ultra-4 5000MWCO spin columns.
9. 5 M NaCl solution.
10. PMSF stock: 200 mM PMSF in methanol.
11. β-Mercaptoethanol.
12. DMSO.
13. 100 mM DTT stock.
14. EDTA 0.5 M, pH 8.0, stock.
15. Mero-87-IAA (www.hahnlab.com).
16. Talon resin (Clontech).
17. Sephadex G-15.
18. Imidazole.
19. Poly-prep chromatography column.

2.2 MeroCBD Characterization

1. GTPase lysis buffer: 30 mM Tris–HCl, pH 7.8 (room temperature), 250 mM NaCl, 5 mM $MgCl_2$, 10 % glycerol. Just before lysis, add PMSF to final concentration of 1 mM, β-mercaptoethanol to 2 mM, and 1 μM GTPγS.
2. GTPase dialysis buffer: 50 mM Tris–HCl, pH 7.6, 50 mM NaCl, 0.5 mM EDTA, pH 7.0, 5 mM $MgCl_2$, 1 mM DTT, 10 nM GDP, 10 % glycerol.

3. Imidazole.
4. Talon resin (Clontech).
5. Amicon Ultra-4 5000 MW CO columns.
6. GTPase measurement buffer: 50 mM NaCl, 50 mM Tris–HCl, 5 mM MgCl, pH 7.6.

2.3 Cell Lysate Preparation

1. 1 M Tris–HCl, pH 7.4.
2. 2 M NaCl.
3. 1 M $MgCl_2$.
4. 10 % Triton X-100.
5. 10× Tris-buffered saline (TBS): 1.37 M NaCl, 250 mM Tris-Base, adjusted to pH 7.4.
6. RAW/LR5 cells derived from RAW 264.7 [5].
7. RPMI 1640 medium with glutamine.
8. 10 cm cell culture dishes.
9. Fractalkine/CX3CL1 (FKN) (R&D Systems); reconstituted at 50 μg/mL with PBS and stored at −80 °C.
10. Cell lysis buffer: 50 mM Tris–HCl, pH 7.4, 500 mM NaCl, 50 mM $MgCl_2$, 1 % Triton X-100. Store at 4 °C. Prepare on the day before use to cool overnight. Cool on ice on the day of stimulation.
11. 1× TBS: 137 mM NaCl, 25 mM Tris-Base; pH 7.4; prepare on the day before use to cool overnight. Cool on ice on the day of stimulation.
12. Protease inhibitor stock: prepare in double-distilled water and store at −20 °C.
    - 1 M Benzamidine.
    - 1 mg/mL Aprotinin.
    - 1 mg/mL Leupeptin.
13. Cell scraper.

2.4 Fluorometry of Lysates

1. Bradford reagent.
2. Spetrofluorometer capable of measuring the fluorescence response of the sensor (i.e., Horiba-Jobin-Yvon Fluorolog system).

3 Methods

3.1 MeroCBD Preparation

1. Transform 100 μL of BL21(DE3) bacteria with 1 μL of plasmid DNA following the manufacturer's protocols. For the plasmid DNA, pET21-CBD-EGFP, or pET21-CBD-MBP can

be used (*see* **Note 2**). These constructs are available from Addgene.com. The bacteria should be plated out onto 4× LB agar plates overnight.

2. Following day, prepare 2× 520 mL of LB broth in 2 L baffled flasks, with 100 μg/mL final concentration of carbenicillin (1:1,000 dilution from the carbenicillin stock solution). Taking the colonies grown from the day before, rinse using this LB medium and place the bacterial suspension into the flask using 2× agar plates worth of colonies per one flask (*see* **Note 3**). Culture the bacteria in the shaker incubator at 37 °C, 225 rpm.
3. When the bacterial OD600 reading reaches 0.9–1.0, protein synthesis is induced. IPTG is added to the cultures to make the final concentration of 0.2 mM from the 1 M stock IPTG solution. The bacterial cultures are then transferred to a room temperature shaker and incubated for additional 6 h to overnight.
4. At the conclusion of the induction period, the bacterial suspension is centrifuged at 6,000×*g* for 10 min at 4 °C. The pellet can be frozen at this point at −80 °C for processing at a later date.
5. On the day of the protein purification, prepare 200 mL of the chilled (4 °C) Talon buffer with 300 mM NaCl (*see* **Note 4**). This will be used from **step 6** on.
6. Taking 50 mL of the Talon buffer, add PMSF to 1 mM final concentration, and β-mercaptoethanol to 2 mM final concentration just prior to resuspending the bacterial pellet (*see* **Note 5**). Here, the bacterial pellet should be weighed. We typically process up to 2 g of bacterial pellet into 30 mL of this buffer. Scale accordingly.
7. When the pellet is fully resuspended, proceed to ultrasonication. Use 4×30 s ultrasonic pulses with 30 s to 1 min rest on ice in between pulses. Users must optimize the ultrasonic amplitudes for their individual equipment and the probe specification. We typically use a ½in. tip probe at 60 % amplitude setting (*see* **Note 6**).
8. When the ultrasonication is complete, collect 100 μL as a sample to run on a PAGE-gel for analysis of the protein expression, label "Crude lysate." The crude lysate is spun at 22,000×*g* for 25 min at 4 °C.
9. While the centrifugation is in progress, prepare the Talon resin. The resin is at 1:1 (v:v) mixture with the storage buffer containing methanol. For 2–3 g bacterial pellet, 2 g of resin is required, thus 4 mL of the resin suspension is pipetted into a 50 mL conical tube. This is then spun at 750×*g* for 3 min at room temperature. The suspension buffer is carefully decanted and the resin is washed twice using 20 mL of the cold Talon buffer without PMSF or β-mercaptoethanol.

10. When the bacterial centrifugation is complete, transfer the supernatant into the tube containing the prepared and washed Talon resin. Be sure to collect 100 μL sample of the supernatant prior to mixing with the Talon resin for later analysis, labeled "Cleared lysate." The pellet should also be resuspended in Talon buffer to the same volume used to resuspend the original bacterial pellet. 100 μL sample of this resuspended pellet should then be kept for later analysis, labeled "Pellet." The cleared lysate–Talon resin suspension should be covered in aluminum foil and placed on a rotating wheel (or a Nutator) for 1 h at room temperature.
11. When the binding reaction is complete, spin at 750 × *g* for 3 min at room temperature. Keep 100 μL sample of the unbound fraction (supernatant) for later analysis, labeled "Unbound." Discard the unbound supernatant. Wash with 20 mL of the Talon buffer twice. At every wash, spin down and keep taking the 100 μL samples for analysis, labeled "Wash1" and "Wash2."
12. The final wash should be done in Talon buffer containing 5 mM Imidazole, for 5 min at room temperature on a rotating wheel (or a Nutator). The imidazole stock solution (1 M) should be made fresh, just prior to use, in the Talon buffer being used here). The third wash is then spun and 100 μL sample of the supernatant is kept, labeled "Wash3."
13. Using 5 mL of 100 mM imidazole in the Talon buffer, the protein is eluted from the Talon resin by incubating for 5 min at room temperature on a rotating wheel (or on a Nutator). The mixture is then spun, and 100 μL of the eluted protein (supernatant) is kept for analysis, labeled "Elution" (*see* **Note 7**). The supernatant is carefully removed from the Talon resin by filtration through a Poly-prep chromatography column (Bio-Rad) to remove any trace of the Talon resin.
14. The eluted protein is dialyzed overnight in 2 L of the cold dialysis buffer at 4 °C for proceeding directly to dye labeling the following day. Otherwise, it can be dialyzed against 2 L of the storage buffer for freezing and storage at –80 °C for up to 1 year.
15. Following the overnight dialysis, the protein solution is collected and concentration is determined (*see* **Note 8**). The concentration is determined by measuring the absorbance at 280 nm. The equation is:

$$[\text{conc}] = \text{absorbance} \times \text{path length} \times \text{dilution factor} / \text{extinction coefficient}$$

The molar extinction coefficient at 280 nm for CBD-EGFP is 28,260 M^{-1} cm^{-1}, and for CBD-MBP is 72,970 M^{-1} cm^{-1}. The target concentration ideal for the dye labeling reaction is 100 μM.

If the protein solution is too dilute, use the Amicon Ultra-4 5000MWCO spin columns to concentrate the protein.

16. Prepare the reactive dye mixture. The crystals of mero-87-IAA should be brought up to room temperature, wrapped in foil, prior to opening the glass vial. Use 30 μL of fresh, anhydrous DMSO to dissolve the dye crystals, then transfer the dye mixture into a 0.2 mL Eppendorf tube, spin in a benchtop microcentrifuge at 18,000 rcf at room temperature for 2 min to pellet any undissolved material. Transfer the supernatant into a fresh 0.2 mL tube. The concentration of this dye mixture must be determined. The dye solution should be diluted in anhydrous DMSO at 1:5,000, and absorbance should be measured at 596 nm. The molar extinction coefficient for mero-87 in DMSO at 596 nm is 135,301 M^{-1} cm^{-1}. The same equation as in **step 15** can be used to determine the concentration. Ideally, the reactive dye solution should be at 10–20 mM concentration.
17. In labeling the protein, the protein to dye ratio and the reaction time must be optimized. As a good starting point, we recommend 1:5 protein to dye ratio, and 2 h reaction at room temperature. The goal here is to approach 100 % reaction efficiency at the single, target cysteine site on the CBD molecule (*see* **Note 9**). Place 300 μL of the purified protein solution (in Dialysis/labeling buffer) into a 2 mL round-bottom Eppendorf tube, calculate correct amount of the reactive dye solution to add directly to this solution in order to obtain the desired protein to dye ratio. The dye solution is added in one quick shot and immediately the mixture is vortexed. This is then wrapped in aluminum foil and placed on a rotating wheel or a Nutator for 2 h at room temperature. At the conclusion of the reaction, the reaction must be quenched by adding 1 μL of β-mercaptoethanol, allowing the mixture to be quenched for 5 min, and then column-purified over Sephadex G-15 column.
18. The Sephadex G-15 column should be made during the dye labeling reaction. Sephadex slurry is equilibrated with fresh Dialysis/labeling buffer, and the slurry should be packed into a 1 cm diameter glass column, approximately 10 cm tall. The quenched reaction mixture should be loaded very carefully so as to maintain as straight and flat a meniscus as possible, which will enable cleaner separation of the labeled protein from the unreacted dyes. The labeled protein will travel faster than the unreacted dyes, so collect the labeled protein fraction when it reaches the bottom of the column. Once the meroCBD is collected, the concentration must be determined and the dye labeling efficiency must be determined.

19. The absorbance should be measured in the Dialysis/labeling buffer (aqueous) to determine the protein concentration, and in DMSO to determine the dye concentration. The mero-87 dye has small amount of absorbance at 280 nm, and this must be subtracted from the measured 280 nm absorbance in the aqueous condition. The correction factor for this is 0.07605. The peak absorbance at 596 nm for the dye is multiplied by the correction factor and then subtracted from the measured 280 nm absorbance to calculate the protein concentration. The dye concentration is in turn calculated from the measurement in DMSO, similar to **step 16**. Once the concentrations are determined, protein to labeled dye ratio can be calculated and the reaction condition can be further optimized if required.
20. The meroCBD is then aliquoted into 0.2 mL tubes (30 μL per tube) and flash-frozen in liquid nitrogen, and then stored at −80 °C until use.

3.2 MeroCBD Characterization

1. Transform 100 μL of BL21(DE3) bacteria with 1 μL of plasmid DNA following the manufacturer's protocols. For the plasmid DNA, pET21-Cdc42wt is used. The bacterial propagation, media conditions, and induction conditions are identical to Subheading 3.1, **steps 1** through **4**.
2. The bacterial lysis and protein purifications steps are similar to Subheading 3.1, **steps 5** through **15**, exception being the GTPase lysis buffer is used, and the overnight dialysis buffer compositions are different. Furthermore, the buffers must contain guanosine phosphates to achieve stability of the purified GTPase. Here, we add 1 μM GTPγS to the lysis buffer, but during the 2× washes of the Talon resin following the protein binding, we exchange to 1 μM GDP, the final wash solution should contain 10 mM imidazole and 1 μM GDP. Following elution, concentration is determined by absorption at 280 nm and the extinction coefficient of 14,650 M^{-1} cm^{-1} is used. The stock protein can then be frozen at −80 °C.
3. To test meroCBD for proper function, we perform an in vitro fluorometry using the purified Cdc42, loaded with GTPγS, GDP or no guanosine phosphate. We resuspend the purified Cdc42 into the GTPase measurement buffer at a final concentration of 1.4 μM, 200 μL per sample (GTPγS, GDP, no guanosine phosphates), together with 1 mM DTT and 10 mM EDTA. To 190 μL of this mixture, add 10 μL of the guanosine phosphate solutions (stock at 200 μM) so that the final concentration in the Cdc42 mixtures is 10 μM. These samples are then incubated at 30 °C for 30 min. The nucleotide loading reaction is quenched by addition of 3 μL of 1 M $MgCl_2$ (final concentration 15 mM); vortex, wait for 5 min at 30 °C, and

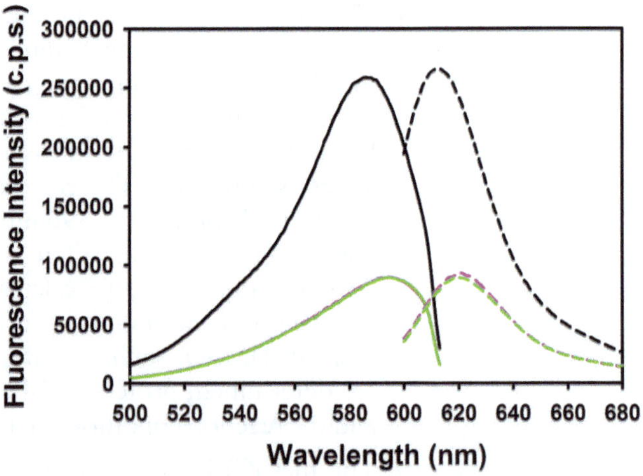

Fig. 1 Excitation and emission spectra of meroCBD in response to GTPgS-loaded (*black*), GDP-loaded (*magenta*) and Apo (*green*) wild-type Cdc42. Excitation spectra are shown with *solid lines* and the emission spectra are shown with *dashed lines*. The protein to dye label ratio for the meroCBD used in this characterization was 69 %

then spin down briefly, and put on ice. The meroCBD is prepared into a separate set of tubes containing 200 μL volume each of the GTPase measurement buffer so that the final concentration of meroCBD will be 300 nM, with 15 mM $MgCl_2$, and placed on ice, shielded from ambient light. Just prior to the fluorometric measurements, the prepared meroCBD suspension and the Cdc42 samples are mixed, and the total volume of 400 μL is loaded into the cuvette and excitation/emission spectra are measured. For the excitation spectral measurement, the emission wavelength is set to 630 nm, and excitation wavelength is scanned from 400 to 623 nm. For the emission spectral scan, the excitation wavelength is set to 596 nm, and the emission wavelength is scanned from 603 to 800 nm (Fig. 1) The peak emission intensities are compared between various guanosine phosphate loaded conditions to determine the total fold-change in fluorescence emission as a function of binding to active versus inactive Cdc42.

3.3 Cell Lysate Preparation

1. Cool cell lysis buffer and 1× TBS on ice.
2. For serum starvation, replace media on RAW/LR5 cells, which were plated the night before in 10 cm dishes at approximately 70 % confluent, with RPMI 1640 for at least 4 h.
3. Pre-warm RPMI 1640 to 37 °C for stimulation. Add FKN at 2× concentration to pre-warmed RPMI 1640 right before use.

4. Add protease inhibitors to cell lysis buffer right before use; working concentrations:
   - 1 mM Benzamidine.
   - 1 μg/mL Aprotinin.
   - 1 μg/mL Leupeptin.
5. Transfer dish to 37 °C water bath.
6. Add equal volume of 2× FKN solution to cells, bringing FKN solution to 1× (50 ng/mL) (e.g., add 5 mL 2× FKN solution to 5 mL serum-starving cells) and swirl gently to mix. Start timer as adding solution. If there are multiple time points, perform one time point at a time.
7. To stop reaction, quickly aspirate media, place dish on ice and add ice-cold 1× TBS right at the time when stimulation ends. It takes ~5 s to aspirate media; start aspirating media while still in water bath ~5 s before stimulation time ends, so 1× TBS can be added immediately as the dish is placed on ice.
8. Immediately aspirate 1× TBS and add ice-cold 0.5 mL of cell lysis buffer (containing protease inhibitors).
9. Scrape cells in dish and transfer lysate to 1.5 mL microcentrifuge tubes. Keep on ice until all time points have been performed.
10. Clear lysates of DNA by centrifuging at 22,000 rcf for 10 min at 4 °C in a microcentrifuge.
11. Transfer supernatant to clean 1.5 mL microcentrifuge. Avoid carrying over the DNA pellet. Keep on ice until performing fluorometry with MeroCBD. Alternatively, lysates can be stored at −80 °C until use.

3.4 Fluorometry of Lysates (See Note 10)

1. Measure the relative concentration differences among the lysates using the Bradford reagent. Here, protein concentration standards are not necessary as long as one operates within the linear range of the detector response of the UV-Visible spectrophotometer. Based on the relative absorbance at 595 nm, determine the correct dilutions necessary to achieve uniform concentrations across all the samples being measured (*see* **Note 11**).
2. Once the concentrations are adjusted to be uniform, aliquot 400 μL of the cell lysate each into Eppendorf tube on ice (*see* **Note 12**). Based on the concentration of the meroCBD solution, calculate how much one would have to add to each sample to achieve 150 nM biosensor final concentration.
3. Just prior to the measurement, mix appropriate volume of meroCBD (may require dilution) with the lysate sample, load into the cuvette, and measure the emission spectral profile. This procedure is repeated for all samples. Wash the cuvette in between each sample to minimize cross-contamination. The representative result is shown in Fig. 2.

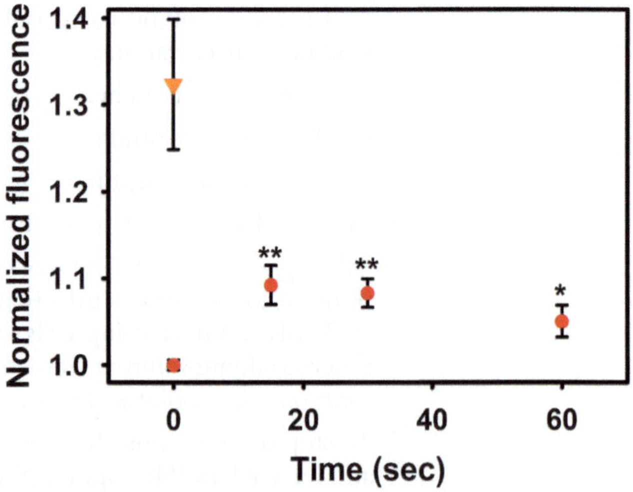

Fig. 2 Fluorometric measurements of lysates of RAW cells stimulated with Fractalkine (50 ng/mL) at indicated times (*red circles*). $^{**}P<0.01$, $^{*}P<0.05$, $n=6$–7. At $T=0$ s, samples of lysates were also pre-equilibrated with 100 μM GTPγS as positive controls (*Orange inverted triangle*)

4 Notes

1. Choice of bacteria is very important. BL21(DE3) must be used. Do not use the BL21(DE3)pLysS, as this will severely reduce the protein yield.
2. pET-CBD-EGFP will produce a fusion protein of the biosensor CBD with EGFP. This will be useful for determining the amount of biosensor pipetted into the cuvette during the fluorometric measurements. If the pipetting of biosensor is done precisely, this step can be omitted and the pET-CBD-MBP can be used. This version does not contain the EGFP. Additionally, EGFP contains two cysteines which will interfere with the stoichiometric labeling of the mero-87-IAA dye, whereas the CBD-MBP only contains one reactive Cysteine in the whole molecule, eliminating this problem.
3. Starting from the plated colonies accelerates the bacterial growth in the flasks as they are still in the log-growth phase. Overnight liquid starter cultures can be used instead, but this will significantly slow down the large-scale bacterial growth as they have reached the plateau-phase in the liquid starter culture.
4. NaCl must be added to the Talon buffer fresh, just prior to use.

5. PMSF is only useful for approximately 30 min after addition into an aqueous solution, so add this just prior to bacterial resuspension.
6. When the ultrasonication starts, the consistency of the bacterial lysis buffer changes from watery, to viscous, to again watery. This indicates the various stages of bacterial lysis. When the liquid becomes viscous, the bacterial outer membrane is lysed but the genomic DNA is not yet sheared. When the viscosity turns to thin again, it indicates that the DNA has been sheared and the lysis is more or less complete.
7. The samples collected throughout the protein purification steps should be run on an SDS-PAGE gel, and Coomassie stained to determine the extent of protein purification.
8. Also collect some amount of the bulk dialysis buffer, as this will be useful for correctly producing the blanking solution during concentration measurement of the dialyzed protein. Trace amount of Imidazole present in the protein sample will affect the absorbance if not properly corrected for by using the bulk dialysis buffer also containing identical amount of imidazole.
9. In the case of CBD-EGFP, prolonged reaction times or increased dye concentrations will result in overlabeling due to the presence of two additional cysteines on EGFP. This effect will reduce the EGFP stability and reduce the fluorescence as well as reducing the fluorescence emission intensity from the mero-87. In the case of CBD-MBP, there is only one Cysteine, but overlabeling is still possible due to reactions with other residues including lysines and histidines if the reaction times/dye concentrations are increased unreasonably.
10. Here, the manner in which pipetting is performed is absolutely critical to remove errors. Use fresh pipette tips all the time, do not pipette into a fresh pipette tip more than one time as the surface wetting and viscosity effects will slightly change the amount being pipetted.
11. Here, choose the lowest concentrated lysate and use that as the baseline. One can always dilute more concentrated lysates to make it the same concentration, never the reverse.
12. The volume of lysates could be more or less than the suggested amount. Use the appropriate cuvettes that are optimized for various volumes.

Acknowledgement

This work was supported by grants GM093121 (D.S., L.H.), T32GM007491 (V.M.), and GM071828 (D.C.).

References

1. Benard V, Bokoch GM (2002) Assay of Cdc42, Rac, and Rho GTPase activation by affinity methods. Methods Enzymol 345:349–359
2. Nalbant P, Hodgson L, Kraynov V et al (2004) Activation of endogenous Cdc42 visualized in living cells. Science 305:1615–1619
3. Toutchkine A, Nguyen DV, Hahn KM (2007) Merocyanine dyes with improved photostability. Org Lett 9:2775–2777
4. Toutchkine A, Kraynov V, Hahn K (2003) Solvent-sensitive dyes to report protein conformational changes in living cells. J Am Chem Soc 125:4132–4145
5. Cox D, Chang P, Zhang Q et al (1997) Requirements for both Rac1 and Cdc42 in membrane ruffling and phagocytosis in leukocytes. J Exp Med 186:1487–1494

Chapter 16

Analysis of the Cell Surface Expression of Cytokine Receptors Using the Surface Protein Biotinylation Method

Mahmud Arif Pavel, Clarissa Lam, Parul Kashyap, Zahra Salehi-Najafabadi, Gurpreet Singh, and Yong Yu

Abstract

Cytokines are pleiotropic, low-molecular-weight proteins that regulate the immune responses to infection and inflammation. They stimulate the immune responses by binding to cytokine receptors on the cell plasma membrane. Thus, knowledge of the expression level of particular cytokine receptors on cell surface is crucial for understanding the cytokine function and regulation. One of the techniques to explore the membrane embedded cytokine receptors is cell surface biotinylation. Biotinylated surface proteins can be rapidly purified through the strong interaction between biotin and streptavidin. Here, we describe the procedure for surface biotinylation and purification of biotinylated cytokine receptors for further downstream analysis.

Key words Cytokine receptor, Biotin, Biotinylation, Streptavidin

1 Introduction

Cytokines are pleiotropic regulatory proteins that are involved in many biological processes that regulate immune and inflammatory responses, cell survival, differentiation, as well as cell growth, migration, and proliferation. Cytokines bind to their receptors present on the cell surface in plasma membrane. Cytokine receptors are transmembrane proteins containing various domains that trigger downstream signaling once the specific cytokine binds to the appropriate receptor [1]. For example, CXCR2, a receptor for interleukin-8 (IL-8, CXCL8), contains a crucial G-protein docking site in the second intracellular loop which is required for downstream signaling of the MAPK pathway, ultimately leading to cell growth [2]. To better understand the interaction between cytokines and their receptors, it is crucial to analyze the expression levels of these receptors on plasma membrane. Cell surface biotinylation is a commonly used technique to fulfill this purpose [3, 4],

Ivana Vancurova (ed.), *Cytokine Bioassays: Methods and Protocols*, Methods in Molecular Biology, vol. 1172, DOI 10.1007/978-1-4939-0928-5_16, © Springer Science+Business Media New York 2014

and it has been effectively applied in cytokine research in recent years [5–11].

In particular, cytokine receptors on cell surface can be modified covalently with biotin and purified through affinity chromatography based on the interaction between biotin and streptavidin. The purified sample can then be detected by SDS-PAGE and Western blotting. The non-covalent binding between biotin and streptavidin is very stable as it can tolerate anomalous pH or temperature changes. Also, the presence of detergent, such as 1–2 % SDS, Triton X-100, or Tween, does not affect the binding between biotin and streptavidin.

Many biotin derivatives have been developed. In this chapter, we discuss the application of the widely-used sulfo-NHS-SS-biotin (sulfosuccinimidyl-2-(biotinamido) ethyl-1,3-dithiopropionate). Sulfo-NHS-SS-biotin reacts with primary amines on the side-chain of lysines and the amino-termini of proteins. The negatively charged sulfo-NHS group makes sulfo-NHS-SS-biotin very easy to dissolve even in cold water and makes it impermeable to cell membrane. The latter minimizes nonspecific labeling of cytosolic proteins, thus increasing the degree of purification of surface proteins. The disulfide bond in its spacer arm can be cleaved by reducing agents, such as dithiothreitol (DTT), which facilitate the release of the biotinylated proteins after purification.

In general, cell-surface proteins are incubated with sulfo-NHS-SS-biotin in aqueous milieu to label all membrane proteins (Fig. 1). Free biotins are then quenched and cells are lysed. Biotinylated cell surface proteins can be further precipitated by binding to streptavidin-coupled resin. Proteins are then eluted and analyzed with different methods. The protocol described here is based on the assumption that SDS-PAGE and Western blotting will be used to detect the precipitated biotinylated membrane receptor. The entire procedure is summarized in Fig. 2. This protocol is not only useful in analyzing cytokine receptor expression but also functional in any other cell surface protein analysis.

2 Materials

2.1 Surface Receptor Biotinylation

1. Cells growing in monolayer in T75 cell culture flasks (or 100 mm cell culture dishes), or cells growing in suspension.
2. Phosphate Buffered Saline (PBS): 10 mM Na_2HPO_4, 1.8 mM KH_2PO_4, 137 mM NaCl, 2.7 mM KCl in deionized distilled water, pH 7.4.
3. Sulfo-NHS-SS-biotin (Sulfosuccinimidyl-2-(biotinamido)ethyl-1, 3-dithiopropionate) solution: Just before use, dissolve sulfo-NHS-SS-biotin (Thermo Scientific) in PBS at 250 μg/mL concentration.
4. Quenching Solution: 1 M Tris–HCl, pH 7.4.
5. Tris Buffered Saline (TBS): 150 mM NaCl, 25 mM Tris–HCl, pH 7.4.

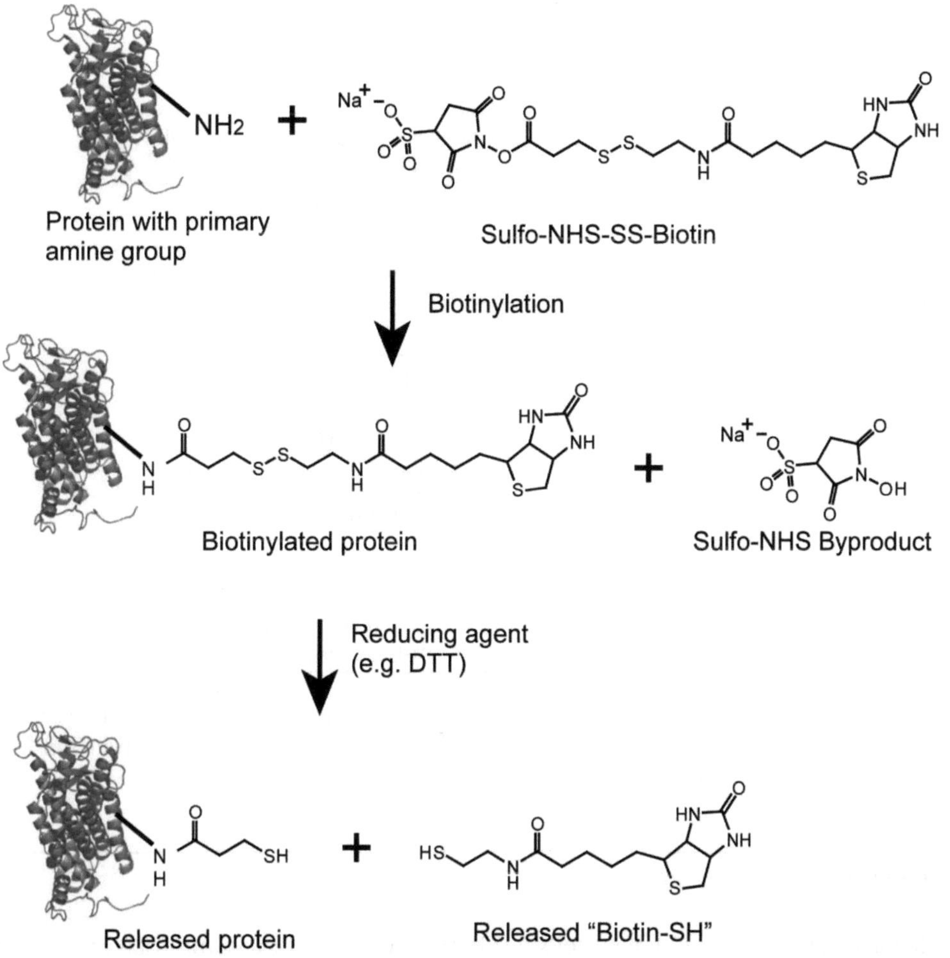

Fig. 1 Schematic illustration of labeling proteins by Sulfo-NHS-SS-Biotin. Sulfo-NHS-SS-biotin contains the biotin part and the sulfo-NHS-SS side chain. Primary amine groups in proteins react with sulfo-NHS-SS-biotin through formation of an amide bond. Reducing agents can be used to cleave the disulfide bond in Sulfo-NHS-SS-biotin to release the labeled proteins

6. Protease inhibitor cocktail (Sigma-Aldrich).
7. Lysis Solution: 100 mM NaCl, 50 mM Tris–HCl, pH 7.4, 10 mM EDTA, 1 % (v/v) Triton X-100. Add protease inhibitor cocktail (Sigma-Aldrich) before use (*see* **Note 1**).

2.2 Precipitation of Biotinylated Proteins with Streptavidin Agarose Resin

1. Streptavidin agarose (Thermo Scientific).
2. Wash Solution: 500 mM NaCl, 25 mM Tris–HCl, pH 7.4, 1 % Triton X-100.
3. SDS-PAGE sample buffer: 62.5 mM Tris–HCl, pH 6.8, 2 % SDS, 10 % glycerol, 0.002 % Bromophenol Blue.
4. Elution Solution: 50 mM DTT (final concentration) is added into the above SDS-PAGE sample buffer just before use.

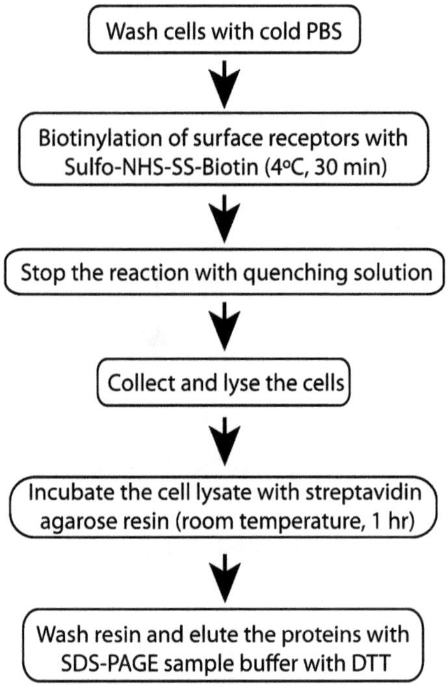

Fig. 2 Procedure summary of the surface protein biotinylation and isolation using streptavidin resin. Isolated proteins can be further analyzed with SDS-PAGE and Western blotting

3 Methods

3.1 Surface Cytokine Receptor Biotinylation

1. Grow cells to 90–95 % confluence in a monolayer in a T75 cell culture flask or on a 100 mm cell culture dish (*see* **Note 2**). For cells growing in suspension, use 1×10^6–10^7 cells for each milliliter of biotin solution prepared in **step 3**.
2. Remove media from the flask/dish and wash the cells twice with 10 mL of ice-cold PBS rapidly (*see* **Note 3**). It is preferred to apply aspiration by Pasteur pipette connected with a vacuum to discard the medium or PBS (or use centrifugation for cells in suspension). Keep the flasks on ice for the following steps.
3. Add 10 mL of an ice-cold sulfo-NHS-SS-biotin solution into the flask/dish (*see* **Note 4**). For cells in suspension, resuspend the cells directly in sulfo-NHS-SS-biotin solution and keep on ice.
4. Place flask on orbital shaker. Let it shake gently for 30 min at 4 °C (*see* **Note 5**).
5. Add 0.5 mL of Quenching Solution into the flask/dish to quench the biotinylation reaction. Incubate for 5 min at 4 °C. Ensure even exposure of the quenching solution by gently tipping the flask back and forth.

6. Detach the monolayer cells by repeatedly pipetting the solutions over them. In case the cells are still firmly attached, gently scrape the cells into solutions and transfer them into 50 mL conical tube. Rinse the flask with 5 mL of ice-cold TBS solution and pool the washing into the same tube of the collected cells.
7. Centrifuge the cells at 500×*g* for 3 min and discard the supernatant.
8. Add 5 mL of TBS solution onto the cell pellet and gently pipette up and down to resuspend the cells. Centrifuge at 500×*g* for 3 min. Discard the supernatant.
9. Repeat **step 8** once (*see* **Note 6**).
10. Add 150 μL of Lysis Solution onto the cell pellet (*see* **Note 7**). Resuspend the cells and transfer the cells in the Lysis Solution to a 1.5 mL microcentrifuge tube.
11. Optional: to improve solubilization efficiency, disrupt the cells by sonicating on ice using five 1-s bursts. Use low power setting to prevent foaming.
12. Incubate the sample with end-over-end mixing in a tube rotator for 1 h at 4 °C, vortexing in medium speed every 10 min for 5 s. Optional: perform additional sonications during incubation.
13. Centrifuge the cell lysate at 10,000×*g* for 15 min at 4 °C. Collect the supernatant (cell lysate) into a new tube.
14. Optional: take out 10 μL of cell lysate for using as a control sample in Western blotting for checking overall expression of target protein.

3.2 Isolation of Biotinylated Proteins with Streptavidin Agarose

1. Take out 120 μL of evenly suspended streptavidin agarose slurry and place it into a 1.5 mL microcentrifuge tube (*see* **Note 8**).
2. Centrifuge for 1 min at 2,000×*g* and discard the supernatant (*see* **Note 9**).
3. Add 300 μL of Washing Buffer onto the resins and mix well by pipetting. Centrifuge at 2,000×*g* for 1 min and discard the supernatant.
4. Repeat **step 3** twice (*see* **Note 10**).
5. Add cell lysate obtained in Subheading 3.1, **step 13**, onto the washed streptavidin agarose resins.
6. Incubate for 1 h at room temperature with end-over-end mixing in a tube rotator. Alternatively, rock back and forth on a rocking platform or orbital shaker (*see* **Note 11**).
7. Centrifuge at 2,000×*g* for 1 min to collect the streptavidin agarose resins.

8. Collect the supernatant into another tube. It can be used as flow-through control.
9. Add protease inhibitor cocktail into 1.2 mL of Wash Solution (*see* **Note 12**).
10. Add 300 μL of Wash Solution onto the resins and incubate for 5 min with end-over-end mixing in a tube rotator at room temperature.
11. Centrifuge at 2,000 × *g* for 1 min and discard the supernatant. Repeat the washing step three times.
12. Add 150 μL of PBS to resuspend the resins and transfer them to a fresh tube (*see* **Note 13**).
13. Centrifuge at 2,000 × *g* for 1 min and discard the supernatant.
14. Add 100 μL of Elution Solution to the precipitated resins. Resuspend the resin by tapping on the tube gently (*see* **Note 14**).
15. Incubate the resins for 60 min at room temperature with end-over-end mixing in a tube rotator (*see* **Note 15**).
16. Centrifuge at 2,000 × *g* for 2 min. Transfer the supernatant into a new tube.
17. The sample is now ready to be detected using SDS-PAGE followed by Western blotting.

4 Notes

1. Certain membrane protein may be better extracted from cell membrane with certain detergent. Thus, optimization of the detergent may be needed to get best extraction yield. Most commonly used detergents include Triton X-100, NP40, and n-Dodecyl β-D-maltoside (DDM). Protease inhibitor should be added immediately before use.
2. Depending on the expression level of target proteins, two or more flasks of cells may be needed for each experiment.
3. A quick wash is recommended because longer contact of cells with PBS may detach the cells from the flask/dish. Do not pipette cold buffer right on top of the cells, instead direct the pipette on the side wall of the dish or on the opposite end of flask surface. For cells growing in suspension, all steps for washing or replacing solution should be done by gentle centrifugation (1,000 × *g*, for 3 min).
4. Dissolve sulfo-NHS-SS-biotin in PBS immediately before adding to the cells. This will help to reduce the hydrolysis of the sulfonate group.

5. Keeping the cells at 4 °C during the biotinylation procedure will help to reduce uptake of sulfo-NHS-SS-biotin by the cells.
6. At this point, cells can be frozen in liquid nitrogen and stored at −80 °C for afterward processing.
7. The volume of Lysis Solution used should be adjusted based on the volume of cell pellet. Use twice as much of Lysis Solution as cell pellet volume.
8. To get better washing efficiency in the following steps, one can use column filter set (such as the Ultrafree-MC centrifugal filter units from Millipore) to hold the resin and perform the washing steps by centrifuging and discarding the flow-through. Use pipette tips with the end cut off to prevent damage to the resins.
9. The resin pellet is very loose. In this and following steps, be very careful when removing the supernatants. Try to remove the supernatant as much as possible while avoiding taking out the resin material. Pipette tips with fine ending such as gel loading tips work well for this purpose.
10. Commercially available streptavidin or similar resins usually use ethanol and sodium azide as preservatives. These preservatives need to be completely removed before use by washing.
11. If the target protein is not stable, binding can be done at 4 °C for 2–4 h or overnight.
12. Some proteins can bind to agarose resin nonspecifically and affect the correct analysis. One can increase the NaCl concentration to 1 M in Wash Solution to obtain better washing efficiency. In this case, larger volume of Wash Solution and longer washing incubation will also help.
13. Transfer to a new tube before elution is necessary since it will avoid eluting the protein nonspecifically bound on the wall of the old tube.
14. Proteins are eluted here with DTT-added to SDS-PAGE sample buffer on the assumption that samples will be analyzed with SDS-PAGE and Western blotting. Other elution solution may be used if SDS-PAGE is not the next analysis method. As long as DTT is added in the Elution Solution, isolated proteins will be cleaved off the streptavidin resins at the disulfide bond site in sulfo-NHS-SS-biotin.
15. Alternatively, the incubation can be done at 37 °C for 30 min or 65 °C for 10 min. Usually, boiling should be avoided since many membrane proteins form aggregates when they are boiled. Protein aggregation results in difficulty of these proteins to migrate into acrylamide gel.

References

1. Allen SJ, Crown SE, Handel TM (2007) Chemokine: receptor structure, interactions, and antagonism. Annu Rev Immunol 25: 87–820
2. Hertzer KM, Donald GW, Hines OJ (2013) CXCR2: a target for pancreatic cancer treatment? Expert Opin Ther Targets 17: 667–680
3. Elia G (2012) Cell surface protein biotinylation for SDS-PAGE analysis. Methods Mol Biol 869:361–372
4. Rybak JN, Scheurer SB, Neri D et al (2004) Purification of biotinylated proteins on streptavidin resin: a protocol for quantitative elution. Proteomics 4:2296–2299
5. Yao Z, Fanslow WC, Seldin MF et al (1995) Herpes virus Saimiri encodes a new cytokine, IL-17, which binds to a novel cytokine receptor. Immunity 3:811–821
6. Nasu J, Mizuno M, Uesu T et al (1998) Cytokine-stimulated release of decay-accelerating factor (DAF;CD55) from HT-29 human intestinal epithelial cells. Clin Exp Immunol 113: 379–385
7. Ozaki H, Ishii K, Horiuchi H et al (1999) Cutting edge: combined treatment of TNF-alpha and IFN-gamma causes redistribution of junctional adhesion molecule in human endothelial cells. J Immunol 163:553–557
8. Partridge EA, Le Roy C, Di Guglielmo GM et al (2004) Regulation of cytokine receptors by Golgi N-glycan processing and endocytosis. Science 306:120–124
9. Espinassous Q, Garcia-de-paco E, Garcia-verdugo I et al (2009) IL-33 enhances lipopolysaccharide-induced inflammatory cytokine production from mouse macrophages by regulating lipopolysaccharide receptor complex. J Immunol 183:1446–1455
10. Adrain C, Zettl M, Christova Y et al (2012) Tumor necrosis factor signaling requires iRhom2 to promote trafficking and activation of TACE. Science 335:225–228
11. Bergström SE, Bergdahl E, Sundqvist KG (2013) A cytokine-controlled mechanism for integrated regulation of T lymphocyte motility, adhesion and activation. Immunology 140: 441–455

Chapter 17

Detection of CXCR2 Cytokine Receptor Surface Expression Using Immunofluorescence

Clarissa Lam, Mahmud Arif Pavel, Parul Kashyap, Zahra Salehi-Najafabadi, Victoria Valentino, and Yong Yu

Abstract

The interleukin-8 (IL-8, CXCL8) chemokine, also known as the neutrophil chemotactic factor, is a cytokine that plays a key role in inflammatory response, cell proliferation, migration, and survival. IL-8 expression is increased not only in inflammatory disorders, but also in many types of cancer, including prostate cancer. IL-8 acts as a ligand for the C-X-C chemokine receptor 2 (CXCR2) protein present on the cell plasma membrane. Binding of the IL-8 ligand to the CXCR2 receptor results in an intracellular signaling pathway mediated by GTP binding proteins coupled to the receptor itself. Knowledge of the CXCR2 expression levels facilitates the understanding of the role and function of IL-8. In this chapter, we describe a protocol that uses the immunofluorescence method and confocal microscopy to analyze the CXCR2 surface expression in human prostate cancer cells. However, this protocol is easily adaptable to analyze the surface expression of other cytokine receptors in different cell types.

Key words Interleukin-8, CXCR2, Cytokine receptors, Immunofluorescence, Prostate cancer PC3 cells

1 Introduction

The CXCR2 protein belongs to a G protein-coupled receptor (GPCR) super family. Human CXCR2 is composed of 360 amino acids and has been found to have about 25–80 % identity with other chemokine receptors [1]. It has seven transmembrane domains and is able to couple with GTP binding proteins to transduce intracellular signals when bound to various cytokines [2]. It has been shown in other chemokine receptors that the extracellular N-terminus plays an important role in ligand binding, while the second intracellular loop contains the G protein docking site [1, 2]. As other GPCRs, CXCR2 may function in higher-order oligomers through dimerization [3, 4]. CXCR2 has also been associated with Rho, Rac, and MAP kinase signaling pathways responsible for cell growth and migration [1]. It was also found to play a role in aging [5].

Ivana Vancurova (ed.), *Cytokine Bioassays: Methods and Protocols*, Methods in Molecular Biology, vol. 1172, DOI 10.1007/978-1-4939-0928-5_17, © Springer Science+Business Media New York 2014

Many chemokine receptors have more than one corresponding chemokine ligands [2]. IL-8, a member of the CXC chemokine family, is one such cytokine that is able to act as a ligand for the CXCR2 receptor. Binding of IL-8 to its receptor activates a specific signaling pathway that plays a role in inflammatory response [6], cell proliferation, and cell survival [7]. Endogenously expressed CXCR2 has been detected in human metastatic prostate cancer cells, where it has been proposed to play a role in IL-8 signaling and, consequently, the progression and proliferation of prostate cancer cells [8].

Immunofluorescence is a versatile method used to visualize antigen molecules in cells. It utilizes fluorophore-coupled antibody to detect the location and relative abundance of target protein molecules. Combined with confocal microscopy, this method is widely used in detecting the cell surface expression of membrane proteins. Employing the immunofluorescence technique to study the cell surface expression of the CXCR2 receptor in prostate cancer cells facilitates the understanding of the IL-8 function and regulation in these cells. Figure 1 illustrates the principle of the immunofluorescence protocol described here.

2 Materials

2.1 Cell Culture

1. PC3 cells obtained from the American Type Culture Collection.
2. RPMI complete medium: RPMI medium supplemented with 10 % fetal bovine serum (FBS), 2 mM glutamine, 1 mM sodium pyruvate, and 1 % penicillin–streptomycin solution. Filter to sterilize using Corning 0.2 μm bottle top filter.
3. Phosphate buffered saline (PBS): to make 1 L of PBS solution, dissolve 8 g of sodium chloride, 0.2 g of potassium chloride, 1.44 g of sodium phosphate dibasic, and 0.24 g of potassium phosphate monobasic anhydrous in 800 mL of distilled water. Adjust pH to 7.2 with HCl or NaOH. Make the final volume up to 1 L with distilled water. 10× or 1× PBS solutions are commercially available from many companies.
4. 24-well tissue culture plates.
5. Glass microscope slides and 12 mm circular coverslips.
6. 0.01 % poly-L-lysine solution: Add 50 mL of sterile tissue culture grade water to 5 mg of poly-L-lysine (*see* **Note 1**).
7. 1 M HCl solution: Slowly add 83 mL of 37 % HCl to 917 mL of deionized water.
8. 95 % ethanol.

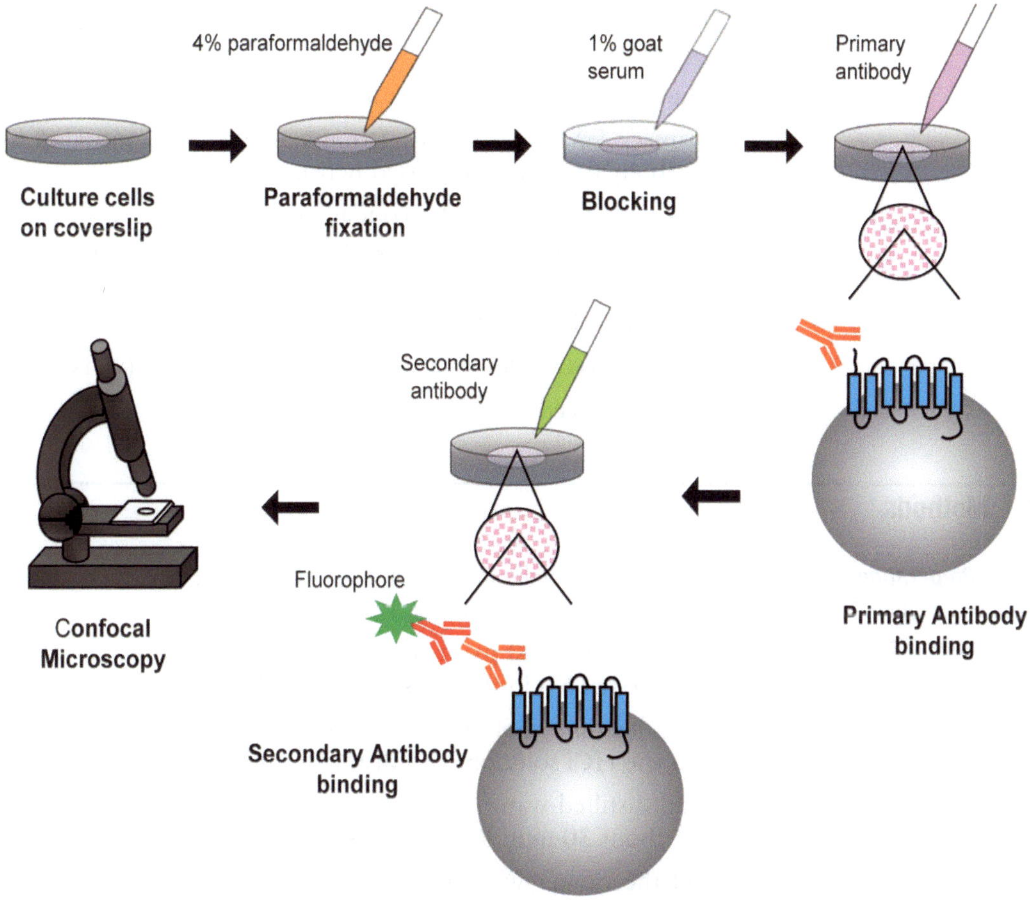

Fig. 1 Schematic illustration of the immunofluorescence method. First, cells are fixed with formaldehyde and blocked with goat serum. Subsequently, CXCR2 is labeled by using specific primary antibody and fluorophore-conjugated secondary antibody

2.2 Immunofluorescence

1. 4 % paraformaldehyde solution: Weigh out 0.4 g of paraformaldehyde and put it in a 15 mL conical Falcon tube. Add 8 mL of distilled water and 10 μL of 1 M NaOH to the tube. Heat the solution to 70 °C and mix until the paraformaldehyde is completely dissolved. Cool the solution down to room temperature and adjust the solution volume to 9 mL with distilled water. Add 1 mL of 10× PBS to make the solution up to a total volume of 10 mL. Adjust pH to 7.0 using diluted HCl. Filter the solution through a 0.2 μm syringe filter. The solution can be aliquoted and frozen for up to 1 month (*see* **Note 2**).
2. 0.2 % Triton X-100: Add 0.1 mL of 20 % Triton X-100 stock solution into 9.9 mL of water and mix well. To make the 20 % stock solution, add 2 mL of 100 % Triton X-100 into 8 mL of water and mix well (*see* **Note 3**).

3. Blocking solution: Mix 0.1 mL of goat serum in 9.9 mL of PBS (*see* **Note 4**).
4. Fluoromount mounting medium (Sigma Aldrich).
5. Primary CXCR2 antibody (mouse monoclonal anti-CXCR2 antibody from Abcam is used here).
6. Fluorophore-conjugated secondary antibody (FTIC-conjugated goat anti-mouse IgG from Jackson Immuno Research Laboratory is used here).
7. Washing solution: PBS + 0.05 % Tween 20.
8. CoverGrip coverslip sealant (Biotium) or nail polish (*see* **Note 5**).
9. Fine-tip forceps.

3 Methods

3.1 Preparation of Acid-Washed Glass Coverslips (See Note 6)

1. Separate coverslips from one another and place them into a glass container containing 1 M HCl solution.
2. Heat the coverslips in the temperature range of 50–60 °C for about 16 h in this loosely covered beaker.
3. After heating, cool the solution to room temperature and rinse out the HCl solution with distilled water.
4. Add distilled water to the beaker and sonicate in water bath for about 30 min. Repeat this step twice (*see* **Note 7**).
5. Fill the beaker with fresh 95 % ethanol. The coverslips may remain in this solution until they are ready to be used.

3.2 Preparation of Poly-L-lysine Coated Coverslips

Perform steps in this section, except for **step 3**, in a sterile cell culture hood.

1. Take out the acid-washed coverslips out of 95 % ethanol and dry them on a sterile petri dish.
2. Place the dried coverslips in a sterile petri dish and cover them with poly-L-lysine solution. Make sure the coverslips are separated from each other.
3. Place the petri dish on an orbital shaker and shake for at least 30 min at room temperature (*see* **Note 8**).
4. Suction out poly-L-lysine solution after the incubation period and rinse the coverslips thoroughly with sterile water (*see* **Note 9**).
5. Place the coverslips into 24-well culture plate, one for each well.
6. Let the coverslips dry completely before moving onto the next step.

3.3 Culturing PC3 Cells on Coverslips

Perform steps in this section in a sterile cell culture hood.

1. Add 500 μL of RPMI complete medium to the wells with coverslips.
2. Plate PC3 cells onto the coverslip (*see* **Note 10**).
3. Place the plate into a CO_2 incubator and incubate at 37 °C.
4. Wait at least a couple of hours for the cells to firmly attach onto the coverslips.

3.4 Paraformaldehyde Fixation and Blocking

1. Suction out the cell culture RMPI medium and rinse the coverslips twice with 500 μL of PBS at room temperature.
2. Add 200 μL of 4 % paraformaldehyde solution onto each coverslip. Gently shake the plate on an orbital shaker for 15 min at room temperature.
3. Suction out the paraformaldehyde solution and wash the cells with 500 μL of PBS by gently shaking for 5 min.
4. Repeat the above washing step twice.
5. Optional: Add 200 μL of 0.2 % Triton X-100 solution onto the coverslip and incubate for 20 min at room temperature. Repeat washing with PBS three times as in **steps 3** and **4** (*see* **Note 11**).
6. Add 200 μL of blocking solution onto each coverslip and block for 30 min at room temperature.

3.5 Incubation with Primary Antibody

1. Dilute the primary antibody to 10 μg/mL in blocking buffer (*see* **Note 12**).
2. Suction out the blocking solution.
3. Add 200 μL of diluted primary antibody into each well (*see* **Note 13**).
4. Incubate for 1 h at room temperature or overnight at 4 °C. Gently shaking the plate during incubation.
5. Suction out the antibody solution from the wells.
6. Wash the cells with 500 μL of Washing solution by gently shaking for 5 min.
7. Repeat the above washing step twice.

3.6 Incubation with Secondary Antibody

1. Dilute the secondary antibody 50 times with blocking buffer (*see* **Note 14**).
2. Add 200 μL of diluted secondary antibody into each well onto the coverslip.
3. Cover the tissue culture plate completely with foil. Keep the plate covered with foil for all of the following steps (*see* **Note 15**).

4. Incubate for 1 h at room temperature. Gently shake the plate during incubation.
5. Suction out the secondary antibody.
6. Wash the cells with 500 μL of Washing solution by gently shaking for 5 min.
7. Repeat the above washing step twice.

3.7 Preparation for Microscopy

1. Drop 10 μL of the mounting medium onto a glass slide.
2. Pick up the coverslips with a fine-tip forceps and invert the coverslips facing down onto the mounting medium.
3. Remove the excess of mounting medium with fiber-free paper, without disturbing the coverslip.
4. Cover the slides with foil and allow the mounting medium to dry for 30–45 min at room temperature.
5. For long-term storage of the coverslips, seal the edges of coverslips with CoverGrip coverslip sealant and allow them to dry for about 3 min. The cells are now ready to be analyzed with confocal microscope.

4 Notes

1. Poly-L-lysine promotes the cell adhesion by increasing electrostatic interaction between cell membrane and the culture surface. Both D- and L-lysine can be used to coat culture surface. Poly-L-lysine is widely used for many cell types. However, certain cells can digest poly-L-lysine. In these cases, poly-D-lysine has to be used. Although higher molecular weight poly-L-lysine (>300,000) provides more attachment sites per molecule, it is more viscous and difficult to work with. Usually poly-L-lysine with molecular weight of 70,000–150,000 provides sufficient attachment site density for normal use.
2. Freshly prepared paraformaldehyde solution works better.
3. When mixing Triton X-100 with water, do not vortex with high speed, otherwise a lot of foam will form. Triton X-100 is very viscous; therefore, cutting the pipette tip to enlarge the opening will make handling Triton X-100 much easier.
4. Blocking solution is used to prevent any nonspecific binding of antibody to proteins that are not being analyzed. 1–2 % bovine serum albumin (BSA) is also widely used as blocking solution.
5. Traditionally, nail polish has been used to seal the edges of mounted coverslips. However, nail polish can leach into aqueous mounting medium and contaminate the sample.

CoverGrip coverslip sealant or similar product is a good replacement for nail polish. To save money, nail polish can still be used if long-term storage is not needed.

6. Washing coverslip with acid helps cells and polyamino acids stick to glass. It also helps to get rid of small particles stuck on the glass, so better DIC pictures can be obtained under microscope.
7. If a sonicator water bath is not available, one may place the coverslips in water-filled container on an orbital platform shaker and shake overnight. Change the water and repeat the shaking once. Use plenty of water to make sure acid residue is completely removed.
8. Alternatively, coverslips can be coated in poly-L-lysine solution at 4 °C for overnight. The authors sometime even store the coverslips in poly-L-lysine solution at 4 °C for future use.
9. For longer storage, allow coverslips to dry on a piece of sterile filter paper then store them in a sterile tissue culture dish. Coated coverslips can be stored for up to a year. After long time storage, coverslips should be rinsed with 100 % ethanol and dried before use.
10. Adjust the number of cells plated to obtain about 70–80 % confluence when doing immunostaining.
11. Triton-X 100 is a commonly used detergent to permeabilize the cell membrane. It should only be used if the goal is to visualize the overall expression of a protein in the cells. If no permeabilization is carried out, only the proteins on the cell surface will be visualized.
12. The working concentration of the primary antibody needs to be optimized for different antibodies.
13. Although it is more convenient, incubation in 24-well plate needs a larger volume of antibody solution so another incubation method may be used to conserve antibody. First place the coverslip, with cell-side facing up, on a piece of parafilm inside of a petri dish, then slowly cover the cells with a few drops of the diluted antibody. 50 μL of antibody solution is usually enough. A piece of water-soaked filter papers can also be placed in the dish to help keeping the coverslips from drying out. Cover the dish with its lid and seal the edge of the dish with parafilm or tape during incubation.
14. The working concentration of the secondary antibody needs to be optimized for different antibodies.
15. Protect fluorescent conjugates from the light. Incubate samples in the dark and cover whenever possible after the secondary antibody is applied.

References

1. Hertzer KM, Donald GW, Hines OJ (2013) CXCR2: a target for pancreatic cancer treatment? Expert Opin Ther Targets 17:667–680
2. Allen SJ, Crown SE, Handel TM (2007) Chemokine:receptor structure, interactions, and antagonism. Annu Rev Immunol 25:87–820
3. Trettel F, Di Bartolomeo S, Lauro C et al (2003) Ligand-independent CXCR2 dimerization. J Biol Chem 278:40980–40988
4. Rodriguez-Frade JM, Mellado M, Martinez-A C (2001) Chemokine receptor dimerization: two are better than one. Trends Immunol 22:612–617
5. Acosta JC, O'loghlen A, Banito A et al (2008) Chemokine signaling via the CXCR2 receptor reinforces senescence. Cell 133:1006–1018
6. Brat DJ, Bellail AC, Van Meir EG (2005) The role of interleukin-8 and its receptors in gliomagenesis and tumoral angiogenesis. Neuro Oncol 7:122–133
7. Vlahopoulos S, Boldogh I, Casola A et al (1999) NFκB-dependent induction of interleukin-8 gene expression by tumor necrosis factor alpha: evidence for an antioxidant sensitive activating pathway distinct from nuclear translocation. Blood 94:1878–1889
8. Wilson C, Wilson T, Johnston PG et al (2008) Interleukin-8 signaling attenuates TRAIL- and chemotherapy-induced apoptosis through transcriptional regulation of c-FLIP in prostate cancer cells. Mol Cancer Ther 7:2649–2661

Chapter 18

Detecting Tie2, an Endothelial Growth Factor Receptor, by Using Immunohistochemistry in Mouse Lungs

Prajna P. Guha, Sascha A. David, and Chandra C. Ghosh

Abstract

Immunohistochemical (IHC) staining is an invaluable, sensitive, and effective method to detect the presence and localization of proteins in the cellular compartment in tissues. The basic concept of IHC is detecting the antigen in tissues by means of specific antibody binding, which is then demonstrated with a colored histochemical reaction that can be observed under a light microscope. The most challenging aspect of IHC techniques is optimizing the precise experimental conditions that are required to get a specific and a strong signal. The critical steps of IHC are specimen acquisition, fixation, permeabilization, detection system, and selection of the antigen specific antibody and its optimization. Here, we elaborate the technique using the endothelial growth factor binding receptor Tie2 in mouse lungs.

Key words Angiopoietins, Antibody, Immunohistochemistry, Tie2

1 Introduction

IHC staining methods use enzyme-labeled antibodies to identify and quantify proteins in tissues. Although this is a simple technique, proper optimization of the steps involved determines its outcome. IHC is now widely used for disease diagnosis and for scientific research as it does not involve sophisticated microscope and the detection signal remains stable for years. This technique however can only detect one antigen as opposed to immunofluorescence (IF) where multiple antigens can be detected simultaneously.

Our laboratory has longstanding interest in understanding the vascular integrity via Tie2-angiopoietin pathway in normal and pathological conditions [1, 2]. The endothelial receptor tyrosine kinases (Tie1 and Tie2), and their ligands, angiopoietins (Angpt), which otherwise maintain the quiescent endothelium, play an important role in acute and chronic inflammation (Fig. 1) [1, 3–7]. The angiopoietins represent an example of growth factors consisting of both

Ivana Vancurova (ed.), *Cytokine Bioassays: Methods and Protocols*, Methods in Molecular Biology, vol. 1172, DOI 10.1007/978-1-4939-0928-5_18, © Springer Science+Business Media New York 2014

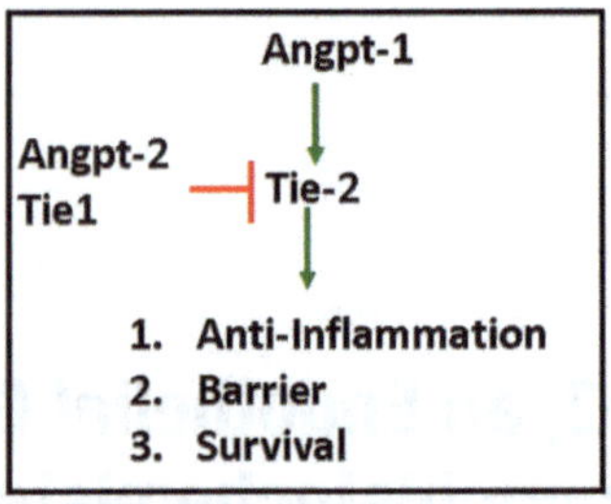

Fig. 1 Tie2 angiopoietin pathway

receptor activator (Angpt-1) and receptor blocker (Angpt-2) [3]. In this chapter, we used IHC to detect Tie2 in mouse lungs which was earlier detected by using IF as described previously [8].

2 Materials

Prepare all solutions in ultrapure water. Disposal of the used chemicals should be done by following Material Safety Data Sheet (MSDS) carefully. The toxic materials should be stored inside the satellite area with proper labeling (date, name of the chemical, and approximate amount).

2.1 Animals

All experiments should be approved by Institutional Animal Care and Use Committee. Use 8-week-old male C57BL/6J (Jackson Laboratories, ME) to collect organs.

2.2 Buffers and Diluents

1. Rinse buffer: 1× Phosphate Buffered Saline (PBS).
2. Quenching, Blocking buffer/Antibody diluents: Quench the endogenous peroxidase by using 3 % H_2O_2. Add 1 ml of 30 % H_2O_2 to 9 ml of water. Block nonspecific proteins by using Protein block-serum free (Dako North America Inc) and 0.1 % Triton X-100 as the blocking buffer, and use the same to dilute primary and secondary antibodies.
3. Periodate-Lysine-Paraformaldehyde (PLP) fixative buffer: Prepare solution A and solution B and mix them on the day of use. Store solution A at −20 °C and prepare solution B on the day of use.

   For solution A: Dissolve 7.3 g L-Lysine-HCl in 200 ml of water. Add enough of 1 M sodium phosphate, dibasic (Na_2HPO_4) to bring the pH to 7.4. To make 100 ml of 1 M Na_2HPO_4, dissolve 14.2 g of Na_2HPO_4 in 100 ml of water. Add 1× PBS to bring the final volume to 400 ml. Aliquot into 18.75 ml and freeze at −20 °C until the day of use.

   For solution B: Add 1 drop (~50 μl) of 10 N sodium hydroxide (NaOH) and 0.4 g of para formaldehyde (PFA) to 5 ml of

1× PBS in 15 ml tube. Dissolve PFA and vortex immediately for 2 min. Prepare 10 N NaOH solution by adding 40 g of NaOH into 100 ml of water.

After mixing solutions A and B, add 51 mg of sodium periodate ($NaIO_4$). Keep the mix on ice. Do not reuse the PLP fixative buffer.

4. Sucrose Buffer: Make 200 ml of 18 % sucrose by dissolving 36 g of sucrose in 200 ml of 1× PBS.
5. The base mold (TissueTek, CA).
6. Optimum Cutting Temperature (OCT) freezing medium (TissueTek, CA).

2.3 Substrates, Chromogens, Endogenous Enzyme Blocking and Counterstain Solutions

1. DAB-hydrogen peroxide substrate solution: In presence of peroxidase, Vector® DAB Substrate (3, 3′-diaminobenzidine) produces a brown reaction product.
2. Mayer's Hematoxylin (Dako, CA).
3. 95 % alcohol: Prepare 95 % alcohol by mixing 5 ml of water and 95 ml of 100 % alcohol.
4. Xylene (Sigma-Aldrich).
5. Xylene mounting media, VectaMount Permanent Mounting Medium (Vector Labs).
6. ABC reagent or avidin–peroxidase: VECTASTAIN Universal Elite ABC kit (Vector Labs).

3 Methods

Paraffin embedding is considered to be the best option for long term storage of tissue samples; however the tissues need to be fixed prior to being embedded in paraffin. Tissues can also be snap-frozen by immersing the sample in liquid nitrogen or in isopentane cooled on dry ice. These samples do not need to be fixed prior to freezing, but unlike paraffin embedded samples, they cannot be stored as long. Frozen tissues can be stored at −80 °C for about a year. Fixing the samples preserves the stabilization and morphology of proteins. Incomplete fixation (under-fixation) causes proteolytic degradation of antigen, while prolonged fixation (over-fixation) results in masking of antigens. We found PLP method of fixation very effective that provides adequate fixation.

The challenge commonly faced with IHC is the nonspecific interactions between the primary and secondary antibodies with the tissues which can be overcome by blocking reagents. Since IHC is an enzyme based chromogenic assay, the endogenous enzyme activity of the tissues need to be blocked as well (quenching). IHC staining that uses horseradish peroxidase (HRP) or alkaline phosphatase (AP) need the quenching step. Organs such as kidney, liver, and heart

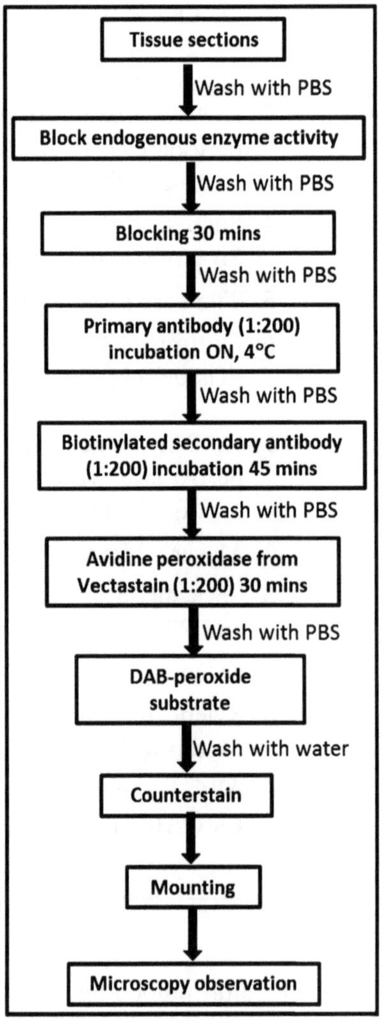

Fig. 2 Flow chart of immunohistochemistry method

have high enzyme activity. These steps are performed prior to antibody incubation.

Counterstaining of immunostained tissues is performed for better visualization of the nucleus and the tissue morphology. However, nuclear staining such as hematoxylin must be avoided when the protein of interest is localized in the cell nuclei (Figs. 2 and 3).

3.1 Tissue Processing, Fixation, Sectioning, Mounting, and Storage

1. Remove tissues, trim and cut in half and quickly wash in ice-cold 1× PBS. Place the tissues in ice-cold PLP fixation buffer on ice or perform the process in a cold room, for 2 h (*see* **Note 1**).
2. Incubate the tissues overnight in ice-cold 18 % sucrose buffer and place it at 4 °C.

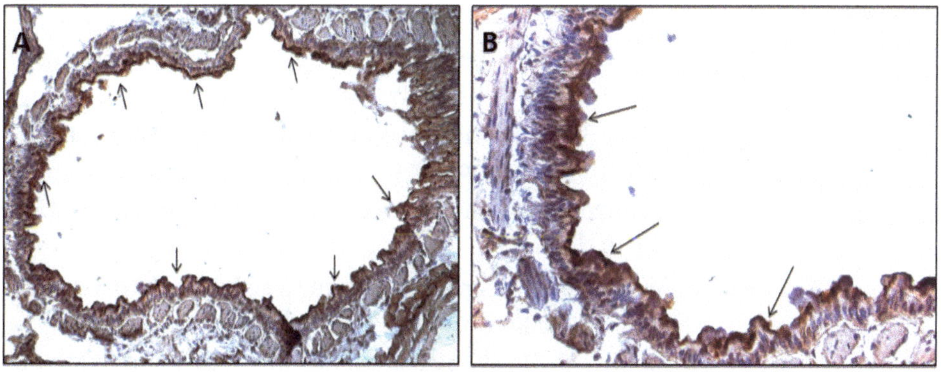

Fig. 3 Tie2 is detected (brown DAB staining) in endothelial cells lining of the alveolar vasculature and in bronchial endothelial cells (*black bar*) as shown (**a**) 20× magnification and (**b**) 40× magnification. Nuclei were counterstained with hematoxylin

3. Label the base mold appropriately and partially fill with OCT freezing medium.
4. Place a beaker with isopentane on dry ice and let it cool adequately.
5. Place the tissue into the mold and then completely immerse the mold carefully in cold isopentane.
6. Allow OCT to solidify and store in −80 °C freezer until sectioning.
7. Equilibrate the block to −20 °C temperature for cryostat. Place the tissue sample in the specimen disk of the cryostat and cut about 6–8 μm sections.
8. Place the sections on Superfrost slides and let them dry for 30 min. At this point, the slides can be stored in a sealed slide box at −80 °C for future staining. When ready to stain, remove the slides from the freezer and equilibrate at room temperature (RT) for about 10 min.

3.2 Quenching/Blocking Endogenous Enzyme Activity

1. Carefully dry slides and surround individual tissue slices with a hydrophobic barrier using a barrier pen, ImmEdge Pen.
2. Wash the slide three times with 1× PBS for 5 min.
3. Block the endogenous peroxidase enzyme activity by incubating the slides with 3 % H_2O_2 for 30 min (*see* **Note 2**).
4. Gently wash the slides three times with 1× PBS for 2 min (drain the slides between washes; DO NOT rinse).
5. Block nonspecific binding with blocking buffer for 30 min at room temperature (RT) (*see* **Note 3**).

3.3 Antibody Incubation and Detection

1. Drain slides gently to remove excess blocking buffer.
2. Incubate with primary antibody (1:200 in blocking buffer) overnight at 4 °C (*see* **Notes 4–8**).
3. Wash 3 times with rinse buffer with 5 min incubation. Drain gently after washes.
4. Incubate with biotinylated secondary antibody for 45 min at RT in dark.
5. Wash the slide three times with rinse buffer with 1 min incubation. Drain gently after each wash.
6. Apply 1:200 ABC reagent or avidin–peroxidase to the biotinylated secondary antibody for 30 min. It amplifies the signal as compared to secondary antibody conjugated with HRP (*see* **Note 4**).
7. Wash three times with rinse buffer.
8. Develop with chromogen, add the prepared substrate DAB-H2O2 and incubate for 5 min (*see* **Notes 9** and **10**).
9. Rinse the sections very gently under running water for 5 min.
10. Drain the slides gently and proceed to counterstaining section.

3.4 Counterstaining Tissue Sections

1. Place the slides in a slide rack and immerse slides in hematoxylin solution for 1 min.
2. Drain the excess hematoxylin dye for 30–60 s and carefully rinse the slides twice in water for 1 min (*see* **Notes 11** and **12**).
3. Wash in 95 % alcohol for 2 min. Drain excess alcohol.
4. Wash in 100 % alcohol for 2 min. Drain excess alcohol.
5. Immerse the slides in xylene for 5 min, three times, and then place the coverslip on the tissue sections using xylene based mounting solution avoiding any bubbles. Dry the slides overnight (*see* **Notes 13** and **14**). The slides are now ready for visualization.

4 Notes

1. Tissue over- or under-fixation can be overcome by decreasing or increasing the fixation durations, respectively.
2. If tissue contains higher levels of endogenous enzymes, increase the quenching time.
3. If there is nonspecific binding of secondary antibody, it can be overcome by using the serum derived from same species as the secondary antibody for blocking reagent. Care must be taken to avoid mouse antibodies used on mouse tissues or tissue must be treated with mouse on mouse blocking reagent prior

to primary antibody incubation. If there are inadequate washes of sections, increase the duration of washes

4. If the antibody concentration is too low or too high, increase or decrease concentrations of primary and/or secondary antibodies by performing serial dilutions and determining the optimized antibody concentration, respectively. Titration of incubation time can be achieved by increasing or decreasing the duration of the primary and/or secondary antibody incubations.
5. There are several monoclonal and polyclonal antibodies available for IHC staining. Monoclonal antibodies are known to give lesser background as compared to polyclonal antisera. Care must be taken while choosing specific antibodies from a whole range of antibodies. Once the antibody is chosen, the optimal concentration needs to be determined by performing serial dilution.
6. Avoid using sodium azide as a preservative as it inhibits enzyme activity.
7. The entire process must be performed in a moist and dark atmosphere. For this, a covered box can be wrapped with aluminum foil and a wet sponge can be used to make the box moist.
8. To evaluate the nonspecific binding and Fc receptor binding, same isotype as the primary antibody must be analyzed in parallel with the primary antibody staining. If an isotype control for an antibody is not available, a negative antibody control must be performed where instead of using the primary antibody, the blocking buffer is used. A positive control is strongly recommended to ensure the antibody performance, where staining is performed on tissues where the protein of interest is known to be present.
9. After adding substrate, color intensity must be visualized under microscope and the incubation must be stopped by washing slides in water. If chromogen substrate incubation is too high, decrease the duration of reaction.
10. Since DAB is toxic and a potential carcinogen, its proper disposal is mandatory. DAB can be inactivated by adding bleach or sodium hypochlorite. The solution will turn black and can then be washed down the drain with plenty of water. But this step needs to be approved by the waste hazard officer.
11. If there is inadequate hematoxylin staining, repeat **step 2** in Subheading 3.3.
12. Choice of counterstain is critical in IHC so that it is compatible with the color of the chromogenic reagent. In this chapter, we have used hematoxylin stain as a counterstain but there are a wide variety of counterstains available that can be looked up for compatibility.

13. Some other slide adhesives (e.g., poly-L-lysine) can be used if sections seem to get detached from the slide.
14. Xylene based mounting media binds strongly to plastic. Hence care must be taken during drying step.

Acknowledgement

The authors would like to thank Harita Dharaneeswaran and Dr. Stanley Artap for useful discussion about the IHC method. The authors also convey their sincere gratitude to Dr. Samir M. Parikh for his constant support and encouragement.

References

1. Parikh SM (2013) Dysregulation of the angiopoietin-Tie-2 axis in sepsis and ARD. Virulence 4:517–524
2. David S, Kumpers P, van Slyke P et al (2013) Mending leaky blood vessels: the angiopoietin-Tie2 pathway in sepsis. J Pharmacol Exp Ther 345:2–6
3. Dumont DJ, Gradwohl G, Fong GH et al (1994) Dominant-negative and targeted null mutations in the endothelial receptor tyrosine kinase, tek, reveal a critical role in vasculogenesis of the embryo. Genes Dev 8: 1897–1909
4. Suri C, Jones PF, Patan S et al (1996) Requisite role of angiopoietin-1, a ligand for the TIE2 receptor, during embryonic angiogenesis. Cell 87:1171–1180
5. Sato TN, Tozawa Y, Deutsch U et al (1995) Distinct roles of the receptor tyrosine kinases Tie-1 and Tie-2 in blood vessel formation. Nature 376:70–74
6. David S, Ghosh CC, Kumpers P et al (2011) Effects of a synthetic PEG-ylated Tie-2 agonist peptide on endotoxemic lung injury and mortality. Am J Physiol Lung Cell Mol Physiol 300:L851–862
7. Kim W, Moon SO, Lee SY et al (2006) COMP-angiopoietin-1 ameliorates renal fibrosis in a unilateral ureteral obstruction model. J Am Soc Nephrol 17:2474–2483
8. Ghosh CC, Mukherjee A, David S et al (2012) Impaired function of the Tie-2 receptor contributes to vascular leakage and lethality in anthrax. Proc Natl Acad Sci 109:10024–10029

Chapter 19

Use of shRNA for Stable Suppression of Chemokine Receptor Expression and Function in Human Cancer Cell Lines

Nicole Salazar, Daniel Muñoz, James Hoy, and Bal L. Lokeshwar

Abstract

In this chapter, we describe a protocol used for stable silencing of chemokine receptor CXCR7 in human cancer cells using shRNA in a lipid transfection setting, previously published by our laboratory. We provide thorough detail and background information about the process of shRNA to clarify the importance of this process. We use CXCR7 shRNA and scrambled sequence shRNA constructs cloned into a pRS plasmid under the control of a U6 promoter for stable expression. Human cancer cells are transfected with shRNA-pRS using Lipofectamine 2000. Cells stably expressing the shRNA are selected from transfected cultures following 2 weeks in medium containing the selection antibiotic puromycin. The emergent cell colonies are evaluated for knockdown of CXCR7 mRNA and protein expression by q-PCR and immunoblotting with rabbit anti-CXCR7 IgG, respectively.

Key words shRNA, Lipofectamine 2000, Stable transfection, Selection, CXCR7

1 Introduction

The discovery that double-stranded RNA was orders of magnitude more potent than single-stranded RNA at gene silencing revolutionized modern biology and opened the door to RNAi [1]. This discovery in 1998 earned Fire and Mello the Nobel Prize in Medicine in 2006 [2]. Lee and Ambros described that conserved RNA species undergo processing through a system known as the RNAi machinery, and that the starting product of this process is a stem-loop or short hairpin RNA precursor [3]. This RNA precursor is produced endogenously in the cell as a long double-stranded non-coding RNA transcript known as pri-miRNA. The pri-miRNA forms a hairpin or stem-loop shaped structure as the RNA anneals with itself due to sense and antisense sequences that flank the loop. This double-stranded precursor is then processed by Drosha in the nucleus and exported to the cytoplasm where it is further processed by Dicer to fragment it into pieces of mature microRNA (miRNA) [4, 5]. These short dsRNA sequences are recognized by the RISC complex.

Ivana Vancurova (ed.), *Cytokine Bioassays: Methods and Protocols*, Methods in Molecular Biology, vol. 1172, DOI 10.1007/978-1-4939-0928-5_19, © Springer Science+Business Media New York 2014

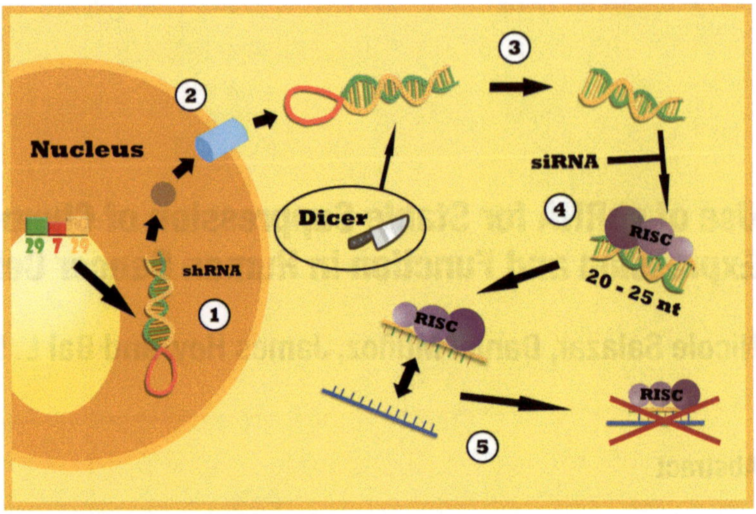

Fig. 1 Schematic illustration of the use of shRNA for stable suppression of chemokine receptor expression and function in human cancer cell lines. (*1*) Pri-miRNA endogenously produced in all mammalian nuclei or shRNA is introduced through transfection. (*2*) Drosha enzyme processes pri-miRNA to pre-miRNA, which is recognized and exported by Exportin V to the cytoplasm. (*3*) Dicer recognizes pre-miRNA and digests it into short oligos of 20–25 nucleotides of dsRNA called miRNA. (*4*) RISC complex can bind to this miRNA or to an introduced set of siRNA delivered through transfection. (*5*) The mi/siRNA-RISC complex then binds to the target mRNA and prevents its translation. Illustrated by Ms. Maite Lopez

The complex combined with the miRNA can recognize and halt targeted mRNA transcripts from being translated. *See* Fig. 1 for a schematic illustration.

Since miRNAs produced by mammalian cells do not have complete complementarity to their targets, it is possible to produce and deliver small interfering RNAs (siRNA) that mimic the function of miRNAs but are designed to have greater specificity to their targets by having complete complementary sequences [6]. One significant drawback to this assay however, is the depletion of siRNA over a few days from delivery. An alternative to this direct delivery method is the development of shRNA and its delivery through a vector-expressing plasmid which contains a selection marker. Expression of the shRNA sequence includes a 29-mer region complementary to the target transcript, followed by a 7-nucleotide loop, followed again, by the antisense sequence of the 29mer region [7]. This produces a dsRNA structure that is similar to the naturally produced pri-miRNAs of the cell and is processed accordingly to its miRNA mimic, the siRNA. This setup allows for the continuous, stable expression of the shRNA for suppression of the target gene [4, 8].

In this protocol we describe an efficient approach to stably silence the chemokine receptor, CXCR7 adapted from the manufacturer's guide to using the transfection reagents. We use RNA interference (RNAi) implemented with short hairpin RNA (shRNA). Vector expressing shRNA can be used to stably suppress gene expression in cell lines. We used a retroviral silencing plasmid (pRS) that contains the puromycin resistance gene obtained from Origene [7]. Our lab has successfully used these shCXCR7s from Origene to stably down regulate CXCR7 expression in breast and prostate cancer cell lines used for functional assays including in-vivo xenograft tumor assays [9]. Origene provides four different shRNA plasmids for the sequence of interest, fully verified by sequencing, thus guaranteeing that at least one of the four constructs provides over 90 % gene expression inhibition. This allows us to use the vectors for sequence modifications and be able to perform RNA rescue experiments with ease. More information about these and lentiviral shRNA vectors can be found at www.origene.com. Origene now also offers a lentiviral vector for CXCR7 shRNA knockdown, which may provide better transfection efficiency.

2 Materials

2.1 Cell Culture

1. MCF7 cells (American Type Culture Collection, Manassas, VA).
2. LNCaP cells (American Type Culture Collection, Manassas, VA).
3. 12-well cluster plates.
4. RNAase and DNAase-free sterile 1.5 microcentrifuge tubes and barrier-filter tips.
5. Trypsin-EDTA.
6. RPMI complete medium: RPMI 1640 with L-glutamine, 10 % fetal bovine serum (FBS), 20 μg/mL of gentamicin sulfate solution.

2.2 Transfection

1. Transfection medium: RPMI 1640 with L-glutamine; 10 % FBS; no gentamicin.
2. Complex medium: Opti-MEM I reduced serum medium (Invitrogen, Gibco, 31985–062). If not available, use serum-free medium, such as RPMI.
3. Plasmid DNA (*see* **Note 1**): HuSH 29mer shRNA construct against CMKOR1 (chemokine receptor CXCR7), Origene.
4. Transfection reagent: Can be lipid based or viral based (*see* **Note 2**). Lipofectamine 2000 or Lipofectamine LTX with Plus reagent (Invitrogen, 52887 and 94756).
5. Rabbit anti-CXCR7 IgG antibody.
6. E.Z.N.A.® RNA Isolation Kit (101319–242), Omega Bio-Tek.

2.3 Stable Clone Selection

1. Complete medium with antibiotics as described in Subheading 2.1.
2. Puromycin stock (1 mg/mL) for kill curve dilutions.
3. Freezing medium: 70 % RPMI, 20 % FBS, 10 % dimethyl sulfoxide (DMSO; cell culture grade). Add gentamicin sulfate to a final concentration of 20 μg/mL.

3 Methods

3.1 Transfection of Human Cancer Cell Lines LNCaP/MCF-7 with shCXCR7

For transfection of human cancer cell lines (LNCaP/MCF-7) with shCXCR7, perform all transfection experiments in a bio-safety level II cell culture hood. If plating on a smaller plate, vary concentrations proportionately according to well area. Example: In a 12-well plate, plate 1.0×10^5 MCF-7 or LNCaP cells per well in antibiotic-free medium 1 day before transfection. Cells should be healthy, well adhered, and 80–90 % confluent at the time of transfection. Wait a maximum of 24 h after plating to transfect the cells (*see* **Note 3**). Refer to Table 1 for an example of calculations.

1. For each transfection sample, prepare the transfection reagent as follows (*see* **Note 4**). Also see example in Table 1.

   TUBE 1:

   (a) Add the required amount of Opti-MEM medium (100 μL final volume minus Lipofectamine volume).

   (b) Add Lipofectamine reagent into Opti-MEM medium.

   (c) Mix/vortex gently.

   TUBE 2:

   (a) Add the required amount of Opti-MEM medium (100 μL final volume minus DNA volume). Add total volume of DNA required for 0.5 μg (in a 12-well plate well).

   (b) Mix/vortex gently.

   (c) Allow mixtures to incubate at room temperature for 5 min.
2. After the 5 min incubation, combine the diluted DNA with the diluted Lipofectamine and incubate 25 min at room temperature. You should have 200 μL of DNA/reagent complex. While waiting for the complex to form, remove old media from cells, and add 800 μL of antibiotic-free media per well, so the final volume in the well is 1 mL after the complex is added.
3. Add the required amount of complex to each well (200 μL per well).
4. Incubate at 37 °C in a CO_2 incubator for 6 h.
5. Replace with antibiotic-free complete media after 6 h of transfection to minimize cell toxicity.

Table 1
Example: for 3 wells of a 12-well dish, we multiply by a 10 % factor, or 1.1 to account for pipetting errors

Vector	Concentration (μg/μL)	Tube 1		Tube 2	
		DNA: (0.5 μg/well) × 3 wells × 1.1 (μL)	DNA dilution: Serum-free medium (100 μL per well × 3 wells × 1.1) (100 × 3 × 1.1) − (DNA)=	Lipofectamine 2000: 2.0 μL per well × 3 wells × 1.1 (μL)	Lipo dilution: Serum-free medium (100 μL per well × 3 wells × 1.1) (100 × 3 × 1.1) − (Lipo)=
Control	0.588	(0.5 × 3 × 1.1)/0.588 = 2.81	327.19	(2 × 3 × 1.1) = 6.6	323.4
shCXCR7	0.441	(0.5 × 3 × 1.1)/0.441 = 3.74	326.26	(2 × 3 × 1.1) = 6.6	323.4

6. Incubate for 24–72 h before measuring gene expression by real-time quantitative polymerase chain reaction (RT-qPCR), western blot (WB), or the observed presence of a green fluorescent protein (GFP)-tag signal if you use shGFP as an expression marker. We find that 24 h after transfection, mRNA levels indicate the efficiency of the reaction, and 48 h after transfection is optimal for protein level assessment by western blot analysis (*see* **Note 5**). The cells can be evaluated for knockdown of CXCR7 mRNA and protein expression by q-PCR and immunoblotting with rabbit anti-CXCR7 IgG antibody, respectively.

3.2 Generating Stable Clones (See Note 6)

1. Seventy-two hours after transfection, begin antibiotic selection of the cells with puromycin. Pick up cells from the well by trypsinization and resuspend them in enough media to aliquot them into a 96-well plate to generate a kill curve (*see* **Note 7**) for the resistance gene; in this case, puromycin.
2. Plate approximately 1×10^4 cells per well in triplicate wells of the 96-well plate or in duplicate if using a 48-well plate. Use complete media. Freeze the remainder of your transfected cells for future selection needs (*see* **Note 8**). You may wait until the cells settle, then change the media to the puromycin containing media, or wait 24 h and change media to selection media.
3. Make a concentration gradient of puromycin ranging at least 6 concentrations. Make 2.0 μg/mL of antibiotic the middle point of the curve, since Origene recommends this concentration as the selection concentration for its plasmid. For example, use the following concentrations: 0, 0.25, 0.5, 1.0, 2.0, 3.0, 5.0, and 7.5 μg/mL of puromycin.
4. Change the media every 3 days and replace with fresh complete media with puromycin dilutions. Check for death of non-transfected cells before every media change. Note the lowest and highest concentrations at which the cells survived. The lowest concentration at which the cells survived is the selection concentration (*see* **Note 9**).
5. After 1–2 weeks, you should be able to see colonies forming that are your stable clones. Select at least three separate clones and transfer to a well of a 6-well plate, while continuing selection. Once cultures are 50 % confluent, split them into wells of a 12-well plate for RNA and protein extraction and passage remaining cells into a T25 flask, using the routine Trypsin-EDTA cell detachment procedure.
6. Assay your stable knockdown efficiency of CXCR7 mRNA and protein expression with RT-qPCR and immunoblotting with rabbit anti-CXCR7 IgG antibody, respectively (*see* **Note 10**).
7. We recommend doing mRNA analysis before protein analysis. If the mRNA results are successful, then proceed to the Western blot analysis.

8. *mRNA analysis*: For mRNA expression analysis of CXCR7, purify RNA and use the isolated RNA as the template. Use CXCR7 mRNA transcript specific primers for the qPCR reaction. You can design the primers using Primer-BLAST from the NCBI Web site and mFOLD (http://mfold.rna.albany.edu/?q=mfold) to ensure no secondary structure interferes with the primer target. The qPCR reaction mix from PerfeCTa SYBR Green FastMix for iQ PPS is as follows:

PerfeCTa SYBR Green FastMix for iQ (2×)	10 μL
Forward primer	0.5 μL
Reverse primer	0.5 μL
Nuclease-free water	4 μL
Total volume	15 μL

The SYBR Green FastMix itself includes dNTPs, a Taq DNA Polymerase, SYBR Green I dye, and stabilizers. Load 5 μL of cDNA sample to complete the reaction. Load each reaction into a low-profile 96-well unskirted PCR plate, and run the reaction in the qPCR System using the programmed cycle shown below:

1	95 °C		30 s
2	95 °C		1 s
3	60 °C		25 s
4	55–95 °C	0.5 °C increments	10 s
	Repeat **steps 2** and **3**, 30–45 times		

We use the PPIA transcript as the reference mRNA. We analyze the results using the Bio-Rad CFX Manager software. Expression levels are calculated by Threshold Cycles (Ct) of a predetermined relative fluorescence unit (RFU). We calculate relative expression by using the difference between the Cts of CXCR7 and PPIA and taking the inverse squared of this difference for comparable values among different cDNA samples ($1/[2\text{^}\{Ct_{test} - Ct_{reference}\}]$).

9. *Protein analysis*: Analyze the expression of specific proteins in treated cells by immunoblotting. We analyze the CXCR7 expression by a 10 % SDS-PAGE, followed by immunoblotting to Immobilon P-(PVDF) membrane. The membrane is blocked with 5 % milk in TBS-T. CXCR7 signal is detected using anti-CXCR7 IgG antibody diluted 1:5,000 in 5 % milk in TBS-T. Signal is detected using ECL. The specific CXCR7 (~42 kDa) band shows up slightly above β-actin and right

below the 55 kDa mark. Relative protein band intensities can be quantified using densitometry. Band densities are normalized to those of β-actin.

4 Notes

1. DNA: Using plasmid DNA with a concentration greater than 0.8 μg/μL gives less toxicity. We recommend using 0.5 μg of DNA when working in a 12-well plate. For western blot and RT-qPCR analysis, 12-well plates provide enough material for analysis (2 wells for RNA and 1 well for protein are sufficient) for analysis of transfection efficiency. Use a 1:2–1:4 ratio of DNA (μg) to Lipo (μL) reagent. If using the Lipofectamine LTX reagent, use a 1:1 ratio of DNA (μL) to PLUS (μL) reagent, as in the manufacturer's instructions.
2. We recommend lipid transfection as it is fast approach that requires a single transfection round of the target cells. Our transfection efficiency is 50–90 % on transient transfections and 50–80 % on stable transfections. Delivery of shRNA via viral vectors requires two steps. At first, a transfection of virus producing cells with lentiviral plasmids, which results in the production of lentiviral particles is required. Those vectors are then isolated from the supernatant of the virus producing cells and are used for transduction/infection of target cells. Thus, delivery of shRNA via a lentiviral system is more labor intensive compared to lipid mediated transfection. The advantage of viral transfection is its high efficiency, therefore, stable selection using puromycin is rapid (3–5 days versus 1-weeks). However, we have seen similar efficiency in delivery of shRNA via viral vectors compared to delivery of shRNA via lipid transfection as evaluated by RT-qPCR and Western blot analysis. For lipid based transfections-our lab has successfully used Lipofectamine 2000, and Lipofectamine LTX transfection reagents.
3. Cells should be at a maximum of 80–90 % confluency. This is because the cells should have enough space to divide, in order to be in the growth phase of the cell cycle, allowing for the plasmid DNA to be most efficiently incorporated.
4. The Lipofectamine LTX manufacturer's protocol accounts for pipetting variations, so there is no need to multiply by an extra factor. If following the Lipofectamine 2000 protocol multiply all volumes by a factor of 1.1 to account for pipetting error and volume loss. We focus on Lipofectamine 2000 use as Lipofectamine LTX is usually recommended for more delicate and hard to transfect cells, such as slow growing primary cell cultures.
5. It is a good idea to use GFP as a marker of successful transfection. Counting the number of fluorescent cells versus the total

number of cells provides a good measure of the transfection efficiency. A flow-cytometry based selection can also be used for rapid isolation and selection of transfectants.

6. When generating stable cell lines, it is important to have a positive control and a negative control. For positive control, the cell line transfected with the empty vector backbone is ideal as it will be used to assess biological effect of the specific gene knockdown versus the vector effect. As a negative control, the un-transfected cell lines are used to ensure the puromycin is killing non-transfected cells.
7. A kill curve is used to determine the minimum optimal antibiotic concentration required to select the cells successfully transfected with the knockdown shRNA from those without it. Selection is seen from 3 days to a week.
8. To ensure the long-term efficiency of the shRNA in the stable cell lines, it is important to freeze stocks from early passages. It is possible after multiple passages that some cells may lose the shRNA gene silencing effect while maintaining resistance to the selection antibiotic. Stocks from earlier passages can be used to replace the older line, so non-expressing resistant cells do not overtake those expressing the shRNA gene. Such a situation can be caught early by occasionally analyzing the stable line via RT-qPCR or WB and ensuring the expression of the shRNA is consistent at later passages.
9. We suggest using the concentration of puromycin at twofold the concentration at which no cells survived. For example, if you see no live cells at 0–2 μg/mL puromycin after 7 days, use 4.0 μg/mL of puromycin for selection.
10. For examples of how the CXCR1, CXCR2, and CXCR7 shRNA can be applied after successful stable down-regulation in vitro and in vivo human cancer cell lines, please see our studies published previously [9–11].

Acknowledgement

This work was funded by grants from the National Institutes of Health (1R01CA61038-14, 1R01 CA 156778–01, VA MERIT Award No 5312–01 and 5312–02, all to BLL). The authors thank Ms. Maite Lopez for illustration in Fig. 1.

References

1. Fire A, Xu S, Montgomery MK et al (1998) Potent and specific genetic interference by double-stranded RNA in *Caenorhabditis elegans*. Nature 391:806–811
2. "The Nobel Prize in Physiology or Medicine 2006". Nobelprize.org. Nobel Media AB (2013) http://www.nobelprize.org/nobel_prizes/medicine/laureates/2006/

3. Lee RC, Ambros V (2001) An extensive class of small RNAs in *Caenorhabditis elegans*. Science 294:862–864
4. Lee Y, Ahn C, Han J et al (2003) The nuclear RNase III Drosha initiates microRNA processing. Nature 425:415–419
5. Hutvágner G, McLachlan J, Pasquinelli AE et al (2001) A cellular function for the RNA-interference enzyme Dicer in the maturation of the *let-7*small temporal RNA. Science 293: 834–838
6. Brennecke A, Stark RB, Russell SM et al (2005) Principles of microRNA-target recognition. PLoS Biol 3:e85
7. Origene HuSH shRNA plasmid panels application guide OriGene Technologies (Rockville, MD)
8. O'Keefe EP (2013) Nucleic acid delivery: lentiviral and retroviral vectors. Labome.com
9. Singh RK, Lokeshwar BL (2011) The IL-8-regulated chemokine receptor CXCR7 stimulates EGFR signaling to promote prostate cancer growth. Cancer Res 71:3268–3277
10. Shamaladevi N, Lyn DA, Escudero DO et al (2009) CXC receptor-1 silencing inhibits androgen-independent prostate cancer. Cancer Res 69:8265–8274
11. Singh RK, Lokeshwar BL (2009) Depletion of intrinsic expression of Interleukin-8 in prostate cancer cells causes cell cycle arrest, spontaneous apoptosis and increases the efficacy of chemotherapeutic drugs. Mol Cancer 8:57. doi:10.1186/1476-4598-8-57

Part III

Analysis of Cytokine Intracellular Protein Levels and Gene Expression

Chapter 20

Intracellular Staining and Detection of Cytokines by Fluorescence-Activated Flow Cytometry

Giulia Freer

Abstract

The detection of cytokines inside cells producing them has made a tremendous impact on the way immune reactivity is measured. Intracellular cytokine staining is the only immunological technique allowing determination of antigen-specific T cell function and phenotype at the same time; for this reason, it is one of the most popular methods to measure antigenicity in the evaluation of vaccine efficacy and in the study of infectious diseases. It is a flow cytometric technique based on staining of intracellular cytokines and cell markers (surface or cytoplasmic) with fluorescent antibodies after short term culture of stimulated immune cells in the presence of a protein secretion inhibitor, followed by fixation and permeabilization. Most experiments involve detection of five to ten different colors but many more can be detected by modern flow cytometers. Here, we discuss our experience using a standard protocol for intracellular cytokine staining.

Key words Intracellular cytokine staining, T lymphocyte response, Antigen-specific immunity, Brefeldin A

Abbreviations

BFA	Brefeldin A
BSA	Bovine serum albumin
FBS	Fetal bovine serum
h	Hour
ICS	Intracellular cytokine staining
IFNγ	Interferon γ
IM	Ionomycin
PBMCs	Peripheral blood mononuclear cells
PMA	Phorbol 12-myristate 13-acetate
RT	Room temperature

Ivana Vancurova (ed.), *Cytokine Bioassays: Methods and Protocols*, Methods in Molecular Biology, vol. 1172, DOI 10.1007/978-1-4939-0928-5_20, © Springer Science+Business Media New York 2014

1 Introduction

Intracellular cytokine staining (ICS) has provided a method to assess what cytokines *and* what cells are responsible for an antigen-specific immune reaction. It is now the gold standard method for the determination of immune reactivity in the preclinical and clinical phases of vaccine evaluation, and it has replaced most traditional methods to quantitate immune reactivity to specific antigens, such as the ^{3}H-thymidine incorporation proliferation assay or the ^{51}Cr release cytotoxicity assay. The main reason for its success is that it allows detection of several properties of the same single cells, thanks to retention of newly synthesized cytokines within antigen-specific cells. Because flow cytometry allows gating on the cells of interest, there is no need for prior enrichment of cell populations to obtain detailed information on subsets thereof.

The standard procedure used by many laboratories is typically performed on peripheral blood mononuclear cells (PBMCs) or heparinized whole blood. PBMCs are obtained from heparinized blood by density gradient centrifugation and resuspended in medium, then they are activated by antigen, peptides, or polyclonal stimuli; the latter are also always used to obtain positive control samples. The cells are allowed to process and present antigen for 2 h at 37 °C. Then, to obtain cytokine retention, a protein secretion inhibitor such as Brefeldin A (BFA) is added to cells; this also accelerates cytokine detection, thereby increasing sensitivity compared to cytokine EIA assays. Incubation is carried out for another 4–8 h. At the end, all the cells are detached from culture vessels by a short incubation in EDTA and washed. At this point, they can be stained with fluorescently labeled antibodies against cell surface markers then fixed, or they can be fixed directly if the antibodies chosen still react with their epitopes after fixation. To stain cytokines retained inside the cells, permeabilization is achieved by adding saponin, a mild detergent. All samples are acquired with a flow cytometer and data are analyzed successively. The procedure can be conveniently interrupted at various steps without significant loss in performance.

Thanks to the ability of modern flow cytometers to analyze a wide variety of fluorochromes at the same time, 5–10 different colors are used in most routine work, although the most sophisticated modern cytometers can detect up to 18 colors [1]. Numerous sites online can be extremely helpful in selecting the wide variety of different combinations of markers and cytokines that can be analyzed [2, 3].

Because the choice of antigens, protein transport blocker, activation conditions and incubation period are critical parameters for the accurate analysis of intracellular cytokines in a given cell population, it is essential to optimize each ICS assay condition empirically [4]. To assist the investigator, many firms have developed reference tables and protocols that can be a starting point [5, 6]. It is very important to treat all samples exactly alike. To help in this, mastermixes are recommended when the same reagents must be added to more samples.

Sensitivity of ICS depends on the level of positive response to a given stimulus compared to background, defined as the number of positive events in the absence of a stimulus. It is also influenced by the number of events acquired, on antibodies/fluorochromes used, cytokines investigated and other factors [7]. To lower background, and increase sensitivity, care must be taken to gate out autofluorescent and dead cells, because they can contribute to background significantly. Cells with low level staining should only be taken into account if viability staining is performed, otherwise they are likely to include dead cells staining nonspecifically [7]. On the other hand, dead cells should be gated out of the population of interest; to this aim, a viability dye can be added to cells before fixation. Fixable viability stains for flow cytometry with different lasers, and compatible with the ICS protocol, are available on market [8]. Propidium iodide, a DNA intercalating agent that emits red fluorescence commonly used to stain dead cells, cannot be used.

ICS is presently the only method that allows simultaneous enumeration and phenotyping of antigen-specific T cells. The continuous development of new reagents and fluorochromes that can be used in highly sophisticated cytometers allows to obtain so much information during the evaluation of candidate vaccines that its suboptimal sensitivity is considered a minor problem. A future challenge of the method might be accurate standardization among laboratories involved in vaccine evaluation so that results can be compared.

Modifications of the original protocol may include determination of proliferation at the same time as cytokine production and cell surface markers. The ability of cells to proliferate, seen as the number of divisions they perform, in terms of halving of the viable fluorescent cell dye carboxyfluorescein diacetate succinimidyl ester in 5–7 days of culture, can be additional information confirming their antigen specificity [9].

2 Materials

2.1 Reagents

1. Complete medium: RPMI 1640, 3 % autologous plasma or 5–10 % fetal bovine serum (FBS), 2 mM L-glutamine, 50 μg/ml gentamycin (*see* **Note 1**).
2. Ficoll-Paque (GE Healthcare Bio-Sciences GmbH) or equivalent.

3. Türk solution: 3 % acetic acid in distilled water, 0.01 % crystal violet, sterile filtered (*see* **Note 2**).
4. Trypan blue solution: Trypan blue, 0.4 %, in PBS, sterile filtered.
5. Positive control stimuli. Phorbol 12-myristate 13-acetate (PMA)/ionomycin (IM): PMA stock solution is 10 mg/ml in ethanol, store at −20 °C in 5 μl aliquots. IM: stock solution is 1 mg/ml in ethanol; store at −20 °C in 5 μl aliquots. Right before adding to cells, add 0.4 ml of RPMI to the IM aliquot (1:100 dilution), dilute PMA 1:100 in RPMI then add 125 μl to the IM solution. Add 5 μl of this mixture to 500 μl of cell suspension. Discard the rest (*see* **Note 3**).
6. Antigen of choice (*see* **Note 4**).
7. Costimulatory antibodies: anti-human CD28, clone 28.2, and anti-Human CD49d, clone 9F10, functional grade purified. Stock concentration 1 mg/ml in medium, 0.5 % human serum. Store at −20 °C until use. Dilute 10 μl of each antibody in 80 μl of complete medium; once diluted, store at 4 °C. Add 10 μl to each test tube (*see* **Note 5**).
8. BFA. Dissolve 5 mg of powder in 1 ml of ethanol (5 mg/ml). Store at −20 °C. Before adding to cells, dilute 1:10 in medium and add 10 μl to each tube (final 10 μg/ml).
9. EDTA, 20 mM, in PBS.
10. (Optional) Fixable viability dye (*see* **Note 6**).
11. FACS buffer: PBS, 1 % BSA, 0,1 % sodium azide.
12. (Only required if whole blood is used as a cell source) OptyLyse B (Beckman Coulter) or FACS Lysing Solution (Becton Dickinson) or equivalent.
13. Fix solution: PBS, 1 % paraformaldehyde. Add 1 g of paraformaldehyde to 100 ml of PBS. Heat to 70 °C under a fume hood while stirring until dissolved. Store at 4 °C. PFA is very toxic and aerates easily. Avoid breathing in powder or fumes.
14. 5 % Saponin: dissolve 5 g in 100 ml of FACS buffer.
15. Permeabilization buffer: add 10 ml of 5 % saponin to 95 ml of FACS buffer (0.5 % saponin final concentration).
16. Fluorescent antibodies (*see* **Note 7**).

2.2 Other Equipment

1. 15 and 50-ml polycarbonate or polypropylene tubes.
2. Falcon conical 12 × 75 mm 5 ml polystyrene or polypropylene tubes (*see* **Note 8**).
3. Table top centrifuge.
4. Flow cytometer.

3 Methods

All procedures should be carried out at room temperature (RT). Mastermixes can be made of Brefeldin A/costimulatory antibodies/peptides, or of saponin and staining antibodies, so that a volume of maximum 10–20 μl is added to each sample.

3.1 Cell Preparation

1. Draw 20 ml of heparinized blood (*see* **Note 9**).
2. Layer 7.5 ml of whole blood over 4 ml of Ficoll-Paque in 3 15-ml conical polystyrene tubes (*see* **Note 10**).
3. Centrifuge tubes at 500 × *g* for 15 min with no brake.
4. Collect plasma (*see* **Note 1**).
5. Collect the PBMC ring into a 50-ml tube (*see* **Note 11**).
6. Fill with RPMI and centrifuge at 400 × *g* for 10 min to wash cells.
7. Resuspend the cell pellet obtained in RPMI and centrifuge again at 300 × *g* for 10 min.
8. Resuspended PBMCs in 10 ml of complete medium.
9. Mix 10 μl of cell suspension in 90 μl of Türk solution and make a total cell count.
10. Mix 10 μl of cell suspension in 90 μl of Trypan blue solution and make a viable cell count.
11. Bring to 4×10^6 cells/ml in complete medium, or freeze if desired (*see* **Note 12**).
12. Aliquot 0.5 ml in Falcon conical 12 × 75 mm polystyrene tubes. Set up a sufficient number of tubes (*see* **Note 13**). Total cells per tube should be between 1.5 and 3×10^6 cells.
13. Leave overnight at 37 °C in a 5 % CO_2 humidified atmosphere (resting) (*see* **Note 14**).

3.2 In Vitro Short-Term Culture of Lymphocytes

1. (Optional) Add 10 μl of costimulatory antibodies anti-CD28/CD49d to all tubes (1 μg/ml final concentration) (*see* **Note 5**).
2. Add antigens to the appropriate tubes (*see* **Note 4**).
3. Add positive stimulus to vials, for example, PMA/IM (*see* **Note 3**).
4. Incubate for 2 h at 37 °C in a 5 % CO_2 humidified (*see* **Note 15**).
5. Add BFA to a final concentration of 10 μg/ml to all vials (*see* **Note 16**).
6. (Optional) Briefly spin the tubes to speed up contact between cells.
7. Incubate for another 4 h (total 6 h) at 37 °C in a 5 % CO_2 humidified atmosphere (*see* **Note 17**).

8. Add 50 μl of 20 mM EDTA in PBS to every tube and vortex cells.
9. Incubate for 15 min.
10. (Optional) Stain with fixable viability dye (*see* **Note 18**).
11. Add 1 ml of Fix solution while constantly vortexing (*see* **Note 19**).
12. Incubate cells in Fix solution for 15 min.
13. Resuspend in FACS buffer and centrifuge at 400 × *g* for 10 min twice.
14. Vortex the pellets and leave them in approximately 100 μl in their capped tubes (*see* **Note 20**).

3.3 Staining for Flow Cytometry

1. Resuspend samples in Permeabilization buffer and centrifuge at 400 × *g* for 10 min.
2. Resuspend in 100 μl of Permeabilization buffer (*see* **Note 21**).
3. Make mastermixes of the antibodies used for staining (one containing the anticytokine antibody and the other for the isotype control) (*see* **Note 22**).
4. Add the appropriate amount of mix to each tube.
5. Incubate in the dark for 30 min. Avoid longer incubation.
6. Resuspend samples in Permeabilization buffer and centrifuge at 400 × *g* for 10 min.
7. Resuspend in FACS buffer, 350–500 μl. Concentration should be 20,000–100,000 cells/ml.

3.4 Flow Cytometry Acquisition and Analysis

Acquisition is the step requiring the most expertise. Instrument setup and optimization vary between different cytometers. The following is a brief outline for a BD flow cytometer and based on [10]:

1. Create an acquisition template for future experiments. You should create a file with an FSC × SSC dot plot, histograms for each fluorescence used, and dot plots of FL1 × FL2, etc. to adjust compensation. Quadrants should be set, no gate should be active while setting the instrument.
2. Set PMT voltage (Instrument gain). This is the first thing to do: once set, PMT voltage should not be changed because compensation is voltage dependent. In setup mode, use a fixed, permeabilized, unstained sample of the cells of interest to find the minimal voltage needed to ensure that each detector has enough gain applied to show dim signals. Adjust the voltage of FSC and SSC photomultiplier tubes until the bulk of cells appear roughly in the lower left quadrant of the FSC × SSC dot plot.
3. Use a fixed, permeabilized, negative control sample of the cells of interest, stained with one fluorescent antibody to adjust the FL1, FL2 etc. voltage (log scale) till the fluorescence peak is between 10^0 and 10^1.

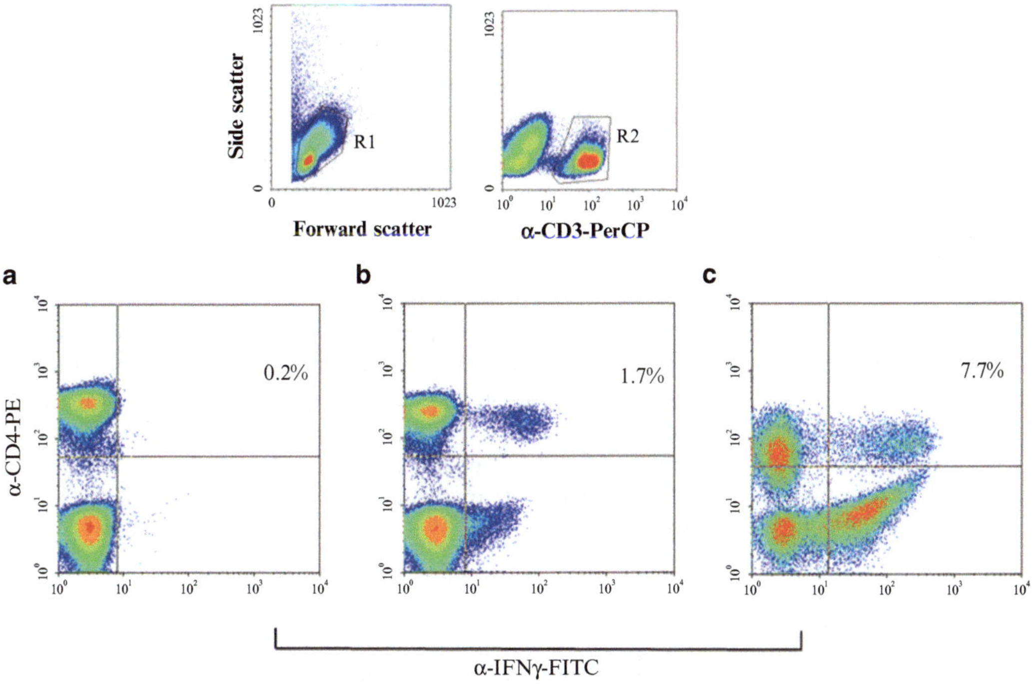

Fig. 1 Gating strategy for three-color ICS. Live cells are gated by SSC × FSC (R1), then $CD3^+$ lymphocytes are selected (R2). Cells gated on R1 + R2 are analyzed for CD4 and IFNγ. The percentage of $CD4^+$ $IFN\gamma^+$ lymphocytes will be obtained by subtracting the percentage of events in the upper right quadrant of the plot from untreated cells (**a**) from the same percentage in the plot of antigen-treated cells. (**b**) In the example, the latter plot was obtained by stimulating human PBMCs with cytomegalovirus-infected cell lysate, while the positive control (**c**) was obtained with PMA/IM

4. Set compensation. Use a positive control sample stained with FL1 antibody and start acquisition in the setup mode. While looking at the FL1 × FL2 dot plot, adjust (FL2 = FL2 − %FL1) so that events positive for FL1 do not exceed a value of 10^1 for FL2. Repeat for each fluorescence channel.
5. Set an FSC threshold below which data will not be stored. A value of 50 is usually sufficient and will help making files smaller.
6. Gate cells of interest: to set up a gate, draw a “Region” with a tool from the tool palette around the cells of interest (*see* **Note 23**). Make sure the gate chosen is suitable for all samples analyzed. In any case, gated and ungated events will be stored in the relevant files, unless otherwise chosen in “Acquisition and Storage” (Fig. 1).
7. Select a place where you want to store your experimental files under “Parameter description,” and name sample identity. It is a good idea to take note of the file number as well.

8. Select the number of events you need to acquire and the events you want to save in a file. In the "Acquisition" window, deselect the "Setup" button and program the instrument to acquire 300.000 events (at least) in the CD3+ gate (*see* **Note 24**).
9. Analyze acquired events by gating on the cells of interest.

4 Notes

1. We use 3 % autologous plasma during cell activation. Plasma is collected in a separate vial after gradient centrifugation of whole blood and stored at 4 °C. However, to avoid potential effects of traces of Ficoll, a separate tube of blood can be obtained for the preparation of autologous serum or plasma. In most protocols in the Literature, 5–10 % heat-inactivated FBS is used.
2. Türk solution is used for leukocyte counting. It will cause red blood cells to lyse. Viability determination cannot be carried out in this solution, but counting human PBMC is easier because all visible cells should be taken for good; we found viability was always >95 % after PBMC isolation.
3. The positive control stimulus used depends on the cytokine of interest: the most common is PMA/IM at final concentrations of 10 and 250 ng/ml, respectively. This stimulus bypasses the T cell receptor and is extremely strong; therefore, no costimuli need to be added to this sample. PMA/IM is toxic for cells even at the concentration recommended for optimal stimulation. Also, PMA/IM tend to lose activity upon freezing–thawing; therefore, all leftovers should be discarded.

   Not all cytokines are produced under PMA/IM: for example, to produce interleukin 12, interferon (IFN) γ and lipopolysaccharide are used to stimulate human PBMCs. For a list of stimuli, manufacturers of labeled antibody (eBioscience, Becton Dickinson, and others) often have useful online guides. Positive controls can be obtained also with immobilized anti CD3 antibody, or with Staphylococcal Enterotoxin A or Staphylococcal Enterotoxin B, that act as superantigens and stimulate T lymphocytes via their Vβ chain [11]. Mitogens, such as pokeweed mitogen, can be also used [12].
4. Several forms of antigens can be chosen and stored as 50× stock solutions at −20 °C. Repeated freezing/thawing is not recommended, although it may be found to be acceptable for defined antigens. Optimum antigen concentration is empirically defined. As a guideline:
   - *Peptides* are often used in pools of overlapping 20-mers designed on the basis of the primary amino-acid sequence

of the protein of interest [13]. Use between 1 and 0.01 μg/ml for each peptide (this concentration may vary a lot with peptide affinity to MHC molecules).

- *Whole recombinant protein* between 1 and 5 μg/ml. Surprisingly, whole recombinant antigens are sometimes found to be less effective than single peptides thereof [14].
- *Inactivated purified virus* between 1 and 5 μg/ml.
- *Complex protein mixtures*, such as infected cell lysates as a source of viral antigens, bacterial extracts etc. at 5–20 μg/ml (this concentration varies with the percentage of the protein of interest in the protein mixture).

5. CD28 is a ligand for CD80 (B7-1) and CD86 (B7-2) and is a potent co-stimulator of T cells; specific clones of anti CD28 antibodies, such as clone 28.2, mimic the stimulatory effect of B7 binding. Human CD49d is an integrin α4 subunit, which, once bound by antibodies, acts as a costimulatory molecule. Many Authors recommend using costimulatory antibodies because they increase cytokine production by low-affinity T cells. However, if used, they must be included in negative control tubes as well, where they can increase the number of positive nonspecific T cells (background), therefore their use is optional.
6. Fixable viability dyes are used to irreversibly label dead cells prior to cryopreservation, fixation and/or permeabilization procedures. Using fixable viability dyes allows exclusion of dead cells from analysis. Every fixable viability dye can be excited by a selected laser wavelength and has a peak excitation that can be detected by specific filters, therefore the choice of such dyes is limited by the FACS used. In turn, the use of such dyes will diminish the number of colors that can be analyzed by one; for example, Viability Dye eFluor® 450 cannot be used on a FACScan because the latter does not have the appropriate (405 nm) laser; the dye can be used on a FACScanto, but not together with Pacific Blue because they emit in the same blue fluorescence channel. A useful fluorophore selection guide to see what color can be used in different FACS instruments can be found online [15].
7. When selecting a panel of fluorescent antibodies for an ICS experiment, it is critical to know some basic fluorochrome properties:
   - *Excitation wave length.* The number and type of fluorochromes that can be used depends, in first instance, on the flow cytometer available. This must have a laser for the fluorochrome chosen (488-nm laser output, the one they all have, for PE, 633-nm laser output for APC, etc.). Old

models such as FACScan only have a 488-nm laser and can only analyze three colors at a time and a limited number of fluorochromes (mainly FITC, PE, and PerCP, in addition to tandem dyes). When using newer-generation cytometers, which can have up to seven different lasers [1], ICS is often optimized for the determination of eight to ten colors [4, 7]; more can be analyzed, although one must bear in mind that emission of one fluorochrome into the spectrum of another needs to be compensated at the expenses of sensitivity.

- *Stability* depends on the buffers used and on the chemical nature of the fluorochrome. Tandem dyes (molecules chemically coupled to transfer excitation from one to the other) are particularly tricky, being sensitive to photobleaching, to prolonged incubation in fixation buffers and prone to uncoupling.
- *Brightness*. The most often used fluorochromes are PE and APC (the latter cannot be used on a FACScan) because these fluorochromes are amongst the brightest ones, whereas FITC is chosen because it is cheaper but it is dimmer and should be used to detect highly expressed molecules (e.g., CD3). The brightest fluorochromes should be coupled to antibodies against the proteins that are less abundant in cells, generally cytokines. Fluorochrome brightness charts are available from many sites online.
- *Fluorescence emission spectrum overlap*: Numerous sites online show fluorochrome excitation and emission charts for fluorochromes online [3]. Emission spectra of the fluorochromes used should not overlap; if they do, compensation is required: this will mean subtracting events that fall in the overlap from the positive event count, thereby decreasing sensitivity. For a great discussion on compensation *see* ref. 16.

8. We use tubes because they allow short term culture, staining and FACS acquisition in a single vessel. Only one wash after each protocol step, done by filling the tube, is enough. If monocytes/macrophages are the cells of interest, polypropylene tubes are recommended, as a number of blood monocytes and activated macrophages adhere to polystyrene. We generally handle a maximum of 30–50 tubes. If more samples are prepared, 96-well polypropylene plates can be used for activation and staining, but more washes will have to be done at the end of each step and, eventually, all samples will have to be transferred to tubes for acquisition [17].
9. The most commonly used cells for ICS are fresh PBMCs obtained from whole venous blood. Blood must be drawn in the presence of heparin and it can be stored 8 h at room temperature (RT)

before processing, if necessary [7]. Although whole blood may be much simpler to use [12], we are not keen on using it because (1) too much antigen is required to obtain the same amount of stimulated cells as with PBMCs; (2) whole blood may lead to underestimate frequencies of antigen-specific cells [18]. If whole blood is used, 0.5–1 ml of heparinized whole blood is added to individual tubes and red blood cells need to be lysed at the end of the 6-h incubation; excellent lysing solutions are available on market; most fix cells at the same time as lysing.

10. Because density depends on temperature, it is important that blood and Ficoll-Paque are at RT. Avoid 50-ml tubes for gradient separation because the yield of PBMCs will be lower. For higher PBMC purity, follow Ficoll-Paque manufacturers' instructions. In this case, plasma will be diluted 1:2.
11. A layer made up of red blood cells/platelets, Ficoll-Paque, a fluffy thin disk of PBMCs and plasma (from bottom to top) should be visible. When autologous plasma is needed, collect the top plasma layer in a separate sterile vial and store at 4 °C. We collect all the PBMC ring and half of the Ficoll-Paque volume, carefully harvesting the cells sticking to the tube wall, to increase yield.
12. At this stage, PBMCs can be frozen for later use: resuspend pellet after second wash by tapping and add 1 ml 90 % FBS, 10 % dimethylsulfoxide in polypropylene cryovials. We usually freeze enough cells for a 10-sample experiment in a single tube, i.e., $20–30 \times 10^6$ cells. Tubes are placed in a freezing unit (e.g.: Nalgene Mr. Frosty, cat. number 5100–0001) at RT, which is then put in a −80 °C freezer for no longer than 3 days. The cells are stored in liquid nitrogen until use. To thaw them, incubate the tubes in a 37 °C water bath and, as soon as they are thawed, immediately place the cells in a 15-ml tube. We then slowly add 14 ml warm RPMI and centrifuge for 10 min at $250 \times g$ [19]. They are counted and aliquoted in tubes as described. Resting is highly recommended (*see* **Note 14**). Reactivity of cryopreserved cells should not be compared with fresh ones; cell recovery and viability should be > 70 % to preserve antigen-specific reactivity.
13. For all samples, one will be stained with the antibody panel selected, the other should be stained with all antibodies except the cytokine antibody, replaced by a matching isotype/fluorochrome (*see* **Note 23**). The following tubes with 0.5 ml of cell suspension are required for each patient PBMC:
    (a) *negative control, two tubes* (medium and costimulating antibodies only).
    (b) *positive control tubes* (medium and an appropriate stimulus for cells/cytokines), two tubes. In addition, one tube of

positive control should be set up for every single antibody (employed as controls to set compensation) and two tubes for negative control.

(c) *antigens of interest, two tubes for each.* It might be a good idea to analyze PBMCs from naïve donors stimulated with the antigens of interest to assess possible nonspecific mitogenic effect of the antigen (specificity control).

14. Fresh PBMC may be left to "rest" in a 37 °C CO_2 incubator until antigen is added the next day. The resting step has been found to improve performance of ICS for fresh and thawed PBMCs [7]. Rested cells are counted again the next day.
15. This incubation allows antigen processing to occur, therefore it is not required for peptides and PMA/IM. It is advisable to make a mastermix of BFA, costimulatory antibodies and antigens that do not require this incubation and add them in a total 10–20 μl.
16. BFA is a fungal metabolite which disrupts the structure and function of the Golgi apparatus thereby inhibiting secretion, allowing cytokines to be retained intracellularly. Monensin may be better in certain circumstances [5]. It is recommended to test which inhibitor to use or, alternatively, use both at 5 μg/ml each.
17. Duration of incubation after addition of BFA depends on the cytokine used as a read out, with IFNγ and tumor necrosis factor α as the most commonly chosen ones. It also depends on the cells studied and stimuli used; retention of cytokines in lymphocytes generally peaks between 4–8 h and up to 12 h (for interleukin 12) [20]. Helpful suggestions as to the incubation times for different cytokines, and other variables, may come from manufacturers of labeled antibody (for example [5]). Protein secretion inhibitors exert toxic effects after incubations lasting longer than 8–10 h. We found that BFA could not be left in cultures for more than 8–10 h; even when BFA concentration was decreased to 3 μg/ml, as suggested by eBioscience, overnight incubation altered cell morphology and decreased the number of positive events.
18. The experimental procedure to stain dead cells varies according to the fixable dye chosen. As an outline:
    - After the EDTA incubation step, before or after staining for surface markers, wash the cells in PBS (without protein or azide) twice.
    - Resuspend in 1 ml PBS.
    - Add 1 μl of the fixable viability dye, or according to manufacturers' instruction, and mix.
    - Incubate for 30 min at 4 °C in the dark.
    - Wash with FACS buffer.
    - Carry on with the fixing step.

19. Many monoclonal antibodies bind native molecules better than fixed ones, therefore many Authors stain unfixed cells with anti-surface marker antibodies, then fix. Certain antibodies will not bind their fixed antigen at all, especially if they are used on species they cross-react with.
20. After fixing, we prefer to store samples in FACS buffer at 4 °C and stain them within 3 days. For longer storage, they can be frozen at −80 °C in FBS, 10 % DMSO. BD FACS lysing solution has also been used to freeze the cells [21].
21. If samples are already in 100 μl FACS buffer, 5 % saponin can be added instead of washing, saving time and loss of cells. Even better, 5 % saponin can be included in the mastermixes of staining antibodies (10 μl/sample).
22. The amount of antibody needed to stain is empirically determined. Use double the lowest quantity that gives maximum staining for the cells number used in 30 min (saturating quantity). We found big differences with the manufacturers' recommended amount, which can be up to ten times higher than required. Using more than needed can increase fluorescence of the negative population. One tube should be stained with matching isotype/fluorochrome instead of the cytokine antibody. Most manufacturers sell antibodies and their matched isotype/fluorochrome controls for this purpose. This control is important to establish the number of nonspecifically stained stimulated PBMCs; it will be essential to establish a negative cutoff value during analysis.
23. For example, a gate based on physical parameters of cells can be drawn in the FSC × SSC dot plot. However, gating on physical parameters of permeabilized cells is of limited use because permeabilization tends to shrink cells and, in addition, a number of positive events may be gated out because activated cells may have greater scatter than resting ones. If multicolor staining is performed, for example CD3-CD4-IFNγ, then the acquisition gate will be set on the broader cell phenotype marker, i.e., CD3. In this case, CD3+ cells will be displayed and events analyzed will be CD4+/IFNγ+. One can also use more than one gate; in the example, first gate all leucocytes by FSC × SSC, then gated them again by CD3 using the function AND when selecting the region (*see* Fig. 1).
24. When analyzing rare cells such as antigen-specific T cells, the number of events acquired in a given gate depends on the frequency expected. A spreadsheet to calculate the minimum sample size with statistical significance has been made available online [3]. We recommend storing all events except those below an FSC threshold; this will give a large file but will allow to examine different populations amongst PBMCs, independently from the acquisition gate, during analysis.

References

1. Preffer F, Dombkowski D (2009) Advances in complex multiparameter flow cytometry technology: applications in stem cell research. Cytometry B 76B:295–314
2. De Rosa SC (2012) Vaccine applications of flow cytometry. Methods 57:383–391
3. http://www.bdbiosciences.com/research/multicolor/tools/index.jsp
4. Lamoreaux L, Roederer M, Koup R (2006) Intracellular cytokine optimization and standard operating procedure. Nat Protoc 1:1507–1516
5. http://www.ebioscience.com/resources/best-protocols/flow-cytometry/flow-cytometry-intracellular-staining-quick-guide.htm
6. http://www.bdbiosciences.com/resources/cytokines/index.jsp#cytoprotocols
7. Horton H, Thomas EP, Stucky JA et al (2007) Optimization and validation of an 8-color intracellular cytokine staining (ICS) assay to quantify antigen-specific T cells induced by vaccination. J Immunol Methods 323:39–54
8. Perfetto SP, Chattopadhyay PK, Lamoreaux L et al (2010) Amine-reactive dyes for dead cell discrimination in fixed samples. Curr Protoc Cytom Chapter 9:Unit 9.34. doi: 10.1002/0471142956.cy0934s53
9. Lyons AB, Parish CR (1994) Determination of lymphocyte division by flow cytometry. J Immunol Methods 171:131–137
10. Maecker HT, Trotter J (2006) Flow cytometry controls, instrument setup, and the determination of positivity. Cytometry A 69:1037–1042
11. Gauduin MC (2006) Intracellular cytokine staining for the characterization and quantitation of antigen-specific T lymphocyte responses. Methods 38:263–273
12. Suni MA, Picker LJ, Maino VC (1998) Detection of antigen-specific T cell cytokine expression in whole blood by flow cytometry. J Immunol Methods 212:89–98
13. Maecker HT, Dunn HS, Suni MA et al (2001) Use of overlapping peptide mixtures as antigens for cytokine flow cytometry. J Immunol Methods 255:27–40
14. Tesfa L, Koch FW, Pankow W et al (2004) Confirmation of *Mycobacterium tuberculosis* infection by flow cytometry after ex vivo incubation of peripheral blood T cells with an ESAT-6-derived peptide pool. Cytometry 60:47–53
15. http://www.lifetechnologies.com/content/dam/LifeTech/migration/en/filelibrary/cell-tissue-analysis/pdfs.par.13383.file.dat/fluorophore-selection-guide-flow-cytometry.pdf
16. http://www.drmr.com/compensation
17. Suni MA, Dunn HS, Orr PL et al (2003) Performance of plate-based cytokine flow cytometry with automated data analysis. BMC Immunol 2:4–9
18. Hoffmeister B, Bunde T, Rudawsky IM et al (2003) Detection of antigen-specific T cells by cytokine flow cytometry: the use of whole blood may underestimate frequencies. Eur J Immunol 33:3484–3492
19. Disis ML, dela Rosa C, Goodell V et al (2006) Maximizing the retention of antigen specific lymphocyte function after cryopreservation. J Immunol Methods 308:13–18
20. Mascher B, Schlenke P, Seyfarth M (1999) Expression and kinetics of cytokines determined by intracellular staining using flow cytometry. J Immunol Methods 223:115–121
21. Nomura LE, deHaro ED, Martin LN et al (2003) Optimal preparation of rhesus macaque blood for cytokine flow cytometric analysis. Cytometry A 53:28–38

Chapter 21

Cytokine Detection by Flow Cytometry

Jian-Ge Qiu, Xiao-Long Mei, Zhe-Sheng Chen, and Zhi Shi

Abstract

Analysis of intracellular cytokines is extremely important in the clinical treatment of numerous diseases. Flow cytometry (FCM) is a highly effective technique that detects intracellular cytokines using specific fluorescence-labeled antibodies. The common steps of this assay include cell collection, fixation, permeabilization, blocking, intracellular staining and analysis by FCM. This technique also allows for analyzing the biological function of cytokines. In this chapter, we describe a modified method to detect the specific intracellular cytokine staining using FCM, with an emphasis on the effects of variables including samples, temperature, buffers, data acquisition, and analysis.

Key words Cytokines, Flow cytometry, Antibody, Intracellular cytokine detection

1 Introduction

The cancer morbidity is getting higher all around the world, and drug resistance remains a major problem in cancer chemotherapy. Our previous findings strongly suggest a close link between cytokines and drug resistance in cancer cells, especially the IL-6 and IL-8 [1]. Cytokines are a type of polypeptide hormones regulating growth and function of various cells [2]. They execute a variety of pleiotropic actions, including cell growth, differentiation, apoptosis, wound healing, signal transduction, and homeostasis [3]. Inappropriate activation of the cytokine network is associated with many diseases. Cytokine dysregulation has been linked to cancer progression, response to chemotherapy, and metastatic status [4–6]. Therefore, as cytokines form a complex network and many are regulated through the same mechanisms, it is critical to profile cytokines expression.

Cytokines in supernatants are customarily measured by ELISA, limiting dilution analysis (LDA), immunohistochemical staining, and other methods [7]. Since the cytokine production by individual cell subsets cannot be determined, these methods, though

Ivana Vancurova (ed.), *Cytokine Bioassays: Methods and Protocols*, Methods in Molecular Biology, vol. 1172,
DOI 10.1007/978-1-4939-0928-5_21, © Springer Science+Business Media New York 2014

useful, provide only an incomplete picture. During the last years, it has become increasingly clear that the production of most cytokines is not confined to one cell type [8, 9]. Thus, a method to detect cytokines at single cell level would be helpful to study the contribution of different cells to cytokine production in heterogeneous cell populations. Followed with the techniques of cell fixation and permeabilization, the identification of respective cytokine producing cells by indirect immunofluorescence microscopy became possible [10, 11]. Owing to the poor discernible ability of lymphocyte phenotype and cytokine production, these methods are helpful but laborious. However, flow cytometry (FCM), a method to detect cytokines at single cell level can be useful to study the contribution of different cells to cytokine production in heterogeneous cell populations [9]. The capability of multi-parameter FCM permits simultaneous detection of two or more cytokines in a single cell with the high throughput inherent of instrument, giving cytokine staining a tremendous advantage over existing signal-cell methods [12, 13].

Fluorescence-labeled cytokine-specific monoclonal antibodies are very useful for the intracellular staining and multiparameter FCM analysis of individual cytokine-producing cells within mixed cell populations [14]. The staining of intracellular cytokines based on the identification of cytokine-specific monoclonal antibodies, and a fixation-permeabilization procedure are compatible with the identification [10, 15, 16]. After treated with carboxylic ionophore monensin for 4–6 h to collect intracellular cytokines, cells were generally digested with tryptase which contained 0.5 % EDTA. If target cytokines embed in the membrane, the process of treated with carboxylic ionophore monensin should be abandon and go to the step of blocking directly. A fixation and permeabilization procedure is necessary subsequently, the stabilization of cell membranes and preservation of intracellular antigenicity of cytokine were achieved with paraformaldehyde fixation, and the fixation allows preservation of cell morphology and intracellular antigenicity, while also enabling cells to withstand permeabilization by detergent [17]. Then cells were treated with detergent (e.g., Triton X-100) to permeate the membrane, which permits the cytokine-specific monoclonal antibodies to penetrate the cell membrane, cytosol, and membranes of the endoplasmic reticulum and Golgi apparatus [18, 19]. With this system, a blocking step is often added prior to intracellular cytokine staining to block the nonspecific binding sites. This procedure reduces the nonspecific staining to the greatest extent and improves the veracity of detection [20, 21]. Next, samples are detected by FCM, and data are analyzed by gating on the cell population of interest. A variety of reagents for the process of staining are used by different laboratories. Below, we describe a protocol based on commercially available reagents.

2 Materials

2.1 Reagents

1. Antibodies: Anti-P-glycoprotein, anti-interleukin-8 (IL-8), mouse IgG, and anti-mouse Ig-PE (*see* **Note 1**).
2. Monensin: 10 mM carboxylic ionophore monensin in ethanol (or carbinol); store at −20 °C. The working concentration is 10 μM.
3. Cell culture medium: DMEM medium, 10 % fetal bovine serum (FBS), tryptase containing 0.5 % EDTA, antibiotic solution (1,000 UI/mL penicillin and 1,000 UI/mL streptomycin).
4. Fixation Buffer: 4 % paraformaldehyde in PBS (*see* **Notes 2–4**).
5. Permeabilization Buffer: 0.3 % Triton X-100 in PBS (*see* **Note 5**).
6. Blocking Buffer: 10 % heat inactivated FBS and 0.3 % Triton X-100 (add when analyzing intracellular cytokines) in PBS.
7. Antibody Dilution Buffer: 0.3 % Triton X-100 (add when analyzing intracellular cytokines) and 3 % heat inactivated FBS in PBS.

2.2 Equipment

1. Cell culture dishes (6 cm).
2. Cell incubator with 5 % carbon dioxide concentration.
3. Low-speed refrigerated centrifuge.
4. Blood counting chamber.
5. Flow cytometer with at least four fluorescence channels.

3 Methods

3.1 Cytokine Gathering

When analyzing cytokines on cell membrane, skip this step, and go directly to Subheading 3.2.

1. Add monensin to cells at the final concentration of 10 μM.
2. Incubate for 4–6 h in cell incubator.

3.2 Sample Collection (See Notes 5 and 6)

1. Discard the supernatant and wash the attached cells gently with PBS buffer.
2. Treat the cells with tryptase containing 0.5 % EDTA, and suspend them in DMEM medium.
3. Collect 1×10^5 to 1×10^6 cells after cell counting using blood counting chamber.

Go to Subheading 3.5 directly if the acquired cytokines are on the surface of membrane.

3.3 Sample Fixation (See Notes 3 and 4)

1. Centrifuge the cell suspension at $200 \times g$ for 5 min, aspirate the supernatant, and resuspend the pellets in PBS by pipetting. Repeat the washing step two times (*see* **Note 7**).
2. Add 1 mL of Fixation Buffer to each tube, resuspend the pellets gently, and incubate at room temperature for 15 min.

3.4 Sample Permeabilization (See Notes 3 and 5)

1. Centrifuge the cell suspension at 200 × *g* for 5 min, aspirate the Fixation Buffer, and resuspend the pellet in PBS by pipetting. Repeat the washing step two times.
2. Add 1 mL of Permeabilization Buffer to each tube, suspend the pellets gently, and incubate at room temperature for 15 min.

3.5 Blocking the Nonspecific Sites

1. Centrifuge the cell suspension at 200 × *g* for 5 min and aspirate the above Permeabilization Buffer.
2. Add 1 mL of Blocking Buffer to each tube, resuspend the pellets, and incubate the cell suspension at room temperature for 1 h.

3.6 Sample Staining (See Note 1)

1. Centrifuge the cell suspension at 200 × *g* for 5 min and aspirate the above Blocking Buffer.
2. Resuspend the pellets in 100 μL of Antibody Dilution Buffer gently, add 1 μL of primary antibody, and incubate at 4 °C for 1 h.
3. Centrifuge the cell suspension at 200 × *g* for 5 min, aspirate the above antibody solution, and wash the pellets three times with Antibody Dilution Buffer.
4. Resuspend the pellet in 100 μL of Antibody Dilution Buffer, add 1 μL of secondary antibody, and incubate at 4 °C for 1 h in the dark.
5. Centrifuge the cell suspension at 200 × *g* for 5 min, wash the pellets twice with Antibody Dilution Buffer, and then resuspend the pellet with 200 μL of PBS. The samples could be analyzed by FCM after passing through the 200 mesh sieve.

3.7 Flow-Cytometric Analysis (See Note 8)

1. Check whether there is enough sheath liquid in the Sheath liquid barrel. If not, turn the gas valves to the "VENT" position, and add sheath liquid to the 4/5 volume of the sheath liquid barrel.
2. Make sure to tighten the lid of the sheath liquid barrel to prevent the gas leak.
3. Forward scatter (FSC) reflects the size of cells and side scatter (SSC) shows the complexity of intracellular components. Set a scatter diagram of FSC–SSC to eliminate debris while still retaining the population, and set a secondary threshold on channels of fluorescently labeled antibody subsequently to show the fluorescence intensity of retaining cell population.
4. Set proper number (e.g., 15,000) of acquisitive cell and source of samples. Number of cell acquisition should depend on the concentration of samples. Run the samples on "low" (e.g., on a BD FACS Calibur™) to improve precise of the detection.

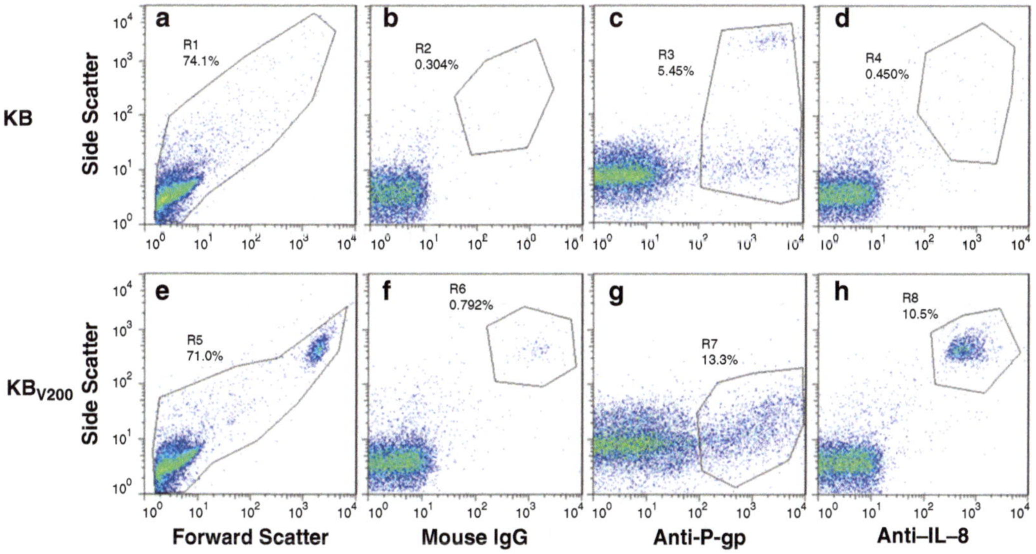

Fig. 1 The expression of P-gp/IL-8 in KB and KB$_{V200}$ cells was analyzed by FCM. The effect regions were created for FSC and SSC (R1 in panel **a** and R5 in panel **e**), isotype control (R2 in panel **b** and R6 in panel **f**), P-gp positive (R3 in panel **c** and R7 in panel **g**), and IL-8 positive (R4 in panel **d** and R8 in panel **h**) in KB and KB$_{V200}$ cells. The P-gp and IL-8 positive rate of KB$_{V200}$ showed significantly higher than those of KB

5. Set the parameters and gate while running the first sample to obtain an optimum figure, save the template and parameters for the latter samples.
6. Acquire data from samples.

3.8 Data Analysis (See Note 9)

In this chapter, a pair of human epidermoid cancer cell lines (KB and KB$_{V200}$) was used as an example to detect membrane surface and intracellular cytokines. KB$_{V200}$, the vincristine-resistant strain of KB cell line, overexpresses the transmembrane P-glycoprotein (P-gp) and the intracellular cytokine IL-8.

1. Analyze data by defining an effective sample region in FSC and SSC (Fig. 1a, e), and a region on the positive samples (Fig. 1c, d, g, h). These regions can be defined by automated gating algorithms. Be sure to size these regions appropriately to include relevant cells.
2. Create a dot plot gated on both of the regions defined in **step 1**, displaying P-gp/IL-8 staining.
3. Using an isotype control sample, draw a region that includes all events of nonspecific positive straining. The region can be a conventional region tethered to the false positive population (Fig. 1b, f).

4. Record the positive percent in the response region for each sample. Subtract the percent positive of isotype sample from each of the acquired samples to obtain the positive population.
5. The percent analysis can be taken as an absolute number of positive cells per milliliter of cell samples by multiplying by the count of P-gp/IL-8 cells per milliliter.

4 Notes

1. Staining controls: Using a correct antibody control should reduce the errors in cytokine detection. The antibody that has the same source, subtype or sub-chains, fluorescent labeled with the first antibody should be used as isotype control antibody for the analysis.
2. Batching of samples: Samples can be stored at −80 °C at the end of **step 2** (Subheading 3.2). Because the samples are fixed, they can be frozen directly in the fixation buffer.
3. Temperature of fixation and permeabilization: The fixation and permeabilization of cells are temperature dependent. The fixation and permeabilization solution should be stored and used at room temperature. If the temperature of solution above or below 20–25 °C, a temperature controlled incubation should be done.
4. Fixation buffer: Low concentration of paraformaldehyde cannot fix the cell completely, which results in cells lysed after permeabilization. 2–4 % paraformaldehyde in PBS could be used as fixation buffer.
5. Permeabilization buffer: Triton X-100 in PBS. NP-40, Tween-20, Saponin, Digitonin, and Leucoperm may also be used and are mild membrane solubilizers. Their percentage (0.1–1 %) depends on the size of cells.
6. Centrifugation speed: Fixation and permeabilization cause increased cell buoyancy. Cells should be collected at 200 × *g* to reduce loss.
7. Washing buffer: PBS could be used as washing buffer before permeabilization, then the washing buffer and incubation buffer of the following step need to add agent of permeabilization (e.g., 0.3 % Triton X-100). Cells have to always stay in the circumstance of permeabilization due to the mobility of phospholipid bilayer, which is beneficial for antibody entering into cell and binding with antigens.
8. Setting of gate: Setting correct gate can ensure veracity of detection results. The gate should wipe out the cell debris. With a proper gate setting, positive and negative control samples can be obtained clearly.

9. Collection number of events: Detection of antigen-specific intracellular cytokine from small population demand requires plenty of effective cell samples, This protocol suggests collecting 15,000 cells, which is sufficient to analyze small positive cell population as little as 0.01 % with reasonable statistical significance.

Acknowledgment

This work was supported by funds from the Chinese National Natural Science Foundation No. 31271444 and No. 81201726 (Z. Shi), the Specialized Research Fund for the Doctoral Program of Higher Education No. 20124401120007 (Z. Shi), the Fundamental Research Fund for the Central University No. 21612407 (Z. Shi), NIH R15 No. 1R15CA143701 (Z.S. Chen), and St. John's University Seed grant No. 579–1110 (Z.S. Chen).

References

1. Shi Z, Yang WM, Chen L-P et al (2012) Enhanced chemosensitization in multidrug-resistant human breast cancer cells by inhibition of IL-6 and IL-8 production. Breast Cancer Res Treat 135:737–747
2. Suni MA, Picker LJ, Maino VC (1998) Detection of antigen-specific T cell cytokine expression in whole blood by flow cytometry. J Immunol Methods 212:89–98
3. Baraldi-Junkins CA, Beck AC, Rothstein G (2000) Hematopoiesis and cytokines: relevance to cancer and aging. Hematol Oncol Clin North Am 14:45–61
4. Fasciani A, D'Ambrogio G, Bocci G et al (2001) Vascular endothelial growth factor and interleukin-8 in ovarian cystic pathology. Fertil Steril 75:1218–1221
5. Bamberger E, Perrett C (2002) Angiogenesis in epithelian ovarian cancer. Mol Pathol 55:348–359
6. Duan Z, Feller AJ, Penson RT et al (1999) Discovery of differentially expressed genes associated with paclitaxel resistance using cDNA array technology: analysis of interleukin (IL) 6, IL-8, and monocyte chemotactic protein 1 in the paclitaxel-resistant phenotype. Clin Cancer Res 5:3445–3453
7. Thorpe R, Wadhwa M, Bird C et al (1992) Detection and measurement of cytokines. Blood Rev 6:133–148
8. Aggarwal B., Puri R. (1995) Common and uncommon features of cytokines and cytokine receptors: an overview. In: Human cytokines: their role in disease and therapy, vol 3. Blackwell Science, Cambridge
9. Jung T, Schauer U, Heusser C et al (1993) Detection of intracellular cytokines by flow cytometry. J Immunol Methods 159:197–207
10. Sander B, Andersson J, Andersson U (1991) Assessment of cytokines by immunofluorescence and the paraformaldehyde-saponin procedure. Immunol Rev 119:65–93
11. Schmid I, Uittenbogaart CH, Giorgi JV (1991) A gentle fixation and permeabilization method for combined cell surface and intracellular staining with improved precision in DNA quantification. Cytometry 12:279–285
12. Foster B., Prussin C., Liu F, et al. (2007) Detection of intracellular cytokines by flow cytometry. Curr Protoc Immunol 6.24. 21–26.24. 21
13. Foster B., Prussin C. (2002) Detection of Intracellular Cytokines by Flow Cytometry. *Curr Protoc Immunol*
14. Prussin C, Metcalfe DD (1995) Detection of intracytoplasmic cytokine using flow cytometry and directly conjugated anti-cytokine antibodies. J Immunol Methods 188:117–128
15. Andersson U, Andersson J (1994) Immunolabeling of cytokine-producing cells in tissues

and in suspension. Cytokine Producing Cells 32–49

16. Sander B, Höidén I, Andersson U et al (1993) Similar frequencies and kinetics of cytokine producing cells in murine peripheral blood and spleen: cytokine detection by immunoassay and intracellular immunostraining. J Immunol Methods 166:201–214
17. Lanier L, Warner N (1981) Paraformaldehyde fixation of hematopoietic cells for quantitative flow cytometry (FACS) analysis. J Immunol Methods 47:25–30
18. O'Brien MC, Bolton WE (1995) Comparison of cell viability probes compatible with fixation and permeabilization for combined surface and intracellular staining in flow cytometry. Cytometry 19:243–255
19. Halldén G, Andersson U, Hed J et al (1989) A new membrane permeabilization method for the detection of intracellular antigens by flow cytometry. J Immunol Methods 124:103–109
20. Carson RT, Vignali DA (1999) Simultaneous quantitation of 15 cytokines using a multiplexed flow cytometric assay. J Immunol Methods 227:41–52
21. Picker LJ, Singh M, Zdraveski Z et al (1995) Direct demonstration of cytokine synthesis heterogeneity among human memory/effector T cells by flow cytometry. Blood 86: 1408–1419

Chapter 22

Analysis of IL-17 Production by Flow Cytometry and ELISPOT Assays

Ling Zhao, Yuan Chou, Yanfang Jiang, Zhenyu Jiang, and Cong-Qiu Chu

Abstract

Interleukin (IL)-17 represents a family of cytokines with six members, namely IL-17A, B, C, D, E, and F. IL-17A and IL-17F are best studied proinflammatory cytokines. $CD4^+$ T helper cells producing IL-17A have been identified as a distinct T helper subset, Th17 cells. IL-17 and Th17 cells are important mediators in tissue inflammation in immune-mediated inflammatory diseases. IL-17 is also produced by other immune cells and plays an important role in host defense against microbial infection. Cell-based assays are sensitive and quantitative, and enable identification of cellular sources of IL-17 production. This chapter describes usage of flow cytometry and ELISPOT assays to quantify IL-17A-producing cells in disease and in vitro experiments to study T cell function.

Key words Interleukin-17, Th17 cells, Inflammation, Flow cytometry, ELISPOT assay

1 Introduction

Interleukin-17 (IL-17) represents a family of cytokines consisting of six members, IL-17A, B, C, D, E, and F. The prototype member, IL-17A (commonly referred to as IL-17), was originally described and cloned in 1993 [1] as CTLA8 (cytotoxic T-lymphocyte-associated protein 8), which shows 58 % homology with an open reading frame of the T lymphotropic herpesvirus samirii. Subsequently five additional family members, IL-17B to IL-17F, were identified by homology-based cloning. Among the IL-17 family members, IL-17F has the highest degree of sequence homology with IL-17A, while IL-17E (also named IL-25) is most distant. In contrast to the other family members, IL-17E induces allergic response and exerts inhibitory effects in IL-17-mediated inflammation [2, 3].

IL-17 is generally considered to be a pro-inflammatory cytokine, which is attributable mainly to the properties of IL-17A and IL-17F. IL-17F is often co-expressed with IL-17A and can form an

Ivana Vancurova (ed.), *Cytokine Bioassays: Methods and Protocols*, Methods in Molecular Biology, vol. 1172,
DOI 10.1007/978-1-4939-0928-5_22, © Springer Science+Business Media New York 2014

IL-17A/F heterodimer [4]. Considerable attention has been paid to IL-17 chiefly due to the identification of Th17 cells, a T helper subset, which is distinctive from Th1 and Th2 cells. It was first noted by Aggarwall et al. that activated/memory $CD4^+$ T cells produce substantial amount of IL-17 upon stimulation by IL-23 [5]. Subsequent studies formally established Th17 cells [6, 7]. Under the so-called Th17 polarizing condition with combination of cytokines, naïve $CD4^+$ T cells can be induced in vitro to differentiate into IL-17-producing Th17 cells. Upon T cell receptor engagement, transforming growth factor (TGF)-β in combination with proinflammatory cytokines, in particular, IL-6, initiates the differentiation of naïve $CD4^+$ T cell towards Th17 [8, 9]. IL-23 perpetuates the expansion and stabilization of Th17 clone. This has been confirmed in vivo. IL-23-deficient animals do not produce stable Th17 cells and are protected from developing experimental inflammatory diseases [10–12]. The importance of Th17 cells in immune-mediated inflammatory disorders in human has been well demonstrated and blocking IL-17 activity is highly efficacious in some of these conditions [3].

On the other hand, to some extent it has been less appreciated that IL-17 plays essential role in host defense against microbial infections. Many cells other than $CD4^+$ T cells produce IL-17A, in particular, those innate immune cells, namely γ/δ T cells, NK cells, NK T cells, monocytes, and dendritic cells which are important in host defense rather than causing inflammation [3, 13]. For instance, IL-17A produced by γ/δ T cells are essential for protection of host again staphylococcus aureus infection in the skin [14]. Deficiency in IL-17 receptor in humans leads to chronic mucocutaneous candidiasis disease [15]. In addition, not all the proinflammatory and host defense properties are shared by IL-17A and IL-17F [15, 16]. It is therefore important to identify the cellular source of IL-17 in diseased state.

IL-17A is primarily a secreted cytokine but the cells producing it can be identified by sensitive cell-based assays. Intracelluar detection of IL-17A with flow cytometry is a sensitive assay and detects IL-17A at single cell level. In combination with cell surface staining, IL-17A-producing cells can be identified and the cell activation status can also be verified. This is particularly useful to detect IL-17A-producing cells in a mixed cell population such as human peripheral blood mononuclear cells (PBMC). The enzyme linked immunospot (ELISPOT) assay is a highly reproducible and sensitive cellular assay which can detect as few as 1 in 100,000 cells producing secreted proteins. ELISPOT assay has been widely used to enumerate cytokine-producing effector T cells in responding to different antigen stimulations [17, 18].

In this chapter, we focus on the detection of IL-17A produced by $CD4^+$ T cells in human PBMC in patients with rheumatoid arthritis (RA) and systemic lupus erythematosus (SLE) [19, 20]

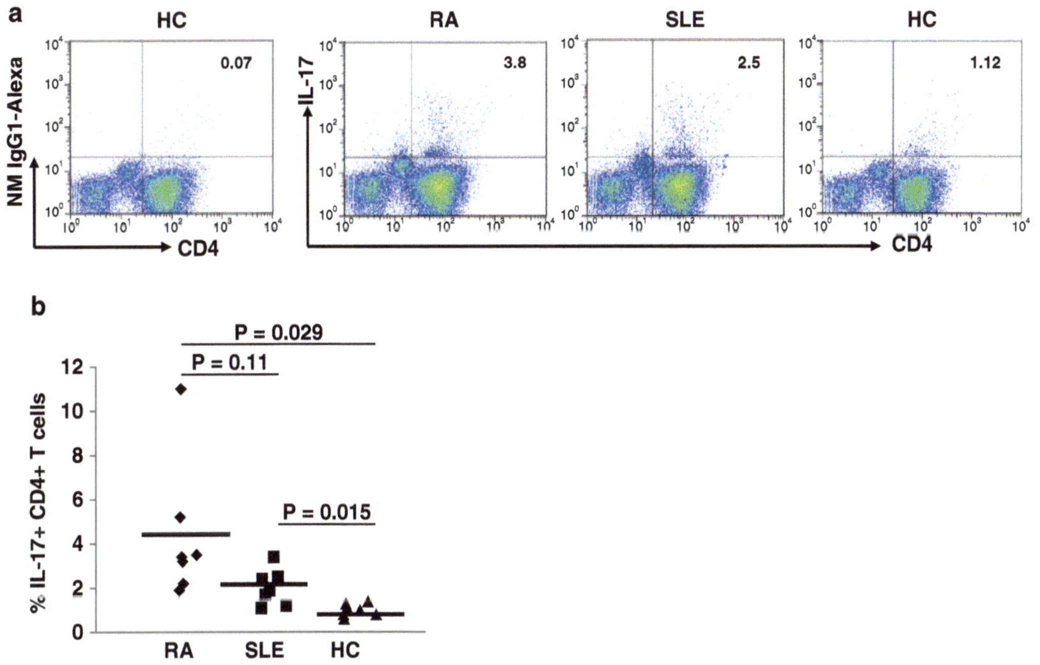

Fig. 1 Analysis of IL-17A-producing CD4⁺ T cells in PBMC by flow cytometry. PBMC from patients with RA, SLE and healthy controls were stained with mouse anti-human CD4-PerCP and mouse anti-human IL-17A-Alexa Fluor as described in Subheading 3.1, and were analyzed using a FACSCanto and Flowjo software. Normal mouse IgG1-Alexa Fluor was used as isotype control for anti-human IL-17A-Alexa Fluor. (**a**) Representative Dot plot of flow cytometry profiles show percent of IL-17A-producing $CD4^+$ T cells in patients with RA, SLE or healthy controls. (**b**) Summary of IL-17A-producing $CD4^+$ T cells in patients with RA, SLE or healthy controls ($n = 7$ in each group). *PBMC* peripheral blood mononuclear cells, *HC* healthy controls, *RA* rheumatoid arthritis, *SLE* systemic lupus erythematosus

(Figs. 1 and 2), on in vitro generated IL-17A-producing $CD4^+$ T cells of mice [21] using flow cytometry (Fig. 3), and on IL-17A production by cloned human Th1 cells in an enzyme linked immunospot (ELISPOT) assay [22] (Fig. 4). Monoclonal antibodies against IL-17F are also available for analysis of IL-17F production and the detection protocol can be adopted from the protocols described in this chapter.

2 Materials

2.1 Materials for Detection of IL-17A-Producing Cells in PBMC

1. Blood collection tube containing sodium heparin (Becton Dickinson) stored at room temperature.
2. Histopaque-1077 (Sigma); store at 4 °C.
3. PMA (phorbol 12-myristate 13-acetate); store aliquots at −20 °C.

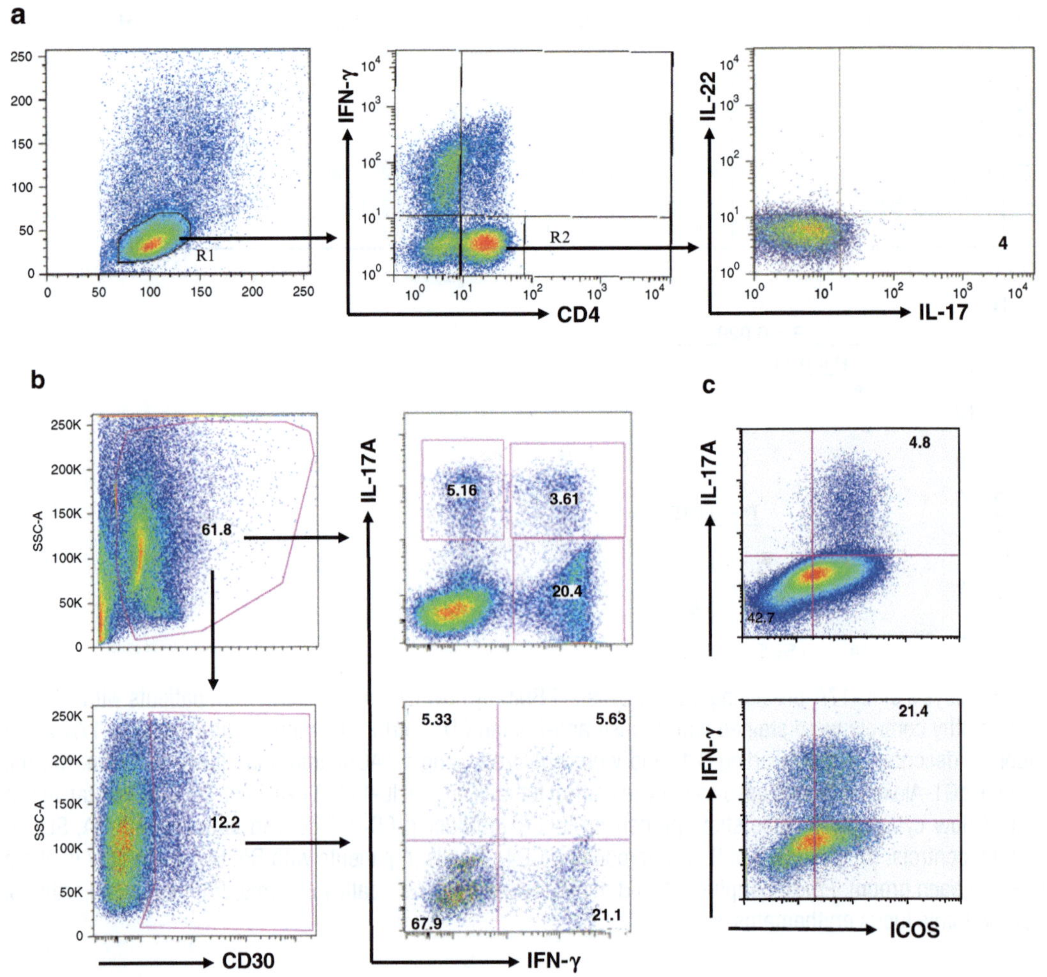

Fig. 2 Analysis of IL-17A-producing cells in different cell populations. (**a**) PBMC from a patient with RA were stained with mouse anti-human CD4-PerCp, mouse anti-human IL-17A-Alexa Fluor, anti-human IFN-γ-FITC, and anti-human IL-22-PE. A live gate was drawn based on cell distribution. CD4+ cell gate was drawn and IFN-γ-producing cells were excluded for further analysis of IL-17A and IL-22 producing cells. (**b**) PBMC from a healthy individual were stimulated with anti-CD3 (1 μg/ml), and anti-CD28 (10 μg/ml) with addition of lipopolysaccharide (100 ng/ml), and were stained with anti-human CD30-APC, IL-17A-Alexa Fluor, anti-human IFN-γ-FITC. IL-17A- and IFN-γ-producing cells were analyzed in total live cell population or in CD30 expression cells. (**c**) Activated PBMC and as in (**b**) were stained with anti-IL-17A, IFN-γ, and anti-ICOS. Analysis was on CD4+ T cells (*see* **Note 8**). *APC* allophycoerythrin, *FITC* fluorescein isothiocyanate, *ICOS* inducible costimulation molecule, *IFN* interferon, *PE* phycoerythrin

4. Ionomycin; store aliquots at −20 °C.
5. Brefeldin A dissolved in DMSO; store aliquots at −20 °C.
6. Mouse anti-human CD4 PerCP (Becton Dickinson); store at 4 °C.
7. Mouse anti-human IFN-γ-fluorescein isothiocyanate (FITC) (Becton Dickinson); store at 4 °C.

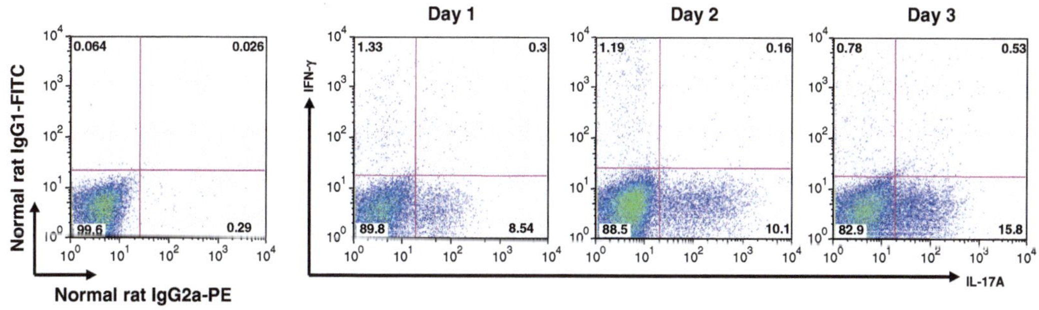

Fig. 3 Th17 cells were differentiated from naïve CD4+ T cells in vitro. Naïve CD4+ T cells were isolated from the spleen of a 6-week-old female C57BL/6 mouse and were stimulated to differentiate to IL-17A-producing cells. Intracellular staining with rat anti-mouse IL-17A-PE and IFN-γ-FITC as described in Subheading 3.2 at day 1, 2, or 3 after culture. Normal rat IgG1-FITC and normal rat IgG2a-PE were used for isotype controls for anti-IFN-γ and anti-IL-17A, respectively. *FITC* fluorescein isothiocyanate, *IFN* interferon, *PE* phycoerythrin

8. Mouse anti-human IL-17A-Alexa Fluor (Becton Dickinson); store at 4 °C.
9. Mouse anti-human IL-22-phycoerythrin (PE) (R&D System); store at 4 °C.
10. Normal mouse IgG1- FITC (Becton Dickinson); store at 4 °C.
11. Normal mouse IgG1-Alexa Fluor (Becton Dickinson); store at 4 °C.
12. Normal mouse IgG1-phycoerythrin (PE) (R&D System); store at 4 °C.
13. Normal mouse IgG (Jackson Immunology); store aliquots at -20 °C.
14. RPMI complete medium: RPMI-1640 supplemented with 10 % fetal calf serum (FCS), 100 U/ml penicillin and 100 U/ml streptomycin; store at 4 °C.
15. 2 % formaldehyde in phosphate buffer saline (PBS) (*see* **Note 1**); store at 4 °C.
16. Permeabilization buffer: 0.5 % saponin in PBS containing 1 % bovine serum albumin (BSA); store at 4 °C.
17. FACS (fluorescence activated cell sorting) buffer: PBS containing 1 % BSA and 0.01 % sodium azide; store at 4 °C.

2.2 Materials for Generation of IL-17A-Producing CD4+ T Cells in Mice

1. Mice: C57BL/6, female, age of 6–8 weeks (The Jackson Laboratory), kept in a specific pathogen-free facility until use.
2. Recombinant mouse transforming growth factor (TGF)-β1 (R&D System); store aliquots at −20 °C.
3. Recombinant mouse IL-6 (eBioscience); store aliquots at −20 °C.

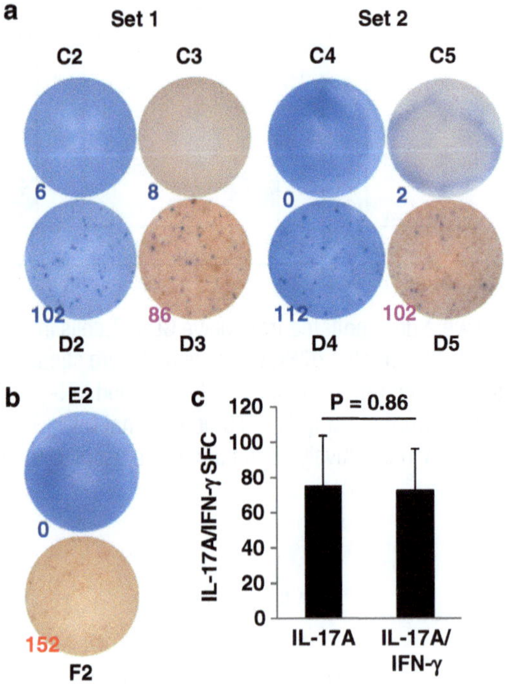

Fig. 4 Superantigen-induced IL-17A production from MOG-specific Th1 clones. MOG_{35-55}-specific cloned Th1 cells were cultured with MOG_{35-55} peptide only or cocultured with superantigen SEB for 24 h in a 96-well PVDF ELISPOT pate as described in Subheading 3.3. Cells producing IL-17A or IFN-γ or both were analyzed by T.C.L. plate reader. (**a**) Wells in row C contain T and NK cell-depleted APC only without cloned T cells show no IL-17A- or IFN-γ-producing cells. Wells in row D contain cloned T cells and T and NK cell-depleted APC. C2, D2, C4, and D4 were coated with anti-IL-17A capture antibody only and developed with anti-IL-17A detection antibody. C3, D3, C5, and D5 were coated with both anti-IL-17A and anti-IFN-γ capture antibodies and developed with both anti-IL-17A and anti-IFN-γ detection antibodies. Numbers in *blue* indicate IL-17A-producing cells and number in *pink* indicate IL-17A and IFN-γ dual producing cells (one representative experiment with duplications). (**b**) Cloned $CD4^+$ T cells were stimulated with MOG_{35-55} peptide only. E2 shows no IL-17A production and F2 shows IFN-γ-producing spots. (**c**) Summary of IL-17A and IL-17A/IFN-γ dual producing SFC as in (**a**) (pooled data of three experiments). *APC* antigen presenting cells, *FITC* fluorescein isothiocyanate, *IFN* interferon, *MOG* myelin olygodendrocyte glycoprotein, *PE* phycoerythrin, *SEB* staphylococcal endotoxin B, *SFC* spot forming cells

4. Recombinant mouse IL-23 (eBioscience); store aliquots at −20 °C.
5. Purified rat anti-mouse CD3 (BioLegend); store aliquots at −20 °C.
6. Purified rat anti-mouse CD28 (BioLegend); store aliquots at −20 °C.

7. Purified rat anti-mouse IFN-γ (BioLegend); store aliquots at −20 °C.
8. Rat anti-mouse IL-17A-PE (BioLegend); store at 4 °C.
9. Rat anti-mouse IFN-γ-FITC (BioLegend); store at 4 °C.
10. Purified anti-mouse CD16/CD32 (BioLegend); store at 4 °C.
11. Anti-mouse CD4-allophycoerythrin (APC) (BioLegend); store at 4 °C.
12. Normal Rat IgG2a-PE (BioLegend); store at 4 °C.
13. Normal rat IgG1-FITC (BioLegend); store at 4 °C.
14. RPMI medium with 2-mercaptoethanol: RPMI-1640 supplemented with 10 % FCS, 100 U/ml penicillin, 100 U/ml streptomycin and 100 μM 2-mercaptoethanol (2-ME); store at 4 °C.

2.3 Materials for ELISPOT Assay to Measure IL-17A-Producing Cells

1. ELISPOT plates with PVDF 0.45 μm filter membrane (Millipore); store at room temperature.
2. Purified mouse anti-human IL-17A antibody (eBioscience). Keep sterile and store at 4 °C.
3. Purified mouse anti-human IFN-γ antibody (BioLegend). Keep sterile, store at 4 °C.
4. FITC-conjugated mouse anti-human IFN-γ antibody (BioLegend); store at 4 °C.
5. Biotinylated mouse anti-human IL-17A antibody (eBioscience); store at 4 °C.
6. Streptavidin-alkaline phosphatase (AP) conjugate (Vector Laboratories); store at 4 °C.
7. Anti-FITC-horse radish peroxidase (HRP) (Southern Biotech); store at 4 °C.
8. Cloned human $CD4^+$ T cells specific to myelin oligodendrocyte glycoprotein (MOG_{35-55}) peptide (previously generated, *see* **Note 2**).
9. Antigen presenting cells (APCs) (T and NK cell-depleted PBMC, *see* **Note 3**).
10. Custom synthesized MOG-$_{35-55}$ peptide (NeoBioscience) dissolved in PBS; store aliquots at −20 °C.
11. Superantigen, staphylococcal endotoxin B (SEB) (Sigma); store aliquots at −20 °C.
12. RPMI medium: RPMI-1640 supplemented with 10 % FCS; store at 4 °C.
13. Sterile PBS; store at 4 °C.
14. PBS/BSA: PBS containing 1 % BSA; keep sterile, store at 4 °C.
15. PBS-Tween: PBS containing 0.05 % Tween-20.

16. Vector Blue alkaline phosphatase substrate kit III (Vector Laboratories); store at 4 °C.
17. 3-amino-9-ehtylcarbazole (AEC) substrate for peroxidase (Vector Laboratories); store at 4 °C.
18. 100 mM acetate buffer, pH 5.0; store at room temperature.
19. Hydrogen peroxide solution.

3 Methods

3.1 Intracellular Staining for Detecting IL-17A-Producing Cells in PBMC

1. Collect 10 ml of venous blood into a heparinized tube and shake well (*see* **Note 4**).
2. Dilute the blood in a 50 ml Falcon tube with PBS. For 10 ml of blood, add 20 ml of PBS and mix well.
3. Draw 13 ml of Histopaque and place the tip of the pipette to the bottom of 50 ml tube containing blood. Slowly release the Histopaque. This will form two layers of solution with Histopaque at the bottom and blood on the top (*see* **Note 5**).
4. Centrifuge the tube at 1,500 × *g* for 20 min at 20 °C.
5. Aspirate the milky white middle layer (containing PBMC) and transfer to a new 50 ml tube. Add 50 ml of PBS and mix well.
6. Centrifuge at 913 × *g* for 7 min at 20 °C.
7. Discard supernatant, PBMC remains at the bottom of tube. Resuspend the cells with 10 ml of PBS, centrifuge at 913 × *g* for 5 min at 20 °C (*see* **Note 6**).
8. Resuspend the PBMC with RPMI complete medium to adjust the cell density to 1×10^6/ml.
9. Culture the PBMC at 1×10^6/ml, 200 μl/well in a 96-well U-bottomed plate, in RPMI complete medium containing PMA at 50 ng/ml, ionomycin 500 ng/ml and brefeldin at 10 μg/ml for 5 h (*see* **Note 7**).
10. Harvest the cells into a 1.5 ml centrifuge tube and add 1 ml of FASC buffer, centrifuge at 200 × *g* in a bench top microcentrifuge for 5 min, discard supernatant.
11. Resuspend the cells with 50 μl of FACS buffer, add mouse anti-CD4-PerCP (*see* **Note 8**) and mix well. Incubate on ice for 15 min.
12. Wash cells by adding 500 μl of FACS buffer, resuspend the cells, and centrifuge at 200 × *g* for 5 min. Repeat once.
13. Fix cells by adding 500 μl of 2 % paraformaldehyde, mix well, and incubate for at least 30 min at room temperature. Centrifuge at 300 × *g* for 5 min, discard supernatant.
14. Permeabilize cells by adding 500 μl of permeabilization buffer, resuspend, and incubate for at least 30 min at room temperature.

Centrifuge at 300×*g* for 5 min, discard supernatant (*see* **Note 9**).

15. Resuspend cells in 50 μl of permeabilization buffer, add 10 μl of normal mouse IgG at final concentration 10 μg/ml, and incubate for 10 min (*see* **Note 10**).
16. Add anti-IL-17A-Alexa Fluor, anti-IFN-γ-FITC, and anti-IL-22 PE, diluted in permeabilization buffer; isotype controls for each antibody to cytokines should be added in separate tubes. Incubate for 30 min at room temperature (*see* **Note 11**).
17. Wash with 500 μl of permeabilization buffer for three times, discard the supernatant.
18. Resuspend the cells in 300 μl of FASC buffer in a FACS tube for flow cytometry analysis (*see* **Note 12**).

3.2 Induction and Detection of Mouse Th17 Cells

1. Isolate naïve CD4⁺ T cells from a mouse spleen (*see* **Note 13**). Naïve $CD4^+$ T cells are defined as $CD4^+/CD62L^{hi}/CD44^{lo}$ and the purity is >96 %.
2. Coat a 96-well flat-bottomed plate with purified anti-mouse CD3 antibody, diluted at 5 μg/ml in sterile PBS, 100 μl/well, incubate at 37 °C for at least 1 h.
3. Wash plate with sterile PBS for two times.
4. Add purified naïve $CD4^+$ T cells at 1×10^6/ml, 200 μl/well in RPMI. Add recombinant mouse IL-6 at 20 ng/ml, TGF-β at 10 ng/ml, IL-23 at 20 ng/ml, purified anti-IFN-γ antibody at 10 μg/ml and anti-CD28 at 0.5 μg/ml. Incubate the cells in a 5 % CO_2 incubator at 37 °C for a total of 3 days (*see* **Note 14**).
5. Add PMA at 50 ng/ml, ionomycin at 500 ng/ml and brefeldin A at 10 μg/ml and incubate at 37 °C for 5 h.
6. Harvest cells into 1.5 tubes for staining for flow cytometry analysis. Cells are washed in PBS once.
7. Resuspend cells in 50 μl of FACS buffer.
8. Add purified anti-mouse CD16/32 antibody at 1 μg/ml, incubate for 15 min on ice (*see* **Note 15**).
9. Wash cells once with FACS buffer as in **step 10** of Subheading 3.1.
10. Add anti-mouse CD4-APC, incubate 15 min on ice.
11. Wash cells in FACS buffer twice.
12. Fix and permeabilize the cells as in **steps 13** and **14** of Subheading 3.1.
13. Add anti-IL-17A-PE and anti-IFN-γ-FITC, or isotype control antibodies diluted in permeabilization buffer and incubate for 30 min.
14. Follow **steps 17** and **18** of Subheading 3.1.

3.3 SEB-Induced IL-17A Production from Cloned Th1 Cells

Fully differentiated Th17 cells can be redirected to produce IFN-γ when stimulated by IL-12. In contrast, Th1 cells exhibit little plasticity, that is, it is difficult to redirect fully differentiated Th1 cells to produce IL-17 [23]. We challenged this dogma to induce IL-17A production by stimulating highly polarized MOG-specific Th1 clones with a superantigen, SEB [22].

1. Coat plates with antibodies to human IL-17 or IFN-γ (in separate wells) diluted in sterile PBS at 10 μg/ml, 100 μl/well (see **Note 16**). For capturing cells producing both IL-17 and IFN-γ, dilute both antibodies in the same sterile PBS solution at the final concentration 10 μg/ml each, and add 100 μl/well.
2. Incubate plates at 4 °C overnight (*see* **Note 17**) in a humidified box.
3. Wash plates with sterile PBS at 200 μl/well three times. Remove PBS by flicking the Plate.
4. Block plates with sterile PBS/BSA, 200 μl/well at room temperature for 1 h.
5. Wash plates as in **step 3** of Subheading 3.3.
6. Add cloned $CD4^+$ T cells at 1×10^5 and APCs at 1×10^5/well with $MOG\text{-}_{35-55}$ at 10 μM with or without SEB at 100 ng/ml in RPMI.
7. Incubate Plates 24 h at 37 °C in a humidified 5 % CO_2 incubator (*see* **Notes 18** and **19**).
8. Wash plates with PBS, 200 μl/well, three times (*see* **Note 20**).
9. Wash plates with PBS-Tween, 200 μl/well, three times.
10. Add biotinylated anti-human IL-17A antibody or FITC-conjugated anti-IFN-γ antibody at 10 μg/ml diluted in PBS-Tween, or both antibodies for detecting both IL-17A- and IFN-γ-producing cells in the same well (*see* **Note 21**).
11. Incubate plates overnight at 4 °C.
12. Wash plates with PBS-Tween for three times.
13. Add streptavidin-AP (diluted 1:1,000) or anti-FITC-HRP (diluted 1:1,000) or both in PBS-BSA and incubate for 1 h at room temperature (*see* **Note 22**).
14. Wash plates with PBS-Tween for three times, followed by washing with PBS for two times.
15. Add AEC at 200 μl/well and wait for color development (red) for 30 min (*see* **Note 23**). Wash with PBS for two times.
16. Add Vector Blue solution 200 μl/well and wait for development (blue) for 30 min (*see* **Note 23**). Wash with PBS for two times.

17. Rinse plates with tap water and air dry them in an upright position in a hood overnight to have completely dry membranes before imaging analysis.
18. Spot forming cells (SFC) are counted by imaging analysis on a Cellular Technology Limited (C.T.L.) ELISPOT reader. This is read by an independent technician at C.T.L. (Fig. 4) (*see* **Note 24**).

4 Notes

1. Perform this procedure in chemical fume hood. Pre-heat PBS to 80 °C and add paraformaldehyde slowly while PBS is being stirred. Keep stirring until paraformaldehyde is dissolved. Slowly cool solution down and store solution at 4 °C. Make fresh solution every 4–6 weeks.
2. $MOG_{35–55}$-specific $CD4^+$ Thl clones were generated from PBMC of a healthy individual by limiting dilution and repeated antigen stimulation as described [24]. These cloned Thl cells produce IFN-γ, but not IL-17A, in response to $MOG_{35–55}$ [22].
3. T and NK cell-depleted APCs: T and NK cells were removed from PBMC by negative selection using mouse anti-human CD3, anti-CD56, and subsequently Dayna-beads-conjugated anti-mouse Ig (Invitrogen) and followed by irradiation at 2,500 rad [22, 24].
4. From each 10 ml blood, expect to obtain $13–30 \times 10^6$ PBMC, depending on individual donors.
5. Alternative method for laying blood on Ficoll-Paque: Add Ficoll-Paque at the bottom of the tube, then lay blood onto the Ficoll-Paque by releasing blood onto the wall of the tube.
6. Count the cell number, and determine the volume of RPMI to add for resuspending the cells to the desired cell density.
7. Brefeldin A is a lactone antibiotic that inhibits transport of cytokines from endoplasmic reticulum to Golgi and leads to cytokine accumulation in the endoplasmic reticulum. The intracellular accumulated cytokines are detected by antibodies. The time for incubation with brefeldin A ranges between 4 and 6 h. Monensin A has also been used for this purpose and its usual concentration is 10 μg/ml. Brefeldin A and monensin A can be combined at 5 μg/ml each.
8. The optimal concentration of anti-CD4-PerCp is predetermined by titration in pilot studies. Staining with antibodies to other cell surface proteins can also be done at this stage (Fig. 2) to evaluate the cell activation status. CD30 and inducible

costimulation (ICOS) molecules are expressed by activated T cells. Figure 2 shows IL-17A-producing cells expressing CD30 and ICOS.

9. After the cells are fixed, they can be resuspended in FACS buffer and stored at 4 °C for up to 1 week. At the time of staining, the cells can be permeabilized as in **step 14** of Subheading 3.1. There are commercial available fix/permeabilization buffers for one step procedure. After fix/permeabilization, cells can be stored at 4 °C for up to 1 week before staining. It is advised to store the fixed/permeabilized cells no longer than 1 week; otherwise the signal for cytokine staining becomes weaker.
10. Normal mouse IgG is added to block the Fc receptor, which can bind to anti-cytokine antibodies and give rise to false positive signal.
11. The optimal concentration of anti-cytokine antibodies is predetermined in pilot studies by titration. It is important to add the same concentration of isotype control antibodies as anticytokine antibodies, respectively.
12. Readers are advised to work with local flow cytometry facility to determine the suitable instrument for acquisition and software for analysis.
13. Spleen cells were prepared from a female C57BL/6 mouse and naïve $CD4^+$ T cells were isolated using a $CD4^+/CD62L^+$ Kit II (Miltenyi Biotec). This is a negative selection for purification of naïve $CD4^+$ T cells. First, a cocktail of biotinylated antibodies to mouse CD8a, CD45R, CD11b, CD25, CD49b, TCRγ/δ, and Ter-119 was added to the spleen cells, followed by magnetic microbeads labeled anti-biotin antibodies. The microbeads labeled cells were removed through a column and magnetic separator. The pre-enriched $CD4^+$ cells were labeled with anti-CD62-microbeads and then separated. Flow cytometry-based cell sorting can also be used to purify naïve $CD4^+$ T cells.
14. Cells can be harvested at day 1 or day 2 of culture as desired (Fig. 3). Intracelluar staining for cytokines can be performed after stimulation with PMA and ionomycin in the presence of brefeldin A. If desirable, further culture for another 3 days can also be done and addition of extra IL-23 is optional.
15. Addition of anti-CD16/CD32 antibodies to block Fc receptor to prevent false positive signal from Fc receptor binding to anti-cytokine antibodies.
16. Each antibody batch should be titrated to find optimal concentration.
17. Four hours of incubation is sufficient if it is done at 37 °C.
18. Incubation for 24 h is sufficient. Longer incubations may produce nondiscrete spots and cause error in spot counting.

19. All the above procedures should be performed in a sterile tissue culture hood.
20. This and all the following procedures are performed in a nonsterile condition.
21. The optimal concentration of detecting antibodies is determined by pilot experiments.
22. For each batch of the conjugate, the optimal dilutions need to be titrated.
23. Prepare AEC solution immediately before use. AEC stock solution is 50 mg/ml in acetate buffer. Dilute AEC stock solution 1:10 in distilled water and add hydrogen peroxide solution at a final concentration of 10 %. Monitor spot color development by watching for the spots in the superantigen or antigen containing wells and the lack of coloration in the control wells. For developing dual colors, develop AEC (red) first, wash with PBS for 2 times, then develop blue color.
24. For independent, objective analysis results, SFC counting and image scan can be done through C.T.L. service.

Acknowledgement

This work was supported by grants from the National Institutes of Health AR-055254 (CQC) and the National Natural Science Foundation of China (No. 30972610) (YJ and ZJ).

References

1. Rouvier E, Luciani MF, Mattei MG et al (1993) CTLA 8, cloned from an activated T cell, bearing AU-rich messenger RNA instability sequences, and homologous to a herpesvirus saimiri gene. J Immunol 150:5445–5456
2. Kolls JK, Linden A (2004) Interleukin-17 family members and inflammation. Immunity 21: 467–476
3. Miossec P, Kolls JK (2012) Targeting IL-17 and TH17 cells in chronic inflammation. Nat Rev Drug Discov 11:763–776
4. Wright JF, Guo Y, Quazi A et al (2007) Identification of an interleukin 17F/17A heterodimer in activated human CD4+ T cells. J Biol Chem 282:13447–13455
5. Aggarwal S, Ghilardi N, Xie MH et al (2003) Interleukin-23 promotes a distinct CD4 T cell activation state characterized by the production of interleukin-17. J Biol Chem 278:1910–1914
6. Harrington LE, Hatton RD, Mangan PR et al (2005) Interleukin 17-producing CD4+ effector T cells develop via a lineage distinct from the T helper type 1 and 2 lineages. Nat Immunol 6:1123–1132
7. Park H, Li Z, Yang XO et al (2005) A distinct lineage of CD4 T cells regulates tissue inflammation by producing interleukin 17. Nat Immunol 6:1133–1141
8. Veldhoen M, Hocking RJ, Atkins CJ et al (2006) TGFbeta in the context of an inflammatory cytokine milieu supports de novo differentiation of IL-17-producing T cells. Immunity 24:179–189
9. Acosta-Rodriguez EV, Napolitani G, Lanzavecchia A et al (2007) Interleukins 1beta and 6 but not transforming growth factor-beta are essential for the differentiation of interleukin 17-producing human T helper cells. Nat Immunol 8:942–949
10. Murphy CA, Langrish CL, Chen Y et al (2003) Divergent pro- and antiinflammatory roles for IL-23 and IL-12 in joint autoimmune inflammation. J Exp Med 198:1951–1957

11. Cua DJ, Sherlock J, Chen Y et al (2003) Interleukin-23 rather than interleukin-12 is the critical cytokine for autoimmune inflammation of the brain. Nature 421:744–748
12. Haines CJ, Chen Y, Blumenschein WM et al (2013) Autoimmune memory T helper 17 cell function and expansion are dependent on interleukin-23. Cell Rep 3:1378–1388
13. Martin B, Hirota K, Cua DJ et al (2009) Interleukin-17-producing gammadelta T cells selectively expand in response to pathogen products and environmental signals. Immunity 31:321–330
14. Cho JS, Pietras EM, Garcia NC et al (2010) IL-17 is essential for host defense against cutaneous *Staphylococcus aureus* infection in mice. J Clin Invest 120:1762–1773
15. Puel A, Cypowyj S, Bustamante J et al (2011) Chronic mucocutaneous candidiasis in humans with inborn errors of interleukin-17 immunity. Science 332:65–68
16. Yang XO, Chang SH, Park H et al (2008) Regulation of inflammatory responses by IL-17F. J Exp Med 205:1063–1075
17. Karlsson AC, Martin JN, Younger SR et al (2003) Comparison of the ELISPOT and cytokine flow cytometry assays for the enumeration of antigen-specific T cells. J Immunol Methods 283:141–153
18. Anthony DD, Lehmann PV (2003) T-cell epitope mapping using the ELISPOT approach. Methods 29:260–269
19. Zhao L, Jiang Z, Jiang Y et al (2013) IL-22(+) CD4(+) T cells in patients with rheumatoid arthritis. Int J Rheum Dis 16:518–526
20. Zhao L, Jiang Z, Jiang Y et al (2013) IL-22+CD4+ T-cells in patients with active systemic lupus erythematosus. Exp Biol Med (Maywood) 238:193–199
21. Morita Y, Ismail DM, Elkon KB et al (2011) Dichotomous response to transforming growth factor beta after T cell receptor activation by naive CD4+ T cells from DBA/1 mice: enhanced retinoic acid receptor-related orphan nuclear receptor gammat expression yet reduced FoxP3 expression. Arthritis Rheum 63:118–126
22. Yomogida K, Chou YK, Chu CQ (2013) Superantigens induce IL-17 production from polarized Th1 clones. Cytokine 63:6–9
23. Zhu J, Paul WE (2010) CD4+ T cell plasticity-Th2 cells join the crowd. Immunity 32:11–13
24. Chou YK, Henderikx P, Vainiene M et al (1991) Specificity of human T cell clones reactive to immunodominant epitopes of myelin basic protein. J Neurosci Res 28:280–290

Chapter 23

Interleukin-1 (IL-1) Immunohistochemistry Assay in Oral Squamous Cell Carcinoma

Takaaki Kamatani

Abstract

This chapter describes an immunohistochemistry method to analyze interleukin-1 (IL-1) in oral squamous cell carcinoma. The described protocol has been optimized for IL-1 detection in formalin-fixed paraffin-embedded oral tissue sections by light microscopy. A few common pitfalls and problems associated with immunohistochemical staining are discussed.

Key words Immunohistochemistry, Interleukin-1, Oral squamous cell carcinoma, Formalin fixed, Paraffin embedded

1 Introduction

The interleukin-1 (IL-1) family comprises 11 cytokines, which play an important role in the regulation of immune and inflammatory responses [1]. The IL-1 family has a conserved beta-trefoil structure [2] and binds to receptors belonging to the IL-1 receptor family [3]. Discovery of the cytokines began with studies on the pathogenesis of fever [4]. The basis for the term interleukin was to streamline the growing number of biological properties attributed to soluble factors from macrophages and lymphocytes. Cells of oral squamous cell carcinoma also express the IL-1 [5]. IL-1alpha and beta are the most studied members [6], because they were discovered first. IL-1 alpha and beta bind to the same receptor molecule, which is called type I IL-1 receptor [3].

Immunohistochemistry is an immunoassay used to localize antigens in tissue sections using labeled antibody as specific reagents through antigen–antibody interactions that are visualized by markers, such as fluorescent dye, enzyme, radioactive element, or colloidal gold [7]. With the expansion and development of the immunohistochemistry technique, enzyme labels have been introduced such as peroxidase and alkaline phosphatase [8].

Ivana Vancurova (ed.), *Cytokine Bioassays: Methods and Protocols*, Methods in Molecular Biology, vol. 1172, DOI 10.1007/978-1-4939-0928-5_23, © Springer Science+Business Media New York 2014

The diaminobenzidine (DAB) molecule is also conjugated to antibodies during the same period [9]. Since immunohistochemistry involves specific antigen–antibody reaction, it has apparent advantage over traditionally used special enzyme staining techniques that identify only a limited number or proteins, enzymes, and tissue structures. This is especially useful for assessing the progression and treatment of disease. Therefore, immunohistochemistry has become a crucial technique and widely used in medical research laboratories as well as clinical diagnosis [10]. There are numerous immunohistochemistry methods that may be used to localize antigens. The selection of a suitable method should be based on parameters such as the type of specimen under investigation and the degree of sensitivity required.

2 Materials

1. 10 % Neutral-Buffered Formalin (NBF): Add 6.5 g anhydrous disodium hydrogen phosphate (Na_2HPO_4), 4.0 g sodium dihydrogen phosphate, monohydrate ($NaH_2PO_4 \cdot H_2O$), and 100 ml of 37 % formaldehyde solution to 900 ml of double distilled water. Adjust pH to 7.0 with 1 N NaOH or 1 M HCl. Adjust volume to 1,000 ml. Store at 4 °C (*see* **Note 1**).
2. 10× Tris-Buffered Saline (TBS): Add 24.2 g Trizma base and 80 g NaCl to 900 ml of double distilled water. Adjust pH to 7.6 with concentrated HCl. Adjust volume to 1,000 ml.
3. TBS containing 0.05 % Tween 20 (TBST): Add 100 ml of 10× TBS and 0.5 ml of Tween 20 to 900 ml of double distilled water. Adjust volume to 1,000 ml.
4. Phosphate-Buffered Saline (PBS): Add 8 g NaCl, 0.2 g KCl, 1.44 g anhydrous disodium hydrogen phosphate (Na_2HPO_4) and 0.24 g potassium phosphate monobasic anhydrous (KH_2PO_4) to 900 ml of double distilled water. Adjust volume to 1,000 ml.
5. Sodium Citrate Buffer: Add 2.9 g trisodium citrate dehydrate to 900 ml of double distilled water. Adjust pH to 6.0 with 1 M HCl. Add 0.5 ml of Tween 20. Adjust volume to 1,000 ml. Store at room temperature, or at 4 °C for storing longer than 3 months.
6. 3 % Hydrogen Peroxide: Add 10 ml of 30 % H_2O_2 to 90 ml of double distilled water.
7. 1 % diaminobenzidine (DAB): Add 0.1 g DAB to 10 ml of double distilled water. Add three to five drops (1 drop = 50 μl) of 10 N HCl and solution turns light brown color. Shake for 10 min and DAB should dissolve completely. Aliquot and store at −20 °C.
8. 0.3 % H_2O_2: Add 1 ml of 30 % H_2O_2 to 99 ml of double distilled water. Mix well. Store at 4 °C, or aliquot and store at −20 °C.

9. DAB Working Solution: Add 250 μl of 1 % DAB to 5 ml of PBS, pH 7.2, and mix well. Add 250 μl of 0.3 % H_2O_2 and mix well. Use the solution within 30 min after preparation (*see* **Note 2**).
10. Ethanol.
11. Xylene.

3 Methods

3.1 Tissue Fixation and Preparation

To ensure the preservation of tissue architecture and cell morphology, prompt and adequate fixation is essential. Inappropriate or prolonged fixation may significantly diminish the antibody binding capability (*see* **Note 3**). There is no one universal fixation that is ideal for the demonstration of all antigens. However, in general, many antigens can be successfully demonstrated in formalin-fixed paraffin-embedded tissue sections with the use of antigen retrieval techniques (*see* **Note 4**). Tissues should be fixed by adding directly to the 10 % NBF for at least 24 h. NBF has a very slow diffusion coefficient so the tissue needs to be no more than 1 cm thick (*see* **Note 5**).

1. Prepare 3–5 μm sections using the microtome and place on clean, positively charged microscope slides (*see* **Note 6**).
2. Heat the slides on tissue-drying oven at 60 °C overnight. Paraffin wax is the most widely used embedding medium for diagnostic histopathology in routine histological laboratories.
3. For deparaffinization, incubate slides 3 times in xylene for 3 min each, followed by 3 min incubation in 100 % ethanol, 3 min incubation in 95 % ethanol, 3 min incubation in 70 % ethanol, and 3 min incubation in 50 % ethanol (*see* **Note 7**).
4. Wash slides 2 times in PBS for 5 min each.
5. Keep the slides in PBS until ready to perform antigen retrieval (*see* **Note 8**).

3.2 Antigen Retrieval

The demonstration of many antigens can be significantly improved by pretreatment with the antigen retrieval reagent that breaks the protein cross-links formed by formalin fixation, and thereby uncovers hidden antigenic sites. There are several methods of antigen retrieval. The techniques involve application of heat for varying lengths of time to formalin-fixed, paraffin-embedded tissue sections in an aqueous solution. The most common method is heat-mediated retrieval in sodium citrate buffer using microwave for IL-1 immunohistochemistry assay in oral squamous cell carcinoma.

1. Add the appropriate 10 mM sodium citrate buffer to the microwaveable vessel and place it inside the microwave.
2. Heat slides in 10 mM sodium citrate buffer, pH 6.0, at 95–100 °C for 20 min.

3. Remove from the microwave and let stand at room temperature in buffer for 20 min.
4. Rinse in TBST for 1 min.
5. If using a domestic microwave, set to full power (*see* **Note 9**).

3.3 Blocking

Background staining may be specific or nonspecific. The main cause of nonspecific background staining is non-immunological binding of the specific immune serum by hydrophobic and electrostatic forces to certain sites within tissue sections. This form of background staining is usually uniform and can be reduced by blocking those sites with normal serum. Endogenous peroxidase activity is found in many tissues and can be detected by reacting fixed tissue sections with DAB substrate. The solution for eliminating endogenous peroxidase activity is by the pretreatment of the tissue section with hydrogen peroxide prior to incubation of primary antibody.

1. Wash slides in 3 % hydrogen peroxide in methanol for 10 min.
2. Wash sections twice in PBS for 5 min each.

3.4 Immunohistochemical Staining

Special controls must be run in order to test the protocol and for the specificity of the antibody being used. A positive tissue control is recommended to ensure that the antibody is performing as expected. It may also be useful to include a negative tissue control in which the protein of interest is not expected to be found.

1. Apply 100 μl per slide of primary IL-1 antibody at recommended concentration diluted in TBS. Incubate 60 min at room temperature or overnight at 4 °C (*see* **Notes 10–14**).
2. Wash slides in TBST 4 times for 5 min each.
3. Apply a 100 μl per slide of conjugated secondary antibody diluted in TBST. Incubate for 30 min at room temperature in the moisture chamber.
4. Wash slides in TBST 4 times for 5 min each.
5. Apply DAB substrate solution. Incubate sections for 1–3 min at room temperature. Adjust the reaction time by microscopic observation.
6. Wash the slide with distilled water at room temperature for 5 min.
7. Apply color development (i.e. hematoxylin) for 10 s.
8. Wash slides in distilled water for 1 min.

3.5 Dehydrate and Mount Slides

1. Wash slides 2 times in 80 % ethanol for 1 min each.
2. Wash slides 2 times in 95 % ethanol for 1 min each.
3. Wash slides 3 times in 100 % ethanol for 1 min each.

4. Wash slides 3 times in xylene for 1 min each.
5. Apply coverslip with sealants.

4 Notes

1. Formaldehyde, a gas highly soluble in water, is typically sold as a saturated aqueous solution of 37 % by mass. This is also referred to as "100 % formalin." 10 % neutral-buffered formalin is the tenfold dilution of this solution, giving a final concentration of formaldehyde of 3.7 %.
2. pH value is important. pH < 7.0 will reduce staining intensity. pH > 7.6 will cause background staining.
3. Inadequate or delayed fixation may give rise to false positive results due to the passive uptake of serum proteins and diffusion of the antigen. Such false positives are common in the center of large tissue blocks or throughout tissues in which fixation was delayed.
4. Certain cell antigens do not survive routine fixation and paraffin embedding. Thus, the use of frozen sections still remains essential for the demonstration of many antigens. However, the disadvantage of frozen section includes poor morphology, poor resolution at higher magnifications, special storage needed, limited retrospective studies, and cutting difficulty. IL-1 can be successfully demonstrated in formalin-fixed paraffin-embedded oral tissue sections.
5. Delayed or inadequate fixation can cause no staining because antigen is denatured or masked during the fixing process. Use less-potent fixative and decrease the fixing time.
6. Tissue sections are best mounted on positively charged or APES (amino-propyl-tri-ethoxy-silane)-coated slides. Otherwise, tissue sections will come off from the slides.
7. Incomplete removal of paraffin can cause poor staining of the section. Remove paraffin thoroughly from the section.
8. Do not allow tissues to dry at any time during the staining procedure. Drying out will cause nonspecific antibody binding and therefore high background staining.
9. Twenty minutes is only a suggested antigen retrieval time. Less than 20 min may leave the antigens under-retrieved, leading to weak staining. More than 20 min may leave them over-retrieved, leading to nonspecific background staining and also increasing the chances of sections dissociating from the slides.
10. Antibodies, especially polyclonal antibodies, are sometimes contaminated with other antibodies due to impure antigen used to immunize host animal.

11. Dilutions of the primary and secondary antibody are listed on the datasheets or are determined by testing a range. Adjust dilutions appropriately from the results obtained.
12. Inadequate incubation with the antibody may cause no staining or only weak staining results. Provide sufficient time for reaction with antibody. In particular, primary antibody should be incubated for longer time.
13. A high room temperature can cause nonspecific staining due to acceleration of enzyme reactions. Keep room temperature at 15–25 °C.
14. A shallow, plastic box with a sealed lid and wet tissue paper in the bottom is an adequate chamber.

References

1. Jensen LE (2010) Targeting the IL-1 family members in skin inflammation. Curr Opin Investig Drugs 11:1211–1220
2. Murzin AG, Lesk AM, Chothia C (1992) beta-Trefoil fold. Patterns of structure and sequence in the Kunitz inhibitors interleukins-1 beta and 1 alpha and fibroblast growth factors. J Mol Biol 223:531–543
3. Sims JE, Gayle MA, Slack JL et al (1993) Interleukin 1 signaling occurs exclusively via the type I receptor. Proc Natl Acad Sci U S A 90:6155–6159
4. Dinarello CA (1994) The interleukin-1 family: 10 years of discovery. FASEB J 8: 1314–1325
5. Kamatani T, Shiogama S, Yoshihama Y et al (2013) Interleukin-1 beta in unstimulated whole saliva is a potential biomarker for oral squamous cell carcinoma. Cytokine 64: 497–502
6. March CJ, Mosley B, Larsen A et al (1985) Cloning, sequence and expression of two distinct human interleukin-1 complementary DNAs. Nature 315:641–647
7. Matos LL, Trufelli DC, de Matos MG et al (2010) Immunohistochemistry as an important tool in biomarkers detection and clinical practice. Biomark Insights 5:9–20
8. Mason DY, Sammons R (1978) Alkaline phosphatase and peroxidase for double immunoenzymatic labeling of cellular constituents. J Clin Pathol 31:454–460
9. Vacca LL, Hewett D, Woodson G (1978) A comparison of methods using diaminobenzidine (DAB) to localize peroxidases in erythrocytes, neutrophils, and peroxidase-antiperoxidase complex. Stain Technol 53:331–336
10. Leong AS, Wright J (1987) The contribution of immunohistochemical staining in tumour diagnosis. Histopathology 11:1295–1305

Chapter 24

Immunofluorescence and Subsequent Confocal Microscopy of Intracellular TNF in Human Neutrophils

Janet Rollins and Veronika Miskolci

Abstract

Immunofluorescence is an important technique required to observe expression, localization and colocalization of proteins within the cell. Here we describe the immunofluorescence and subsequent confocal microscopy technique of tumor necrosis factor-α (TNF) in human neutrophils (polymorphonuclear leukocytes; PMN). The qualitative technique can be used to observe the expression pattern changes from resting to stimulated leukocytes. Colocalization with other cytokines, proteins, or organelles can be observed. This immunofluorescence technique can be done in 1–2 days.

Key words Cytokines, Immunofluorescence, Leukocytes, Confocal microscopy, Neutrophils, TNF

1 Introduction

Immunofluorescence is a powerful technique that allows for the detection of a wide variety of antigens including cytokines. The technique also allows for the visualization of antigen expression patterns at various stages of the immune responses. When employing this technique, one can not only look for changes in subcellular location but also determine if the antigen is interacting with other proteins. This paper describes the technique of indirect immunofluorescence, which will yield a more robust signal than direct immunofluorescence. Many antibodies are commercially available against various cytokines from different organisms so a wide variety of research and clinical applications can be achieved by this technique. Planning the experiment for your particular conditions is a key factor of this technique. The primary antibody, which will bind to the antigen it was raised against, must be carefully chosen and tested at different conditions. The secondary antibody that has the fluorophore covalently attached will amplify the signal from the primary antibody since there are multiple sites that the anti-IgG can bind to. Matching the secondary antibody's organism to the animal the primary antibody is crucial along with picking different and nonoverlapping fluorophores.

Ivana Vancurova (ed.), *Cytokine Bioassays: Methods and Protocols*, Methods in Molecular Biology, vol. 1172,
DOI 10.1007/978-1-4939-0928-5_24, © Springer Science+Business Media New York 2014

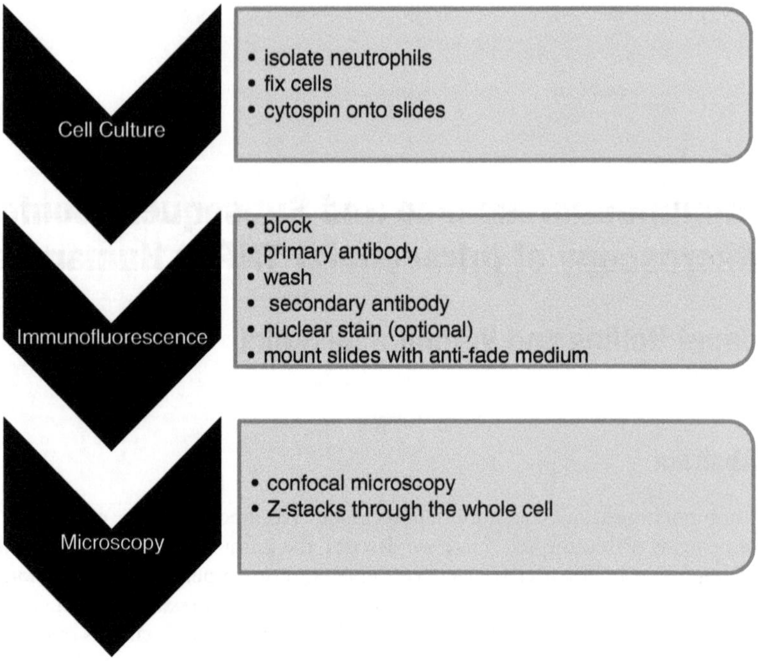

Fig. 1 Schematic illustration of the overall procedure

Confocal microscopy is the desired form of microscopy when visualizing the immunofluorescent samples. The laser gives you the precise excitation and the out of focus light gets eliminated from your image unlike epi-fluorescence that includes all the light in the image. This can lead to false positives in colocalization studies plus the filters of the wide-field scope are much broader that the precise laser wavelengths. This can lead to bleed-through between two channels. Confocal microscopy also allows you to scan at various depths through the cell, which can give you more precise cellular localization data that might have been overlooked using a wide-field microscope. Figure 1 illustrates the individual steps of this technique.

2 Materials

All solutions should be prepared with ultrapure water.

2.1 Preparation of Cells

1. PBS: Dilute 10× PBS, pH 7.4, 1:10 with water. Many solutions will use PBS for dilutions, so make plenty (*see* **Note 1**).
2. 2 % FCS-PBS: Add 2 ml of fetal calf serum (FCS) into 98 ml of PBS.
3. 1 % BSA-PBS: Add 1 g of bovine serum albumin (BSA) into 100 ml of PBS.

4. Permeabilization solution PBST: 0.1 % Triton-X 100 in PBS. Add 200 μl of Triton-X 100 to 200 ml of PBS (*see* **Note 2**).
5. 4 % Paraformaldehyde fixative: Add 4 g of paraformaldehyde to 100 ml of water. Warm solution gently and spin until dissolved (*see* **Note 3**).
6. Microscope slides: Use precleaned frosted slides.
7. Cytospinner.

2.2 Immunofluorescence

1. Blocking solution PBST/BSA: Weigh out 6 g of BSA and bring up the volume to 200 ml with PBST (*see* **Note 4**).
2. Primary antibodies: Dilute with blocking solution PBST/BSA (*see* **Note 5**).
3. Disposable pipettes: Use if washes are done on the slides.
4. Humid chamber: There are commercially available humid chambers where the slides stick lightly to a rack and this makes washing very simple. If that is not an option, then a humid chamber can be made by lining multiple layers of paper towels to a bottom of a plastic container and wetting the paper towels. Make sure the surface is flat so the slides are level. The chamber can be made dark by covering it with aluminum foil.
5. Pap pen (Sigma-Aldrich).
6. Secondary antibodies: Dilute with blocking solution (*see* **Note 6**).
7. Nuclear stain: Propidium iodide (PI) or 4′,6-diamidino-2-phenylindole (DAPI), 5 μg/ml (*see* **Note 7**).

2.3 Mounting and Visualization of Slides

1. Kim-wipes.
2. Mounting medium: Mowiol 10 % (w/v, Calbiochem), 25 % glycerol, 100 mM Tris–HCl, pH 8.5. Add 2.4 g of Mowiol 4-88 to 6 g of glycerol. Stir to mix. Add 6 ml of H_2O and leave for several hours at room temperature. Add 12 ml of 0.2 M Tris–Cl, pH 8.5, and heat to 50 °C for 10 min with occasional mixing. DABCO 1,4-diazabicyclo[2.2.2]octane 2.5 % can be added for prevention of fading (*see* **Note 8**).
3. 1.5 coverslips: Most modern microscope objective lenses are optimized for glass that is 0.17 mm thick, and no. 1.5 coverslips correspond most closely to this figure.

3 Methods

3.1 Isolation of Human Neutrophils

1. Obtain fresh blood from healthy adult human volunteers in heparinized preservative-free tubes [1, 2] (*see* **Note 9**).
2. Isolate neutrophils by Ficoll centrifugation under endotoxin-free conditions [3].

3. Stimulate neutrophils with lipopolysaccharide (LPS), 100 ng/ml, for 3 h.
4. Fix the cells with 2 % paraformaldehyde (*see* **Note 10**).

3.2 Cytospin onto Microscope Slides (See Note 11)

1. Wash cells in cold 2 % FCS-PBS twice and dilute in 100 μl of cold 1 % BSA-PBS. Be sure to keep all samples on ice.
2. Place slides and filters into appropriate slots in the cytospin with the cardboard filters facing the center of the cytospin. In the event that there are few cells available, aliquot about 100 μl of cold 1 % BSA-PBS into each of the wells and spin for 1–2 min. This will serve to wet the filter and allow more cells to reach the slide. Also, be sure that each filter and slide pair is flushed with each other and that the hole in the filter is in proper position so that cells will be able to reach the slide.
3. Quickly aliquot 100 μl of each sample into the appropriate wells of the cytospin. Be careful not to confuse the slides so that the samples are not aliquoted into the wrong wells. Marking the slides with pencil is helpful and less likely to smear.
4. Carefully place the lid of the cytospin over the samples and spin at maximum speed for 1–3 min.
5. Remove the filters from their slides without contacting the smears on the slides.
6. Examine each slide under the microscope to be sure that the cells have adhered properly. The cells should appear to have normal morphology and should be lying flat on the slide. For staining purposes, the cells should also be in a flat layer on the slide.

3.3 Permeabilization and Blocking

1. If fixation was already done (*see* **Note 11**), wash the slides in PBST, 3× for 20 min each.
2. When washing the slides, use a plastic disposable pipette to place a quick rinse, and then drain wash solution. Try not to let the slides dry out between washes (*see* **Note 12**).
3. Block for 45–60 min with PBST/BSA at room temperature, or overnight at 4 °C (*see* **Note 13**).

3.4 Primary Antibody

1. Prepare your primary antibody dilutions by diluting with PBST/BSA (*see* **Note 14**).
2. Drain off blocking solution; do not wash the slides.
3. Wipe off excess blocking solution on the exterior of the slide being careful not to touch the cells.
4. Using a PAP pen, draw a circle around the cells. This provides a hydrophobic well for small amounts of antibody solution to make contact with the cells.
5. Immediately place the antibody solution on the appropriate slide (*see* **Note 15**).

6. Control slide should get PBST/BSA instead of the primary antibody. It will be subsequently treated the same as the other slides.
7. Incubate overnight at 4 °C or 2 h at room temperature in a moist chamber.

3.5 Secondary Antibody

1. If primary antibody was incubated overnight, allow slides to come to room temperature.
2. Wash slides with PBST, 3 × 10 min.
3. Prepare secondary antibody by diluting it with PBST/BSA (*see* **Notes 16** and **17**).
4. Add the secondary antibody to slides and incubate in a moist chamber at room temperature in the dark for 1 h.

3.6 Washes, and DAPI or PI Staining

1. Wash 1 × 10 min with PBST followed by 2 × 10 min washes with PBS.
2. Make DAPI or PI dilution (*see* **Note 18**).
3. Incubate slides with DAPI or PI for 10 min.
4. Wash 2 × 5 min with PBS.

3.7 Mount

1. Dry off excess PBS and wick off on the exterior where the cells are using a Kim-wipe.
2. Place 15–20 μl of Mowiol mounting medium or other commercially available anti-fade mounting medium on the slide; be careful to avoid bubbles.
3. Carefully place 1 cover-slip on the slide. If you plan to view the slides the same day, then you need to seal with nail polish to avoid movement at the microscope.
4. If slides are to be viewed within the next few days, nail polish is not necessary since the Mowiol (and some other commercially available media) will harden. This is the preferred way to view since there is no movement of the cover-slip across the cells.

3.8 Confocal Microscopy

1. Find the best area to view on the microscope and view image on the screen.
2. Set up the parameters recommended for your confocal microscope.
3. Set the upper and lower parameters of the Z-stack in order to view the entire cell.
4. When capturing your images, use the highest resolution because you never know which image will be the one you use for publication.
5. When looking for colocalization of a cytokine with another protein, a single layer of a series must determine if the two colocalize, not a projection (Fig. 2).

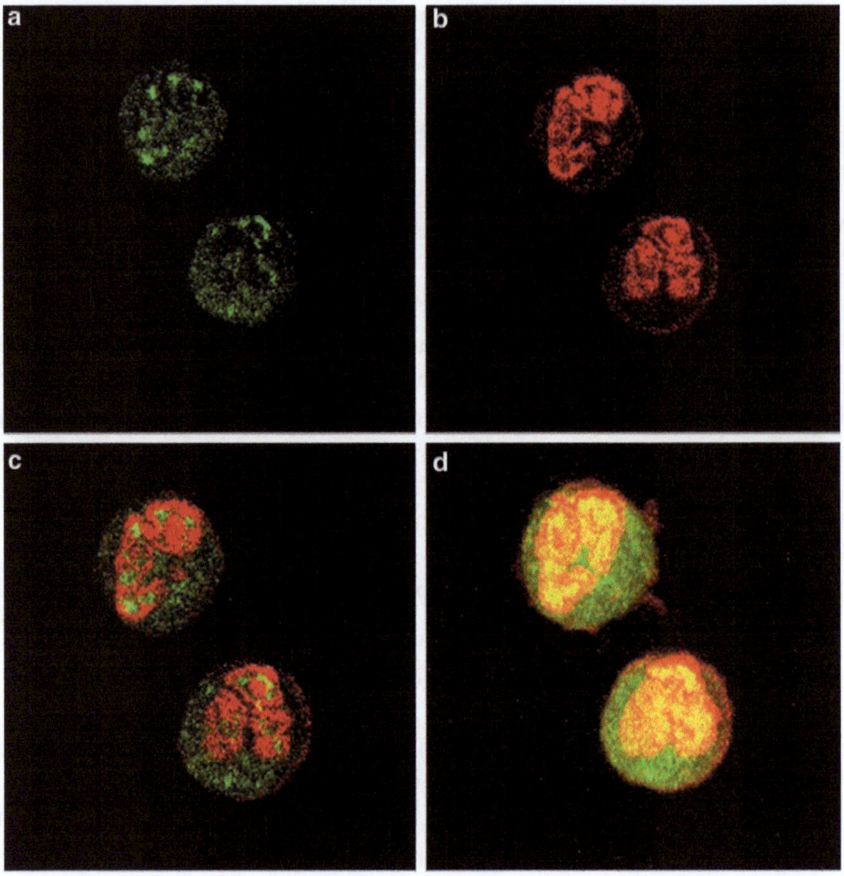

Fig. 2 Confocal microscopy of TNF in human neutrophils. Anti-TNFα in *green* and nucleus stained with propidium iodide in *red*. (**a**) and (**b**) are single scans of each channel. (**c**) is overlay of (**b**) and (**c**). (**d**) is an overlay projection of all the layers scanned. Note there appears to be much more colocalization (*yellow*) of the TNF with the nucleus in the projection

4 Notes

1. Commercially available 10× PBS is the fastest, easiest, and most accurate way to make PBS solution.
2. Triton-X 100 is a viscous solution, so take care in pipetting it. Give time for the solution to fully go into the pipette and wipe off any residual on the pipette tip.
3. It is much more convenient to use ampules containing 16 % paraformaldehyde (Electron Microscopy Sciences). Open the ampule, aliquot 250 μl into 1.5 ml vials, and store at −20 °C for about 6 months. Make the diluted solution fresh before using. If making 4 % solution, add 750 μl of PBS to vial, and mix.
4. Other blocking solutions can be used, such as 10 % goat serum or bovine serum [3, 4].

5. Only about 50–100 μl per slide is required, so when you make the dilutions, take that into consideration as not to waste precious antibody. For example, if you have 5 slides, only 400 μl of antibody is required since the control slide will have no primary antibody.
6. Store in the dark at 4 °C until ready to use.
7. Store in the dark at 4 °C until ready to use.
8. Stirring vigorously with a magnetic stirrer helps dissolve the Mowiol. Aliquot into vials and store at −20 °C.
9. When using human subjects, make sure to get informed written consent and approval from your Institutional Review Board.
10. Many protocols call for 4 % paraformaldehyde for fixation. The percentages can vary depending upon the time of fixation. We have used 2 % for 2 h and even 0.5 % overnight and the cells were preserved properly.
11. An alternate procedure would be to cytospin first, then fixation on the slide.
12. Washes, permeabilization, and blocking can either be done on the slide in a plastic box (to avoid the sample from drying out) or in coplin jars, which will guarantee that the samples will not dry out.
13. If incubating at 4 °C overnight, be sure to put the slide in a moist chamber or in coplin jar. The length of incubation and the type of blocking solution should be experimentally determined by varying the types of agents, such as goat serum, bovine serum, or bovine serum albumin.
14. The primary antibody concentration should also be experimentally determined. Some signals are very weak and fleeting and require a higher concentration. We used 1:10, 1:50, and 1:100 dilutions of primary antibodies [3, 4].
15. When checking for colocalization of two proteins, the antibodies must be raised in different animals, such as rabbit and mouse [5].
16. Usually start at 1:500 dilution for the secondary antibody but test higher and lower concentrations. The control no primary antibody slide will determine if there is nonspecific binding of the secondary antibody even at higher concentrations.
17. Secondary antibodies should match the animal that the antibody was raised in, i.e., if the primary antibody was produced in rabbit, the secondary antibody should be anti-rabbit IgG conjugated to a fluorophore. If you are using two primary antibodies and looking for colocalization, the secondary antibody fluorophores should be far apart on the spectrum to avoid any false positive by bleeding through the channels at the confocal.
18. If staining with PI, pretreatment with RNase is necessary. There are other far-red stains commercially available, such as Draq 5 or To-Pro 3. DAPI can only be used if your confocal has a UV laser.

References

1. Vancurova I, Miskolci V, Davidson D (2001) NFκB activation in TNF-stimulated neutrophils is mediated by protein kinase Cδ. Correlation to nuclear IκBα. J Biol Chem 276:19746–19752
2. Castro-Alcaraz S, Miskolci V, Kalasapudi B et al (2002) NFκB regulation in human neutrophils by nuclear IκBα: correlation to apoptosis. J Immunol 169:3947–3953
3. Miskolci V, Ghosh C, Rollins J et al (2006) TNFα release from peripheral blood leukocytes depends on a CRM1-mediated nuclear export. Biochem Biophys Res Commun 351:354–360
4. Miskolci V, Rollins J, Vu H et al (2007) NFκB is persistently activated in continuously stimulated human neutrophils. Mol Med 13:134–142
5. Riepl B, Grässel S, Wiest R et al (2010) Tumor necrosis factor and norepinephrine lower the levels of human neutrophil peptides 1-3 secretion by mixed synovial tissue cultures in osteoarthritis and rheumatoid arthritis. Arthritis Res Ther 12:R110. doi:10.1186/ar3044

Chapter 25

Evaluating Cytoplasmic and Nuclear Levels of Inflammatory Cytokines in Cancer Cells by Western Blotting

Himavanth R. Gatla, Bipradeb Singha, Valerie Persaud, and Ivana Vancurova

Abstract

Increased expression and cellular release of inflammatory cytokines, interleukin 8 (IL-8; CXCL8), and high mobility group box-1 (HMGB1) are associated with increased cell proliferation, angiogenesis, and metastasis during cancer progression. In prostate and ovarian cancer cells, increased levels of IL-8 and HMGB1 correlate with poor prognosis. We have recently shown that proteasome inhibition by bortezomib (BZ) specifically increases IL-8 release from metastatic prostate and ovarian cancer cells. In this chapter, we describe a protocol to analyze the cytoplasmic and nuclear levels of IL-8 and HMGB1 in prostate and ovarian cancer cells by western blotting. IL-8 is localized in the cytoplasm in both cell types, and its protein levels are significantly increased by BZ. In contrast, HMGB1 is localized in the nucleus, and BZ increases its nuclear levels only in ovarian cancer cells. The protocol includes isolation of cytoplasmic and nuclear extracts, followed by SDS electrophoresis and western blotting, and can be easily modified to analyze the cytoplasmic and nuclear cytokine levels in other cell types.

Key words Bortezomib, HMGB1, IL-8, Nuclear localization, Ovarian cancer, Prostate cancer

1 Introduction

Increased expression of the pro-inflammatory proteins, interleukin-8 (IL-8; CXCL8) and high-mobility group box-1 (HMGB1), contributes to cancer progression through their induction of tumor cell proliferation, survival, and migration. In addition, tumor-derived IL-8 and HMGB1 promote angiogenesis, induce neutrophil recruitment, and activate neutrophils and the tumor-associated macrophages to release more IL-8 and HMGB1, which further amplifies the pro-survival, pro-angiogenic, and metastatic effect. IL-8 and HMGB1 expression is increased in many types of advanced cancers, including the metastatic prostate and ovarian cancers, and correlates with poor prognosis [1–21].

Ivana Vancurova (ed.), *Cytokine Bioassays: Methods and Protocols*, Methods in Molecular Biology, vol. 1172,
DOI 10.1007/978-1-4939-0928-5_25, © Springer Science+Business Media New York 2014

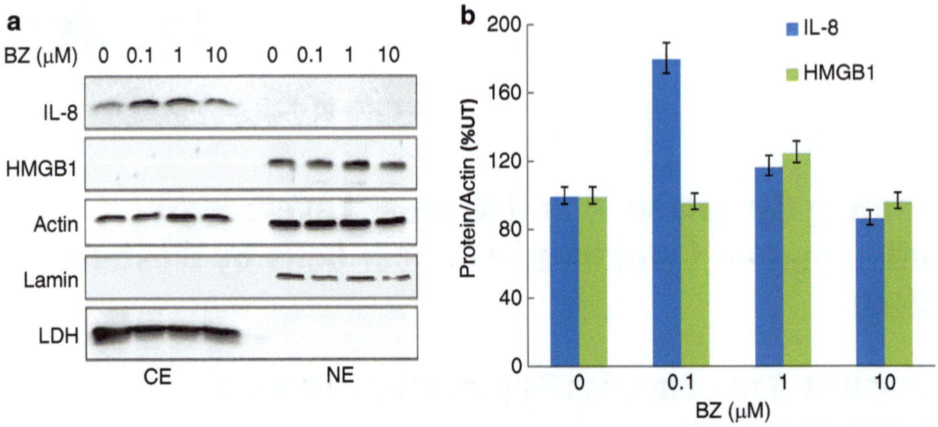

Fig. 1 Analysis of IL-8 and HMGB1 cytoplasmic and nuclear levels in BZ-treated prostate cancer cells. (**a**) Western blotting of cytoplasmic (CE) and nuclear extracts (NE) prepared from PC3 cells treated with increasing concentrations of BZ for 24 h, and analyzed by using IL-8 and HMGB1 antibodies. To confirm equal protein loading, the membrane was stripped and reprobed with actin antibody. Nuclear contamination in the cytoplasmic fraction was assessed by using lamin B-specific antibody. The presence of cytoplasmic proteins in nuclear fraction was evaluated by reprobing the membrane with lactate dehydrogenase (LDH) antibody. Each lane corresponds to approximately 5×10^4 cells. (**b**) Densitometric evaluation of IL-8 and HMGB1 levels in BZ-treated PC3 cells, illustrated in Fig. 1a. The IL-8 and HMGB1 bands were scanned and their densities were normalized to the densities of actin used as a loading control. The values for untreated cells were arbitrarily set to 100 %, and the other values are presented relative to these values. The data represent the means of three experiments ± SE

We have recently shown that proteasome inhibition by bortezomib (BZ) specifically increases IL-8 mRNA expression and protein release from metastatic prostate and ovarian cancer cells [22, 23]. In this chapter, we have analyzed the intracellular protein levels and nuclear-cytoplasmic distribution of IL-8 and HMGB1 in BZ-treated prostate and ovarian cancer cells. In both cell types, IL-8 is localized in the cytoplasm, and its protein levels are significantly increased by 0.1 μM BZ (Figs. 1 and 2), which approximately corresponds to the clinically used BZ concentrations. However, the BZ-mediated increase of IL-8 protein levels is considerably lower compared to the IL-8 mRNA expression in prostate and ovarian cancer cells [22, 23]. This is likely caused by the rapid IL-8 protein release from cells [22]. The cytoplasmic localization of IL-8 in ovarian and prostate cancer cells is in agreement with the previously described cytoplasmic localization of this cytokine [24].

In contrast to IL-8, the cellular localization of HMGB1 in prostate and ovarian cells is exclusively nuclear. Interestingly, while BZ did not significantly increase the HMGB1 nuclear levels in prostate cancer cells (Fig. 1), it increased the nuclear HMGB1 levels in ovarian cancer cells (Fig. 2). While the intracellular HMGB1 localization in most cell types is predominantly in the nucleus, where it associates with chromatin, HMGB1 can also localize in

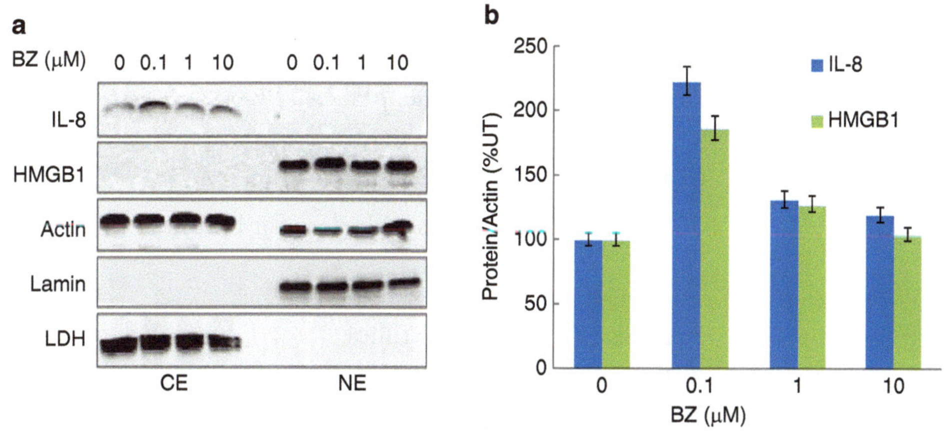

Fig. 2 Analysis of IL-8 and HMGB1 cytoplasmic and nuclear levels in BZ-treated ovarian cancer cells. (**a**) Western blotting of cytoplasmic (CE) and nuclear extracts (NE) prepared from OVCAR3 cells treated with increasing concentrations of BZ for 24 h, and analyzed by using IL-8 and HMGB1 antibodies. To confirm equal protein loading, the membrane was stripped and reprobed with actin antibody. Nuclear contamination in the cytoplasmic fraction was assessed by using lamin B-specific antibody. The presence of cytoplasmic proteins in nuclear fraction was evaluated by reprobing the membrane with lactate dehydrogenase (LDH) antibody. Each lane corresponds to approximately 5×10^4 cells. (**b**) Densitometric evaluation of IL-8 and HMGB1 levels in BZ-treated OVCAR3 cells, illustrated in Fig. 2a. The IL-8 and HMGB1 bands were scanned and their densities were normalized to the densities of actin used as a loading control. The values for untreated cells were arbitrarily set to 100 %, and the other values are presented relative to these values. The data represent the means of three experiments ± SE

the cytoplasm and be either released from cells in response to pro-inflammatory stimuli, or degraded by autophagy [17–21].

The protocol described below includes the isolation of cytoplasmic and nuclear extracts from BZ-treated prostate and ovarian cancer cells, followed by SDS electrophoresis, western blotting, and densitometric evaluation of the nuclear and cytoplasmic IL-8 and HMGB1 levels.

2 Materials

2.1 Cell Culture

1. OVCAR3 and PC3 cells (American Type Culture Collection, Manassas, VA).
2. RPMI complete medium (CM): RPMI 1640 medium supplemented with 10 % fetal bovine serum (FBS), 2 mM glutamine, 1 mM sodium pyruvate, and 1 % penicillin-streptomycin solution.
3. Bortezomib stock solution: Dissolve BZ in dimethylsulfoxide (DMSO) to a final concentration of 10 mM. Store at −20 °C.
4. 75 cm² culture flasks.

5. 0.25 % trypsin-EDTA.
6. 15 mL centrifuge tubes.
7. Phosphate buffer saline, pH 7.4 (PBS).
8. Trypan Blue solution.
9. 1.5 mL microcentrifuge tubes.
10. Standard 6-well plates with clear flat bottom.

2.2 Protein Extraction

1. 1.5 mL microcentrifuge tubes.
2. 2 mL microcentrifuge tubes.
3. Phosphate buffer saline, pH 7.4 (PBS).
4. Relaxation buffer (RB): To 50 mL of distilled water, add 1 mL of 1 M HEPES, pH 7.5, 0.25 mL of 4 M KCl, 75 μL of 4 M NaCl, 0.3 mL of 1 M $MgCl_2$, 0.2 mL of 0.5 M EDTA and 0.2 mL of 0.5 M EGTA. Make up the volume to 100 mL using distilled water and store at 4 °C.
5. 1 M dithiothreitol (DTT): Dissolve 154 mg of DTT (Cleland's reagent) in 1 mL of distilled water and store at −20 °C (*see* **Note 1**).
6. 100 mM phenylmethylsulfonylfluoride (PMSF): Dissolve 17 mg of PMSF in 1 mL of absolute ethanol and store at −20 °C (*see* **Note 2**).
7. Protease inhibitor cocktail (PIC) for mammalian cell extracts containing pepstatin A, bestatin, leupeptin, aprotinin, 4-(2-aminoethyl)-benzenesulfonyl fluoride (AEBSF), and *trans*-epoxysuccinyl-L-leucylamido(4-guanidino)butane (E-64) (*see* **Note 3**).
8. Cytoplasmic extract relaxation buffer (CERB): To 1 mL of RB, add 2 μL of 1 M DTT, 20 μL of 100 mM PMSF and 70 μL of PIC (*see* **Note 4**).
9. Nuclear Wash Buffer (NWB): To 1.5 mL of RB, add 3 μL of 1 M DTT and 30 μL of 100 mM PMSF (*see* **Note 5**).
10. 10 % Igepal CA-630: Add 1 mL of 100 % Igepal CA-630 to 9 mL of distilled water. Invert several times and store at room temperature (*see* **Note 6**).
11. 5× sample buffer (5×SB): To 6 mL of distilled water, add 2 mL of 0.5 M Tris–HCl, pH 6.8, 3.2 mL of 50 % glycerol, 3.2 mL of 10 % SDS, 0.8 mL of 1 % (w/v) bromophenol blue, and 0.8 mL of 2-mercaptoethanol (ME). Add ME just before use, and store at room temperature (*see* **Note 7**).
12. 2× sample buffer (2×SB) (*see* **Note 8**).

2.3 SDS-PAGE

1. 70 % ethanol: Add 300 mL of distilled water to 700 mL of ethanol.
2. 10 % ammonium persulfate (APS): Dissolve 100 mg of APS in 1 mL of distilled water (*see* **Note 9**).

3. 10 % (w/v) sodium dodecyl sulfate (SDS): Dissolve 10 g of SDS in 90 mL of distilled water and make up the volume to 100 mL. Store at room temperature.
4. 30 % acrylamide/*bis* solution: Dissolve 29.2 g of acrylamide and 0.8 g of *N′N′*-bis-methylene-acrylamide in 50 mL of distilled water, and make up the volume to 100 mL. Filter and store in dark at 4 °C (*see* **Note 10**).
5. Tetramethylethylenediamine (TEMED).
6. 1.5 M Tris–HCl, pH 8.8: Dissolve 18.15 g of Tris in 50 mL of distilled water, adjust to pH 8.8 using 1 M HCl, and make up the volume to 100 mL. Store at 4 °C.
7. 0.5 M Tris–HCl, pH 6.8: Dissolve 6 g of Tris in 50 mL of distilled water, adjust to pH 6.8 using 1 M HCl, and make up the volume to 100 mL. Store at 4 °C.
8. Glass plates, casting frames and stand, plastic combs, clamping frame, and electrode assembly.
9. 10× running buffer (10×RB): Dissolve 30.3 g of Tris, 144 g of glycine and 10 g of SDS in 600 mL of distilled water. Make up the volume to 1 L. Store at 4 °C.
10. Kaleidoscope prestained markers.

2.4 Western Blotting and Detection

1. Transfer Buffer (TB): Dissolve 3.03 g of Tris and 14.4 g of glycine in 500 mL of distilled water. Add 200 mL of methanol and make up the volume to 1 L using distilled water.
2. Nitrocellulose membrane.
3. Extra-thick filter paper.
4. Semi-dry transfer apparatus.
5. 10×TBS: Dissolve 81.8 g of NaCl and 3.04 g of $MgCl_2$ in 500 mL of distilled water, and add 100 mL of 1 M Tris–HCl, pH 7.5. Make up the volume to 1 L using distilled water and store at 4 °C.
6. TBST: Add 2.5 mL of 20 % Tween-20 to 100 mL of 10× TBS, and adjust the volume to 1 L with distilled water. Store at 4 °C.
7. TBSTM blocking solution: Dissolve 5 g of nonfat dry milk in 100 mL of TBST and store at 4 °C. Filter. Prepare fresh each day.
8. Small containers for incubating the membranes with antibodies.
9. IL-8, HMGB1, actin, lactate dehydrogenase (LDH), and lamin antibodies.
10. Horseradish peroxidase (HRP)-labeled secondary IgG antibodies.
11. ECL western blotting detection reagents and chemiluminescence imaging system.

12. Stripping buffer: Dissolve 7.5 g of glycine, 0.5 g of SDS and 5 mL of Tween-20 in 400 mL of distilled water, adjust the pH to 2.2 with 1 M HCl, and make up the volume to 500 mL. Prepare fresh just before use.

3 Methods

In this section, we describe the protocol for analysis of cytoplasmic and nuclear levels of IL-8 and HMGB1 in BZ-treated prostate and ovarian cancer cells. The main steps of the protocol are: (1) PC3 and OVCAR3 cell culture and incubation with BZ; (2) preparation of cytoplasmic (CE) and nuclear protein extracts (NE); (3) SDS-PAGE, and (4) western blotting, detection and analysis (Fig. 3). This protocol can be easily modified and used for analysis of different intracellular proteins in different cell types. On average, it can be accomplished within 4 days.

3.1 Cell Culture

1. Grow PC3 and OVCAR3 cells in CM medium until they reach about 90 % confluency (*see* **Note 11**).
2. Discard the media. Trypsinize the cells by adding 4 mL of 0.25 % trypsin-EDTA and 15-min incubation at 37 °C in a 5 % CO_2 humidified atmosphere (*see* **Note 12**).
3. Neutralize the trypsin by adding 8 mL of CM, transfer the cell suspension to a 15 mL centrifuge tube, and centrifuge at 130 × g for 5 min at room temperature.
4. Discard the supernatants and resuspend the cell pellets in 3 mL of CM.

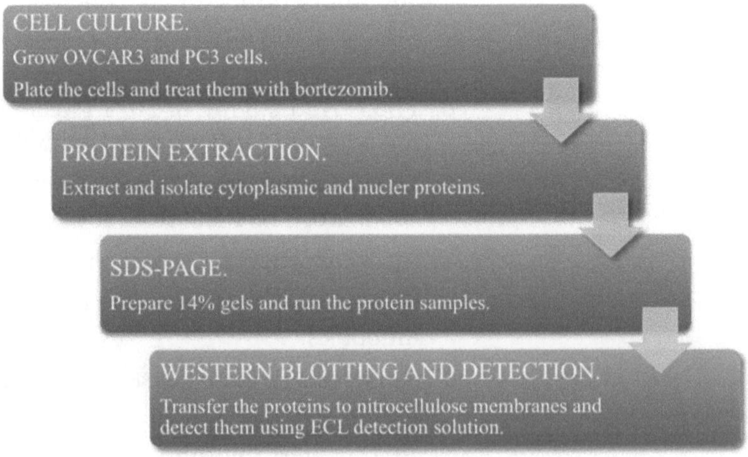

Fig. 3 Schematic representation of the steps involved in the protocol

5. To count the cells, transfer 50 μL of the above cell suspension into a 1.5 mL microcentrifuge tube, add 50 μL of 1×PBS and 100 μL of Trypan Blue solution. Mix well (*see* **Note 13**).
6. Add 10 μL of the Trypan Blue cell suspension obtained from **step 5** to each chamber of the hemocytometer.
7. Count the number of viable cells, and calculate the cell concentration using the following formula: Cell concentration = average cell count in four squares $\times 4 \times 10^4$ cells/mL.
8. Dilute the cells suspension obtained from **step 4** to a final concentration of 0.5×10^6 cells/mL using CM.
9. Mix well, and plate 2 mL of the above cell suspension into each well of a 6 well plate, so that each well has 1×10^6 cells.
10. Incubate the plate 24 h at 37 °C in a 5 % CO_2 humidified atmosphere tissue culture incubator, to allow cell attachment.
11. After 24 h, discard the medium, and add 2 mL of new CM containing BZ of the required concentration (0, 0.1, 1, and 10 μM) into each well.
12. Incubate cells for 24 h at 37 °C in 5 % CO_2 humidified atmosphere.

3.2 Protein Extraction

All following steps are performed on ice or at 4 °C, unless otherwise stated.

1. Following cell incubation, collect 2 mL of unattached cell suspension from each well, and transfer it into two labeled 1.5 mL microcentrifuge tubes; each tube will get 1 mL of the cell suspension. Centrifuge at 1,700 × *g* for 5 min (*see* **Note 14**).
2. Detach the adherent cells still in the wells by adding 800 μL of 0.25 % trypsin-EDTA, followed by 10 min incubation at 37 °C in 5 % CO_2 humidified atmosphere.
3. After 10 min, neutralize trypsin by adding 700 μL of CM into each well.
4. Carefully discard supernatants from the centrifuged 1.5 mL microcentrifuge tubes (from **step 1**), and resuspend the cell pellets in 1.5 mL of the trypsinized cell solution from **step 3**. Centrifuge at 1,700 × *g* for 5 min.
5. Discard supernatants, and wash the cells using 1 mL of ice-cold 1× PBS. Centrifuge at 1,700 × *g* for 5 min.
6. Carefully discard supernatants, resuspend the pellets in 150 μL of ice-cold CERB, and let stand on ice for 15 min (*see* **Note 15**).
7. Start boiling water in a 1 L beaker (to be used later, in **steps 10** and **14**).
8. At the end of the 15 min cell incubation on ice (from **step 6**), add 8 μL of 10 % Igepal CA-630 to each tube, vortex vigorously for 10 s and let stand on ice for 10 s (*see* **Note 16**).

9. Centrifuge the permeabilized cells at 200 × *g* for 5 min. After centrifuging, keep tubes on ice.
10. Transfer the cytoplasmic supernatants (approximately 150 μL) into new 1.5 mL microcentrifuge tubes, add 50 μL of 5×SB and boil for 7 min (*see* **Note 17**).
11. After boiling, centrifuge tubes at 1,700 × *g* for 5 min at room temperature.
12. Transfer supernatants into new 1.5 mL tubes, label as cytoplasmic extract (CE) and store at −70 °C.
13. Wash the nuclear pellets from **step 9** with 250 μL of ice-cold NWB, and centrifuge at 1,700 × *g* for 5 min.
14. Discard supernatants, and resuspend the nuclear pellets in 50 μL of 2×SB. Vortex well, and boil immediately for 7 min.
15. Centrifuge at 6,700 × *g* for 5 min at room temperature. Transfer supernatants into new 1.5 mL microcentrifuge tubes. Label as nuclear extract (NE), and store at −70 °C.

3.3 SDS-PAGE

1. Clean electrophoresis glass plates and combs using 70 % ethanol.
2. Place the plates in the casting frame and fasten it to the casting stand (*see* **Note 18**).
3. To prepare two 14 % resolving gels, mix in a beaker: 2.7 mL of distilled water, 4.7 mL of 30 % acrylamide/*bis*, 2.5 mL of 1.5 M Tris–HCl, pH 8.8, and 0.1 mL of 10 % SDS.
4. Swirl the ingredients, and just before casting the gels, add 5 μL of TEMED and 50 μL of 10 % APS to induce polymerization. Mix well again.
5. Immediately after adding TEMED and APS, pour the gels using a micropipette. Carefully overlay the gels with distilled water to even the gel tops and remove air bubbles.
6. Allow the resolving gels to polymerize for 30 min.
7. Once the gels are polymerized, remove water on the top by inverting the casting stand and using Kim wipes.
8. Prepare stacking gels (4 %) by mixing 6.1 mL of distilled water, 1.3 mL of 30 % acrylamide/*bis*, 2.5 mL of 0.5 M Tris–HCl, pH 6.8, and 0.1 mL of 10 % SDS.
9. Swirl the ingredients, and add 10 μL of TEMED and 50 μL of 10 % APS.
10. Add combs to gels in the casting frame, and fill the remaining spaces by adding the stacking gel solution using a micropipette (*see* **Note 19**).
11. Allow stacking gels to polymerize for 40 min.
12. Meanwhile, thaw the frozen protein samples by boiling them for 7 min.

13. Place the plates in the running module and assemble them in the gel tank.
14. Fill the inner chamber with 1×RB, and gently remove the combs.
15. Clean the wells with 1×RB using a syringe.
16. Using gel loading tips, add the CE and NE samples, as well as prestained markers into the wells (*see* **Note 20**).
17. Fill the tank with 1×RB.
18. Run gels at 120 V until the blue dye reaches the bottom of the gel (*see* **Note 21**).

3.4 Western Blotting and Detection

1. Dissemble plates, discard the stacking gels, and soak the resolving gels each in 400 mL of TB for 10 min.
2. Soak two filter papers and one nitrocellulose membrane per gel in TB for 10 min.
3. Clean the anode platform of semi-dry transfer apparatus with 70 % ethanol and assemble the sandwich in the order from bottom to top (*see* **Note 22**):
   (a) Filter paper (top)
   (b) Gel
   (c) Nitrocellulose membrane
   (d) Filter paper (bottom)
4. Place lid on top, and run the transfer according to the parameters below:
   (a) One 14 % gel: 18 V, 3 A, 25 min
   (b) Two 14 % gels: 20 V, 3 A, 40 min
5. After transfer, disassemble the sandwich, and cut the nitrocellulose membrane with transferred proteins to the size of the blocking container, to minimize the amount of needed antibody (*see* **Note 23**).
6. Transfer membranes to containers for incubation with antibodies, and add blocking solution (TBSTM) to each container (*see* **Notes 24** and **25**).
7. Incubate membranes in the blocking solution TBSTM for 2 h at room temperature, with gentle rocking (*see* **Note 26**).
8. After 2 h, remove the blocking solution, and add primary antibody diluted in TBSTM solution. Each primary antibody requires different dilution. For IL-8 and HMGB1, we use 1:100 dilutions.
9. Incubate membranes with primary antibodies 2 h at room temperature, followed by overnight incubation at 4 °C, with gentle rocking.

10. After incubation with primary antibody, remove the antibody solution, and wash the membranes six times in 50 mL of TBST, 5 min each wash.
11. Transfer membranes into new blocking containers, and add secondary HRP-labeled antibody diluted 1:2,000 in TBSTM (*see* **Note 27**).
12. Incubate on a rocking platform for 1 h at room temperature.
13. Remove the secondary antibody solution, and wash the membranes six times with TBST buffer, 5 min each wash.
14. Mix required amounts of ECL solution A and solution B in a 15 mL centrifuge tube as shown below (*see* **Note 28**):
    (a) 1 membrane: 2 mL of solution A and 50 μL of solution B.
    (b) 2 membranes: 3 mL of solution A and 75 μL of solution B.
15. Carefully blot the membranes to remove any remaining TBST, place them on parafilm, and add enough ECL solution A + B to cover the surface of the membranes. Incubate at room temperature for 5 min.
16. Cover the membranes with plastic wrap and visualize proteins using chemiluminesence imaging system.
17. To probe the membranes with other antibodies, incubate them twice in 50 mL of stripping buffer (10 min each wash), followed by two 5 min washes in TBST.
18. Block the membranes for 2 h in 5 % TBSTM at room temperature with gentle rocking.
19. Incubate the membranes with primary and secondary antibodies as described above.

4 Notes

1. DTT, being a strong reducing agent, is used to prevent formation of intramolecular and intermolecular disulphide bonds in proteins.
2. PMSF, being unstable in water, should be mixed in absolute alcohol (ethanol, methanol, or isopropanol).
3. PIC is a broad-spectrum protease inhibitor, inhibiting serine, cysteine, and aspartic proteases, and amino peptidases.
4. Prepare CERB buffer fresh just before each experiment, by adding DTT, PMSF, and PIC to 1 mL of RB. Keep it on ice.
5. The NWB buffer is the same as the CERB buffer, but it does not contain PIC. It should be prepared fresh just before use, and kept on ice.

6. Igepal CA-630 is a nonionic, nondenaturing detergent, which is chemically indistinguishable from Nonidet P-40. It permeabilizes the plasma membrane.
7. 5×SB (without 2-mercaptoethanol) can be prepared and stored at room temperature in dark. Before each experiment, use the required amount of 5×SB, and add 2-mercaptoethanol.
8. Prepare required volumes of 2×SB by diluting 5×SB with distilled water.
9. Always prepare fresh APS. Old APS results in incomplete catalysis and long polymerization time.
10. Unpolymerized acrylamide is toxic and can be easily absorbed by skin. Proper gloves should be worn while handling the gels.
11. It is optimum to grow the cells until they reach about 90 % confluency.
12. Do not "over-thaw" the trypsin-EDTA solution; remove it from water bath right after it melts. Increasing temperature of trypsin decreases its activity.
13. Make sure to mix the cells evenly by pipetting up and down. Uneven mixing results in incorrect cell counts.
14. It is advisable to centrifuge all tubes in one position—hinges either up or down, so that the pellets can be easily seen.
15. CERB is a hypotonic solution; cells will swell at the end of this step.
16. Cell exposure to 10 % Igepal CA-630 for longer times results in the disruption of nuclear membrane, and loss of nuclear proteins.
17. Boiling the protein samples in sample buffer (2×SB or 5×SB) denatures proteins, so that the samples can be stored for longer periods. It is better to aliquot the samples in smaller volumes, so that they are not repeatedly thawed.
18. Make sure that the plates are leak-proof by aligning them correctly. Otherwise, the gel solution might leak.
19. Avoid air bubbles, which could result in improper well formation.
20. Load markers in different positions in the two gels, so that they can be used to differentiate the gels.
21. Keep checking the gel run. Fall in the level of the running buffer halts the run.
22. Avoid air bubbles, especially between the gel and the membrane. Bubbles result in improper transfer of proteins.
23. To avoid cutting out an important piece of membrane with transferred proteins, cut along the outer borders of the markers and the blue dye on the bottom of the gel.

24. For blocking, use either TBSTM, or 5 % BSA in TBST, which works better for most phosphoproteins.
25. The membrane has to be completely covered with the blocking solution. Typically, 6 mL of blocking solution will be sufficient for one membrane in a small container.
26. Western blotting can be interrupted at this point, by blocking overnight at 4 °C.
27. Always use the secondary antibody raised against animal species in which the primary antibody was raised.
28. Since ECL solutions are light sensitive, keep the solution mixture in dark.

Acknowledgement

This work was supported by NIH grant CA173452 to I. Vancurova.

References

1. Xu L, Fidler IJ (2000) Interleukin 8: an autocrine growth factor for human ovarian cancer. Oncol Res 12:97–106
2. Waugh DJ, Wilson C (2008) The interleukin-8 pathway in cancer. Clin Cancer Res 14: 6735–6741
3. Balkwill F, Mantovani A (2001) Inflammation and cancer: back to Virchow? Lancet 357: 539–545
4. Balkwill F (2004) Cancer and the chemokine network. Nat Rev Cancer 4:540–550
5. Mantovani A, Allavena P, Sica A et al (2008) Cancer-related inflammation. Nature 454:436–444
6. Araki S, Omori Y, Lyn D et al (2007) Interleukin-8 is a molecular determinant of androgen independence and progression in prostate cancer. Cancer Res 67:6854–6862
7. Singh RK, Lokeshwar BL (2011) The IL-8-regulated chemokine receptor CXCR7 stimulates EGFR signaling to promote prostate cancer growth. Cancer Res 71:3268–3277
8. Chen J, Xi B, Zhao Y et al (2012) High-mobility group protein B1 (HMGB1) is a novel biomarker for human ovarian cancer. Gynecol Oncol 126:109–117
9. Scaffidi P, Misteli T, Bianchi ME (2002) Release of chromatin protein HMGB1 by necrotic cells triggers inflammation. Nature 418:191–195
10. Lotze MT, Tracey KJ (2005) High-mobility group box 1 protein (HMGB1): nuclear weapon in the immune arsenal. Nat Rev Immunol 5:331–342
11. Bell CW, Jiang W, Reich CF III et al (2006) The extracellular release of HMGB1 during apoptotic cell death. Am J Physiol Cell Physiol 291:1318–1325
12. Sha Y, Zmijewski J, Xu Z et al (2008) HMGB1 develops enhanced proinflammatory activity by binding to cytokines. J Immunol 180:2531–2537
13. Tang D, Kang R, Cheh CW et al (2010) HMGB1 release and redox regulates autophagy and apoptosis in cancer cells. Oncogene 29:5299–5310
14. Tang D, Kang R, Livesey KM et al (2010) Endogenous HMGB1 regulates autophagy. J Cell Biol 190:881–892
15. Gnanasekar M, Kalyanasundaram R, Zheng G et al (2013) HMGB1: a promising therapeutic target for prostate cancer. Prostate Cancer 2013:1–8
16. Chen J, Liu X, Zhang J et al (2012) Targeting HMGB1 inhibits ovarian cancer growth and metastasis by lentivirus-mediated RNA interference. J Cell Physiol 227:3629–3638
17. Lu B, Nakamura T, Inouye K et al (2012) Novel role of PKR in inflammasome activation and HMGB1 release. Nature 488:670–674
18. Lu B, Wang H, Andersson U et al (2013) Regulation of HMGB1 release by inflammasomes. Protein Cell 4:163–167

19. Yang H, Antoine DJ, Andersson U et al (2013) The many faces of HMGB1: molecular structure-functional activity in inflammation, apoptosis, and chemotaxis. J Leukoc Biol 93:865–873
20. Guo ZS, Liu Z, Bartlett DL et al (2013) Life after death: targeting high mobility group box 1 in emergent cancer therapies. Am J Cancer Res 3:1–20
21. Li W, Zhu S, Li J et al (2011) EGCG stimulates autophagy and reduces cytoplasmic HMGB1 levels in endotoxin-stimulated macrophages. Biochem Pharmacol 81:1152–1163
22. Manna S, Singha B, Phyo SA et al (2013) Proteasome inhibition by bortezomib increases IL-8 expression in androgen-independent prostate cancer cells: the role of IKKα. J Immunol 191:2837–2846
23. Singha B, Gatla H, Manna S et al (2014) Proteasome inhibition increases recruitment of IKKβ, S536P-p65 and transcription factor EGR1 to interleukin-8 (IL-8) promoter, resulting in increased IL-8 production in ovarian cancer cells. J Biol Chem 289: 2687–2700
24. Miskolci V, Ghosh CC, Rollins J et al (2006) TNFα release from peripheral blood leukocytes depends on a CRM1-mediated nuclear export. Biochem Biophys Res Commun 351:354–360

Chapter 26

Western Analysis of Intracellular Interleukin-8 in Human Mononuclear Leukocytes

Veronika Miskolci, Louis Hodgson, Dianne Cox, and Ivana Vancurova

Abstract

Most cytokines are stored in the cytoplasm until their release into the extracellular environment; however, some cytokines have been reported to localize in the nucleus. Traditional whole cell extract preparation does not provide information about the intracellular localization of cytokines. Here, we describe how to prepare cytoplasmic and nuclear extracts that can be analyzed by immunoblotting. While in this chapter we use this method to analyze intracellular localization of interleukin-8 (IL-8) in human mononuclear leukocytes, this protocol is adaptable to any cell type or protein of interest.

Key words Cytokines, Interleukin-8, Intracellular localization, Cytoplasmic extract, Nuclear extract, Immunoblotting

1 Introduction

Interleukin-8 (IL-8) is a potent pro-inflammatory chemokine. Based on its structure it belongs to the CXC chemokine class and is also denoted as CXCL8. The expression of IL-8 is mainly regulated by NFκB transcription factor and it is synthesized as 99-amino acid precursor that is subsequently processed to a 72-amino acid mature form. While it is produced by many cell types, leukocytes are the major source of secreted IL-8 [1, 2]. Secretion of IL-8 can be stimulated by many different signals, including bacterial lipopolysaccharide (LPS) or other pro-inflammatory signals such as Tumor Necrosis Factor (TNFα). Extracellular IL-8 binds G-protein-coupled receptors CXCR1 and CXCR2 leading to the activation of multiple downstream signaling pathways including pathways that regulate cell proliferation and survival [2]. IL-8 plays a critical role in the immune response of the host by mediating the recruitment of neutrophils and monocytes to sites of inflammation. However, the unresolved secretion of IL-8 can lead to the development of chronic inflammatory diseases such as acute respiratory distress syndrome [3, 4]. IL-8 expression has also been

Ivana Vancurova (ed.), *Cytokine Bioassays: Methods and Protocols*, Methods in Molecular Biology, vol. 1172,
DOI 10.1007/978-1-4939-0928-5_26, © Springer Science+Business Media New York 2014

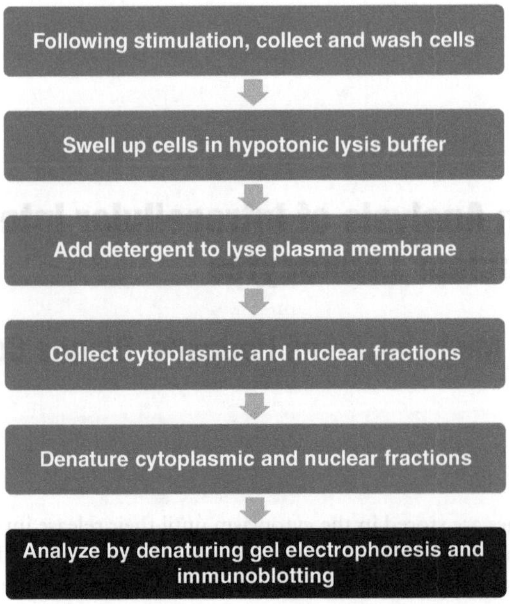

Fig. 1 Schematic outline of preparing cytoplasmic and nuclear extracts

shown to be associated with the progression of cancer [2], therefore the regulation of IL-8 is an important area of research.

There are a variety of methods to study and measure secreted cytokines as described in other chapters of this book. This chapter describes the detection of intracellular cytokines by immunoblotting. Unlike immunofluorescence microscopy techniques that detect intracellular cytokines in their native forms, immunoblotting uses denaturing conditions. Most cytokines are localized in the cytoplasm before secretion, however some exhibit nuclear localization, such as HMGB1, IL-1α, and TNFα [5–7]. Whole cell extract preparations represent total protein expression levels, while cytoplasmic and nuclear extracts can provide information about localization in these cellular compartments. We have previously shown cytoplasmic localization of IL-8 by subcellular fractionation [7]. Here we describe how to prepare cytoplasmic and nuclear extracts, and analyze IL-8 expression in LPS-stimulated human peripheral blood mononuclear cells (PBMCs) by immunoblotting. The preparation of cytoplasmic and nuclear extract is outlined in Fig. 1; it involves the sequential lysing of plasma and nuclear membranes in order to separate the two cellular compartments. First, cells are incubated in hypotonic lysis buffer without detergent to swell up the cells. The plasma membrane of the swollen cells is then lysed by adding nonionic, nondenaturing detergent and the nuclei are pelleted. Cytoplasmic fraction (supernatant) is collected, and both the cytoplasmic fraction and the nuclei are denatured in sample buffer. The denatured proteins can be resolved on denaturing polyacrylamide gel, transferred to nitrocellulose or PVDF

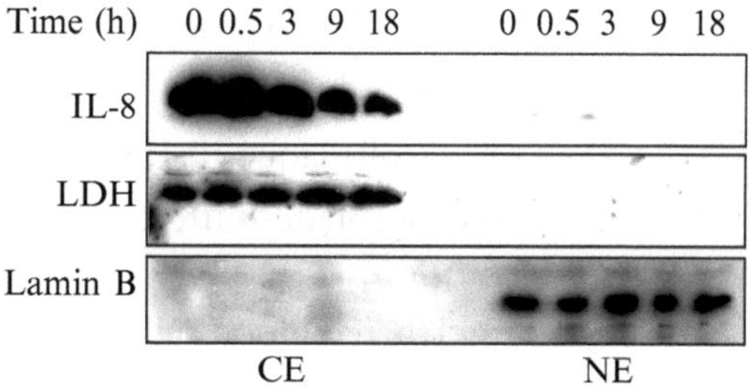

Fig. 2 Intracellular localization of IL-8 in LPS-stimulated human PBMC. Western blot analysis of cytoplasmic (CE) and nuclear extracts (NE) prepared from human PBMC (5×10^6) stimulated with LPS (100 ng/mL) over time as indicated. CE and NE were resolved on 14 % denaturing polyacrylamide gels and transferred to nitrocellulose membrane. IL-8 expression was analyzed by using polyclonal IL-8 antibody (Santa Cruz, sc-7922). The purity of CE and NE was monitored by LDH and lamin B antibodies, respectively. Each lane contains approximately 5×10^5 cells

membrane, and protein expression analyzed by immunoblotting. This chapter focuses on the analysis of intracellular IL-8 in human PBMCs (Fig. 2); however, this protocol can be adapted to any cell type to analyze any intracellular protein.

2 Materials

2.1 Stock Solutions

All solutions are prepared in double-distilled water and stored at room temperature unless noted otherwise.

1. 1 M Hepes, pH 7.5, store at 4 °C.
2. 1 M KCl.
3. 1 M NaCl.
4. 1 M $MgCl_2$.
5. 0.5 M EGTA.
6. 0.5 M EDTA, pH 8.0: To prepare 100 mL, add 20.81 g of endotoxin-tested EDTA (MW 416.2) to ~50 mL of cell-culture grade water. Add NaOH pellets as necessary to start dissolving EDTA. EDTA will not dissolve until the pH is close to 8.0. Adjust pH to 8.0, bring to final volume. Filter. Store at room temperature.
7. 0.5 M Tris–HCl, pH 6.8.
8. 50 % glycerol.
9. 10 % SDS.
10. 1 % (w/v) Bromophenol blue.
11. 10 % NP-40 (Nonidet P-40, can be substituted by IGEPAL CA-630).

2.2 Isolation of Peripheral Blood Mononuclear Cells (PBMCs)

1. 15 mL polypropylene centrifuge tubes, sterile and pyrogen-free.
2. Transfer pipette, sterile and pyrogen-free.
3. Ficoll-Paque PLUS (endotoxin-tested). Store at 4 °C.
4. Phosphate-buffered saline without Ca^{2+} and Mg^{2+} (PBS), pH 7.4, sterile.
5. 10× red blood cell (RBC) lysis buffer: 1.55 M NH_4Cl, 100 mM $NaHCO_3$, 1 mM EDTA. To prepare 100 mL of 10× stock, dissolve 8.291 g NH_4Cl and 0.8401 g of $NaHCO_3$ (endotoxin-tested) in cell-culture grade water. Add 1 mL of 0.5 M EDTA, pH 8.0. Adjust pH to 7.4. Bring to 100 mL and filter sterilize. Store at 4 °C. Use at 1× to lyse red blood cells.
6. Culture medium: RPMI 1640 supplemented with heat-inactivated fetal bovine serum (FBS) at 10 %. Penicillin-streptomycin at 1 % is optional.
7. Wright/Giemsa stain.

2.3 Stimulation of PBMCs

1. 1.5 mL microcentrifuge tubes, sterile and pyrogen-free.
2. Lipopolysaccharide reconstituted at 1 mg/mL with PBS and stored at −20 °C.

2.4 Preparation of Cytoplasmic and Nuclear Extracts

1. Hypotonic lysis buffer (detergent-free): 10 mM Hepes, pH 7.5, 10 mM KCl, 3 mM NaCl, 3 mM $MgCl_2$, 1 mM EDTA, 1 mM EGTA. Filter. Store at 4 °C. Add inhibitors just before use.
2. Protease inhibitors stock concentration: 1 M dithiothreitol (DTT; 500×), 100 mM phenylmethylsulfonyl fluoride (PMSF) prepared in absolute ethanol (50×), 20 mM sodium orthovanadata (20×), protease inhibitor cocktail for mammalian cell extract (10×).
3. 5× sample buffer: 62.5 mM Tris–HCl, pH 6.8, 10 % (v/v) glycerol, 2 % (w/v) SDS, 0.01 % (w/v) bromophenol blue, 5 % (v/v) β-mercaptoethanol. β-mercaptoethanol is added right before use. Stock without β-mercaptoethanol can be stored at room temperature.

3 Methods

3.1 Isolation of PBMCs

This section describes the procedure to isolate PBMCs from freshly collected, heparinized whole blood (*see* **Note 1**). PBMCs are mixture of lymphocytes and monocytes (*see* **Note 2**). It is not necessary to perform this procedure in cell culture hood unless cells will be cultured for longer than 24 h; however, it is critical to minimize pre-activation of cells prior to stimulation. To prevent pre-activation, use all materials that come into contact with cells, including solutions and plastic ware, certified endotoxin-tested/pyrone-free; check to use cell-culture grade (endotoxin-tested) powdered chemicals in preparing stock solutions. In addition, proceed in a timely fashion and keep isolated PBMCs cold until stimulation.

3.1.1 Ficoll Gradient Sedimentation

1. Add 5 mL of Ficoll to 15 mL tube.
2. Carefully layer 8 mL of undiluted heparinized blood over 5 mL of Ficoll using a pyrogen-free transfer pipette (*see* **Note 3**).
3. Centrifuge at 400 × *g* for 40 min. For the remaining of the procedure, keep cells and all solutions cold on ice. This is to minimize pre-activation of cells.
4. Collect the interphase layer (between serum and Ficoll layers) using a transfer pipette into a new 15 mL tube (*see* **Note 4**).
5. Bring up to 10 mL with cold PBS.
6. Invert few times to rinse.
7. Spin cells at 300 × *g* for 10 min.
8. Discard supernatant.

3.1.2 RBC Lysis (Optional, See Note 5)

1. Add 10 mL of cold 1× RBC lysis buffer to cell pellet.
2. Resuspend cell pellet using a transfer pipette.
3. Gently mix by inverting tube 10–20 times (*see* **Note 6**).
4. Centrifuge at 300 × *g* 10 min.
5. Discard supernatant.

3.1.3 Washing

1. Add 10 mL of cold PBS to cell pellet.
2. Resuspend cell pellet using a transfer pipette.
3. Invert few times to rinse cells.
4. Centrifuge at 200 × *g* for 10 min (*see* **Note 7**).
5. Discard supernatant.
6. Repeat washing twice more (*see* **Note 8**).

3.1.4 Resuspending Cells in Culture Medium

1. After the last wash, discard supernatant, resuspend cell pellet in residual fluid and pool cells if there are multiple 15 mL tubes.
2. Bring to 5 mL with cold culture medium.
3. Count cells; use trypan blue to determine viability. Adjust volume for final cell concentration of 5×10^6 cells/mL (or as optimal for cell type used).
4. To determine PBMC purity by differential morphology, smear ~100 μL of cell suspension onto a glass slide and air dry. Stain with Wright/Giemsa stain according to manufacturer's instructions. Count at least 100 cells. Expected purity is ~95 %.

3.2 Stimulation of PBMCs

1. Add 1 mL of cell suspension to 1.5 mL pyrogen-free test tubes (*see* **Note 9**).
2. Transfer tubes to 37 °C water bath. Equilibrate tubes for 15 min to bring to 37 °C.
3. Add LPS to tubes at a final concentration of 100 ng/mL. Invert tubes to mix and start timing stimulation and place back into water bath.
4. Incubate until desired times.
5. Place tubes on ice to cool (*see* **Note 10**).

3.3 Preparation of Cytoplasmic (CE) and Nuclear Extracts (NE)

Keep cells and solutions cold on ice at all times. All centrifugation steps are performed at 4 °C. Volumes can be scaled to number of cells (e.g., for 2.5×10^6 cells use half the volume at each step).

1. Centrifuge cells for 3 min at $300 \times g$ at 4 °C in a microcentrifuge.
2. Aspirate supernatant. Wash cells by adding 1 mL of cold PBS and resuspend cell pellet by pipetting up and down. Centrifuge as in **step 1** (*see* **Note 11**)
3. While the cells are centrifuging add protease inhibitors to hypotonic lysis buffer (*see* **Note 12**).
4. Aspirate supernatant and add 150 μL of hypotonic lysis buffer (with protease inhibitors). Resuspend cell pellet gently by pipetting up and down. Keep on ice for 15 min (*see* **Note 13**).
5. Add 8 μL of 10 % NP-40, at a final concentration of 0.5 %. Vortex vigorously for 10 s. Keep on ice for 10 s (*see* **Note 14**).
6. Centrifuge for 5 min at $845 \times g$ at 4 °C in a microcentrifuge (*see* **Note 15**). Place tubes back on ice. Keep both the supernatant and the pellet.
7. Transfer supernatant to new 1.5 mL microcentrifuge tube. This is the cytoplasmic extract (CE) fraction.
8. The pellet is the nuclear fraction. Wash nuclear pellet by adding 300 μL of hypotonic lysis buffer (with protease inhibitors) to the nuclear pellet and resuspend gently by pipetting up and down. Centrifuge 5 min at $2400 \times g$ in a microcentrifuge.
9. Save 5 μL of the cytoplasmic fraction to measure protein concentration.
10. Add 36 μL of 5× sample buffer to the remaining cytoplasmic fraction and heat immediately for 5 min at 95 °C (*see* **Note 16**). Spin and store CE at −20 °C.
11. Aspirate supernatant from the washed nuclear pellet. Resuspend the nuclear pellets in 50 μL of 2× sample buffer. Vortex well. Heat immediately for 5 min at 95 °C. Centrifuge 5 min at $9400 \times g$ in a microcentrifuge. Transfer supernatant (NE) to a new tube and store at −20 °C.

12. Determine protein concentration of cytoplasmic fraction. If your laboratory does not have a routine method already to measure protein concentration, there are commercially available kits such as Coomassie Plus.

3.4 Immunoblotting

1. Due to the small size of cytokines (IL-8 is ~8 kDa), use 14 or 15 % polyacrylamide gel.
2. Load equal protein amounts per lane; volumes are calculated based on protein concentrations. The recommended volumes for loading are 20 μL of CE and 5 μL of NE, equivalent of $\sim 5 \times 10^5$ cells.
3. Proteins can be transferred onto nitrocellulose or PVDF membrane.
4. Monitor purity of cytoplasmic and nuclear fractions by analyzing lactate dehydrogenase (LDH; cytoplasmic-specific) and lamin B (nuclear-specific) expression, as shown in Fig. 2.
5. Detection can be done by preferred method; chemiluminescence was used in Fig. 2.

4 Notes

1. The procedure to collect whole blood must be reviewed and approved by the respective Institutional Review Board. Start procedure to isolate PBMCs immediately after blood was collected. Keep whole blood at room temperature until isolation.
2. Monocytes and lymphocytes can be separated using additional methods [8].
3. When layering whole blood over Ficoll, make sure not to break the interface between Ficoll and blood; mixing Ficoll and whole blood will destroy the gradient and prevent successful PBMC isolation.
4. Ficoll gradient creates four distinct layers: the top yellow layer contains serum, followed by the clear Ficoll layer in the middle, and red layer at the bottom containing RBCs and granulocytes. There is a thin, cloudy layer between the serum and Ficoll layers. It is possible to directly collect the interphase using a transfer pipette. Alternatively, serum layer may be removed first. Take care to minimize collecting the Ficoll layer as this layer can be a source of non-PBMC cells.
5. RBC lysis is optional. There are usually little contaminating RBCs in the interphase layer collected from adult blood; however, the interphase is usually heavily contaminated from a cord blood source.

6. This step is time sensitive. Time in RBC lysis buffer should be minimized to prevent lysis of non-RBCs.
7. The slower speed allows removal of contaminating platelets.
8. At this point the cell suspension can be used separate monocytes or lymphocytes [8].
9. Alternatively, cells can be stimulated in cell culture dishes. If preparing whole cell extracts, cells may be directly lysed in the cell culture dish after stimulation. However, for preparing cytoplasmic and nuclear extracts, cells have to be collected into 1.5 mL centrifuge tubes first.
10. Alternatively, sample can be quick-spun for 10 s and lysed immediately. The time to quick-spin and aspirate supernatant is part of the incubation time, and the hypotonic lysis buffer is added exactly at the end of the incubation time point.
11. This step is optional.
12. Inhibitors should be added right before use as some protease inhibitors have short half-lives once in aqueous solution; e.g., PMSF has a half-life of ~30 min. If the time points are more than 30 min apart we recommend adding inhibitors freshly at each time point. Each cell type has its own unique combination of proteases. The inhibitors listed here are minimal required; hypotonic lysis buffer can be supplemented with additional protease inhibitors and optimized for working concentrations as necessary depending on the cell type used.
13. During the incubation in hypotonic lysis buffer cells will swell up; do not exceed more than 30 min at this step as it may result in partial lysis of cells and loss of cytoplasmic material.
14. This step is time sensitive. Do not exceed the indicated times as NP-40 may start lysing the nucleus as well, resulting in the cross-contamination of the cytoplasmic fraction.
15. The speed used to pellet the nuclei may have to be optimized for other cell types, e.g., 200 × *g* in a microcentrifuge is sufficient for human neutrophils.
16. If a heat block is not available, tubes may be suspended in boiling water.

Acknowledgment

V.M. is supported by Cellular, Molecular Biology and Genetics training grant T32GM007491.

References

1. Remick DG (2005) Interleukin-8. Crit Care Med 33:S466–S467
2. Waugh DJ, Wilson C (2008) The interleukin-8 pathway in cancer. Clin Cancer Res 14: 6735–6741
3. Puneet P, Moochhala S, Bhatia M (2005) Chemokines in acute respiratory distress syndrome. Am J Physiol Lung Cell Mol Physiol 288:L3–L15
4. Harada A, Sekido N, Akahoshi T et al (1994) Essential involvement of interleukin-8 (IL-8) in acute inflammation. J Leukoc Biol 56: 559–564
5. Wang H, Yang H, Czura CJ et al (2001) HMGB1 as a late mediator of lethal systemic inflammation. Am J Respir Crit Care Med 164:1768–1773
6. Wessendorf JH, Garfinkel S, Zhan X et al (1993) Identification of a nuclear localization sequence within the structure of the human interleukin-1 alpha precursor. J Biol Chem 268: 22100–22104
7. Miskolci V, Ghosh CC, Rollins J et al (2006) TNFalpha release from peripheral blood leukocytes depends on a CRM1-mediated nuclear export. Biochem Biophys Res Commun 351: 354–360
8. Wahl LM, Wahl SM, Smythies LE et al (2006) Isolation of human monocyte populations. Curr Protoc Immunol Chapter 7: Unit 7.6A

Chapter 27

Quantitative Analysis of Bortezomib-Induced IL-8 Gene Expression in Ovarian Cancer Cells

Bipradeb Singha, Sai A. Phyo, Himavanth R. Gatla, and Ivana Vancurova

Abstract

Interleukin-8 (IL-8), originally discovered as the neutrophil chemoattractant and inducer of leukocyte-mediated inflammation, contributes to cancer progression through its induction of tumor cell proliferation, survival, and migration. IL-8 expression is increased in many types of advanced cancers, including ovarian cancer, and correlates with poor prognosis.

Bortezomib (BZ) is the first FDA-approved proteasome inhibitor that has shown remarkable antitumor activity in multiple myeloma and other hematological malignancies. In solid tumors, including ovarian carcinoma, BZ has been less effective as a single agent; however, the mechanisms remain unknown. We have recently shown that in ovarian cancer cells, BZ greatly increases IL-8 expression, while expression of other NFκB-regulated cytokines, IL-6 and TNF, is unchanged. In this chapter, we describe a protocol that uses real-time qRT-PCR to quantitatively analyze mRNA levels of IL-8 and IL-6 in BZ-treated ovarian cancer cells. The protocol can be easily modified and used for analysis of other cytokines in different cell types.

Key words Bortezomib, Interleukin-8, Interleukin-6, NFκB, Ovarian cancer, Proteasome inhibition

1 Introduction

IL-8, a member of the CXC cytokine family, is a neutrophil chemotactic and activating factor. It was first described as a pro-inflammatory cytokine produced by immune cells in response to stimulation [1–5]. However, IL-8 also contributes to cancer progression through its induction of tumor cell proliferation, survival, and migration. In addition, tumor-derived IL-8 activates endothelial cells to promote angiogenesis, induces neutrophil recruitment, and activates neutrophils and the tumor-associated macrophages to release more IL-8, which further amplifies the pro-survival, pro-angiogenic, and metastatic effect. IL-8 expression is increased in many types of advanced cancers and correlates with poor prognosis [6–15].

Ivana Vancurova (ed.), *Cytokine Bioassays: Methods and Protocols*, Methods in Molecular Biology, vol. 1172, DOI 10.1007/978-1-4939-0928-5_27, © Springer Science+Business Media New York 2014

In addition to IL-8, many tumors produce high levels of IL-6, a pro-inflammatory cytokine that is also regulated at the transcriptional level by NFκB [16–20].

In most unstimulated cells, NFκB resides in the cytoplasm in an inactive complex with its inhibitor, IκBα [21–24]. After cell stimulation, IκBα is phosphorylated, ubiquitinated, and degraded by the proteasome [25–29]. This releases NFκB proteins, which then translocate to the nucleus and stimulate transcription of target genes, including IL-8 and IL-6. In ovarian cancer, the NFκB activity is constitutively increased, and conveys a poor outcome [30, 31]. Bortezomib (BZ; PS-341, Velcade), the first FDA-approved proteasome inhibitor, has been originally developed as an inhibitor of the constitutive NFκB activity, and has been remarkably effective in the treatment of multiple myeloma and other hematological malignancies [32–35]. In solid tumors, including ovarian carcinoma, BZ has been less effective; however, the mechanisms remain elusive. Our recent studies have indicated that one of the underlying mechanisms consists of the increased IL-8 expression induced by proteasome inhibition in prostate and ovarian cancer cells [36, 37].

In this chapter, we describe a protocol that uses real-time qRT-PCR to analyze the mRNA expression levels of IL-8 and IL-6 in ovarian cancer cells treated with BZ. The main points of the protocol are: (1) Cell culture of ovarian cancer OVCAR 3 cells and incubation with BZ; (2) Extraction of total RNA; (3) Real-time qRT-PCR using primers for human IL-8 and IL-6 genes. The assay can be accomplished within 3 or 4 days, depending on the time of cell incubation with BZ.

2 Materials

2.1 Cell Culture

1. Human ovarian cancer OVCAR3 cells (American Type Culture Collection).
2. RPMI complete medium: RPMI supplemented with 20 % fetal bovine serum (FBS), 2 mM glutamine, 1 mM sodium pyruvate and 1 % penicillin–streptomycin solution.
3. Bortezomib stock solution: Dissolve bortezomib in dimethylsulfoxide (DMSO) to a final concentration of 10 mM. Store at −20 °C (*see* **Note 1**).
4. 75 cm^2 culture flasks.
5. Tissue culture-treated 6-well plates.
6. Hemocytometer and cover slips.
7. Trypan Blue dye for cell counting (*see* **Note 2**).

2.2 RNA Extraction

1. Phosphate buffered saline (PBS), pH 7.4.
2. 1.5 mL micro-centrifuge tubes.
3. 0.25 % trypsin-EDTA.
4. RNeasy mini-kit (Qiagen).
5. β-mercaptoethanol.
6. 70 % ethanol.
7. RNAse-free DNAse.
8. RNAse-free water.
9. RNA secure reagent.
10. Quartz cuvettes.

2.3 qRT-PCR

1. iScript™ SYBR® Green (Biorad).
2. Nuclease-free water.
3. iScript™ Reverse Transcriptase.
4. iCycler iQ™ PCR Plates, 96 wells.
5. iCycler iQ™ Optical tape.

3 Methods

In this section, we describe the protocol for quantitative RT-PCR analysis of IL-8 and IL-6 mRNA levels in BZ-treated ovarian cancer cells. However, this protocol can be easily modified and used for analysis of other genes in different cell types. Figure 1 represents a schematic illustration of the protocol, starting from cell culture to real-time RT-PCR. Figure 2 illustrates the mRNA expression levels of IL-8 and IL-6 in OVCAR3 cells treated different times with 1 μM BZ. Table 1 explains the calculation of IL-8 and IL-6 mRNA levels in OVCAR3 cells.

3.1 Cell Culture

1. Grow ovarian cancer OVCAR3 cells to a concentration of 0.5×10^6 cells/mL in RPMI complete medium. You will need about 22 mL of the cell suspension.
2. Set up the experiment in five 6-well cell culture plates and label them as 0, 3, 6, 24, and 48 h. Use two wells in each plate; one well will have untreated cells and the other well will have cells treated with 1 μM BZ. Add 2 mL of the cell suspension to each well. Incubate the plates for 24 h in tissue culture incubator (37 °C, 5 % CO_2, humidified atmosphere) to allow the cells to attach to the surface of the culture plate.
3. After 24 h, discard the media and add BZ to a final concentration of 1 μM in 2 mL of fresh media.

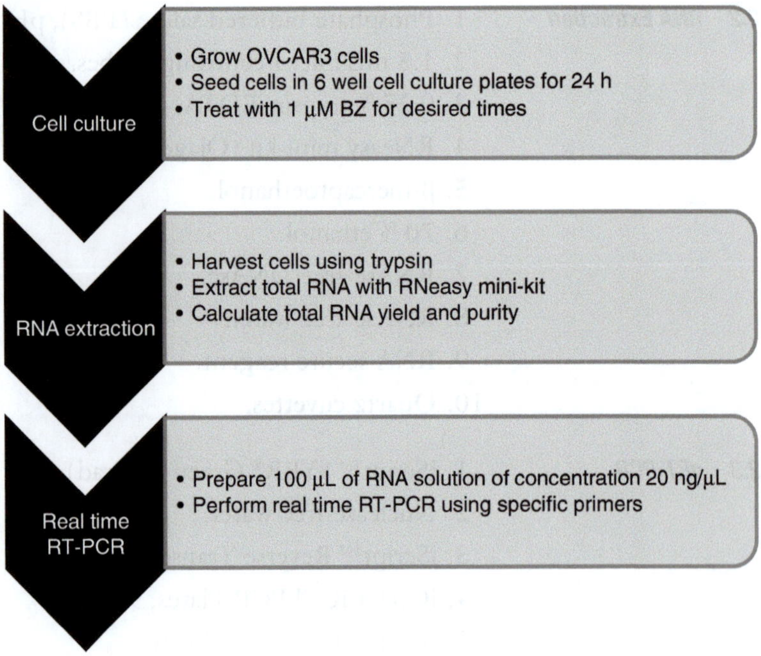

Fig. 1 Schematic illustration of the protocol

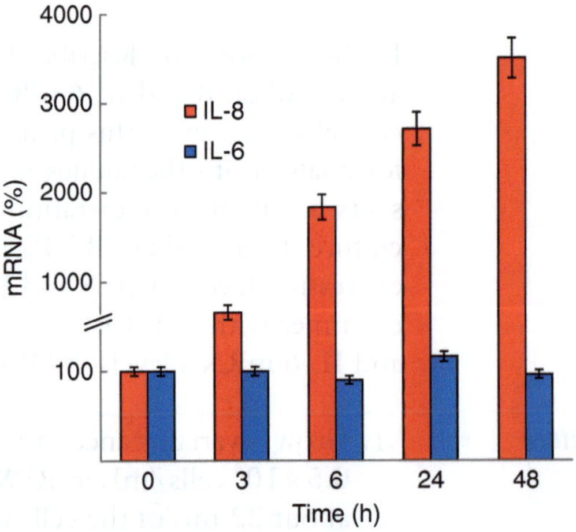

Fig. 2 Real-time RT-PCR analysis of IL-8 and IL-6 mRNA levels measured in OVCAR3 cells incubated with 1 µM BZ for up to 48 h. The values represent the mean ± SE of four experiments and are expressed as the % of values compared to $T = 0$ h

3.2 Extraction of RNA

3.2.1 Cell Harvesting

1. At desired times, remove media from each well into two 1.5 mL tubes to obtain the detached cells. Centrifuged at 2,000 × *g* for 5 min at 4 °C. Discard the supernatant, and keep the cell pellets (detached cells) on ice.
2. Wash the remaining attached cells in wells with 1 mL of 1× PBS. After removing PBS, add 700 µL of 0.25 % trypsin-EDTA

Table 1
Calculation of IL-8 and IL-6 mRNA levels in OVCAR3 cells from Ct values for IL-8, IL-6, and actin obtained after qRT-PCR

BZ (μM)	*T* (h)	Ct values for IL-8	Ct values for actin[a]	Actin-IL8	$2^{(Actin-IL8)}$	IL8 mRNA (%)
1	0	28.45	18.93	-9.52	0.0014	100.00
1	3	26.55	18.70	-7.86	0.0043	316.02
1	6	23.96	18.66	-5.30	0.0255	1,863.6
1	24	24.47	19.70	-4.77	0.0367	2,681.6
1	48	25.74	21.38	-4.37	0.0485	3,550.6
BZ (μM)	***T* (h)**	**Ct values for IL-6**	**Ct values for actin[a]**	**Actin-IL6**	$2^{(Actin-IL6)}$	**IL6 mRNA (%)**
1	0	29.73	18.98	-10.75	0.0006	100.00
1	3	29.56	18.80	-10.76	0.0006	99.31
1	6	29.73	18.64	-11.09	0.0005	79.28
1	24	30.24	19.82	-10.42	0.0007	126.14
1	48	32.29	21.49	-10.80	0.0006	96.60

[a]The increasing Ct values for actin during cell incubation with BZ are caused by the BZ-induced cell killing, resulting in reduced amounts of total extracted RNA

to each well. Incubate at 37 °C until the cells start to detach from the surface. Neutralize trypsin by adding equal amount of complete media (*see* **Note 3**).

3. Collect cells in 1.5 mL tubes and centrifuge at 2,000 × *g* for 5 min at 4 °C. Discard supernatant and resuspend the cell pellets in 1 mL of 1× PBS. In this step, the cell pellets obtained from **step 1** can be also added to the tubes. Centrifuge tubes at 2,000 × *g* for 5 min at 4 °C. Discard supernatant and keep the cell pellets on ice.

3.2.2 Total RNA Extraction (See ***Note 4****)*

1. Disrupt cells obtained from **step 3** (Subheading 3.2.1) by adding 350 μL of RLT buffer (from RNeasy mini kit) containing β-mercaptoethanol (*see* **Note 5**). Keep cells on ice, and vortex vigorously (*see* **Note 6**). Homogenize the lysates by pipetting up and down.
2. Add 350 μL of 70 % ethanol to the homogenized lysates and mix well by pipetting (*see* **Note** 7).
3. Transfer 700 μL of each sample obtained from **step 2** (above; *see* **Note 8**) into each RNeasy spin column placed in a 2 mL collection tube.
4. Centrifuge at 8,000 × *g* for 15 s at room temperature. At this step, RNA binds to membrane of the spin column, while most other components pass through. Discard the flow-through.

5. Add 350 μL of buffer RW1 to each RNeasy column and centrifuge at 8,000 × *g* for 15 s at room temperature. This is the wash step. Discard the flow-through.
6. For each RNeasy column, mix 10 μL of DNAse stock solution with 70 μL of RDD buffer, and pipette the entire volume onto the RNeasy column. Leave the column at room temperature for 15 min (*see* **Note 9**).
7. After 15 min, add 350 μL of buffer RW1 to each RNeasy column and centrifuge at 8,000 × *g* at room temperature for 15 s. This is the second wash step. Discard the flow-through and the collection tube.
8. Place the RNeasy column in a new 2 mL collection tube and add 500 μL of RPE buffer to each column. Centrifuge at 8,000 × *g* for 15 s. Discard the flow-through, and reuse the collection tube for the next step.
9. Add 500 μL of RPE buffer to each column and centrifuge for 2 min at 8,000 × *g* (*see* **Note 10**).
10. Transfer the column to a new 1.5 mL collection tube. Add 50 μL of RNase-free water into the membrane. Incubate for 2 min at room temperature and then centrifuge at 8,000 × *g* for 1 min in a micro-centrifuge at room temperature.
11. Add 2 μL of RNA secure reagent to the 50 μL eluate, and incubate 10 min at 60 °C in a water bath.

3.2.3 RNA Yield and Purity Calculation

1. Take 5 μL of extracted RNA and add it to 495 μL of RNase-free water.
2. Determine absorbance (Abs) at 260 and 280 nm against water in a 0.5 mL quartz cuvette.
3. For calculating the yield of RNA, use the following formula: $Abs_{260} \times 40$ ng/μL × 100 (dilution factor)
4. For calculating purity of RNA, find the ratio of Abs_{260} over Abs_{280}. Ideally, the ratio should be within the range of 1.8 and 2.1 (*see* **Note 11**).
5. Store the extracted RNA at −80 °C (*see* **Note 12**).

3.3 Real-Time qRT-PCR

1. From the extracted total RNA, prepare 100 μL of RNA solution with a concentration of 20 ng/μL. From this diluted RNA sample, take 5 μL for each PCR reaction; each PCR reaction will thus contain 100 ng of total RNA as a template.
2. For each PCR reaction, prepare a master mix solution containing 12.5 μL of SYBR green, 6 μL of RNase-free water, 0.5 μL of reverse transcriptase and 1 μL of specific gene primer (*see* **Note 13**).
3. Add a total of 25 μL (20 μL of master mix + 5 μL of diluted RNA template) into each well of a 96 well PCR plate. Seal the

plate carefully using an optical sealing tape, and continue with the real-time PCR reaction in the thermal cycler (*see* **Note 14**).

4. For calculating the mRNA relative expression values, subtract the Ct value of IL-8/IL-6 from the Ct value of actin for each time point (*see* **Note 15**). Usually, the Ct values for actin are lower compared to the Ct values of studied genes, resulting in a negative value after subtracting the Ct values of IL-8/IL-6, from the Ct value of actin. Since after each cycle of PCR, the amount of DNA doubles compared to DNA in the previous cycle, calculate the value of 2^N (where N = Ct value of actin − Ct value of IL-8/IL-6) for each time point. Consider the 2^N value obtained for $T = 0$ h as 100 %. Calculate mRNA levels for other time points using the formula below:

$$\left(\frac{2^N \text{ value for each time point with BZ treatment}}{2^N \text{ value for 0h BZ treatment}}\right) \times 100.$$

4 Notes

1. Bortezomib (MW 384.2) is a potent, specific, and reversible inhibitor of the 26S proteasome. It can be purchased from Sigma, Biomol, Santa Cruz Biotechnology, or ChemieTek. Prepare a stock concentration of 10 mM BZ in DMSO, and store at −80 °C. The final working concentration of BZ is 1 nM to 1 μM, depending on the cell type. Control untreated cells should always receive an equal volume of DMSO.
2. Cell counting using Trypan Blue should be quick since Trypan Blue is toxic to cells. Longer cell incubations with Trypan Blue before the cell counting may lead to lower cell viability, and thus error in the counting.
3. Neutralize trypsin with complete media as soon as the cells start detaching. Longer incubation of cells with trypsin is toxic to the cells.
4. During RNA extraction, do not to talk, sneeze, or cough near the working place since the RNA sample may become contaminated with saliva, which contains ribonuclease that can degrade the extracted RNA. Work quickly, and use gloves all time.
5. Add 10 μL of β-mercaptoethanol per 1 mL of RLT buffer before use. The RLT buffer containing β-mercaptoethanol can be stored at room temperature for 1 month.
6. Vortex the sample vigorously, at least for 15 s, at maximum speed. Ineffective vortexing may lead to a lower yield in total RNA.

7. Mix the disrupted cell suspension in RLT buffer and 70 % ethanol. Incomplete homogenization leads to reduced RNA yields and can clog the column.
8. Do not worry about some precipitate that may form during this step; it will not interfere with the assay.
9. Make sure to pipette the DNase/RDD buffer solution exactly on the membrane, so that it does not stick to the inner walls of the column. Extracted RNA sample may have genomic contamination if this step is not done effectively.
10. Dry the column completely by centrifugation, since the residual ethanol may interfere with the PCR. After centrifugation, it is recommended to blot the residual ethanol from the tip of the column using Kim-wipes.
11. Do not use RNA sample if the Abs_{260} over Abs_{280} ratio of the extracted RNA is below 1.8; this would mean that the RNA sample is contaminated with protein.
12. Aliquot the extracted RNA into two equal fractions, and store them at −80 °C. When using the RNA for real-time RT-PCR, use one aliquot at a time. This will ensure that the other fraction does not have to go through the freeze–thaw cycle.
13. Prepare the master mix for all PCR reactions, rather than preparing it individually for each reaction. Then add the specific primers and vortex it gently so that it is mixed thoroughly. Do not mix by pipetting, since SYBR green has a tendency to stick to the inner walls of the pipette tip, thus leading to the loss of the master mix.
14. Add the RNA template (5 μL) to the wells of the PCR plate before adding the master mix (20 μL).
15. You can use actin, GAPDH, or other housekeeping gene to normalize for changes of the studied gene(s).

Acknowledgement

This work was supported by NIH grant CA173452 to I. Vancurova.

References

1. Oppenheim JJ, Zachariae CO, Mukaida N et al (1991) Properties of the novel proinflammatory supergene "intercrine" cytokine family. Annu Rev Immunol 9:617–648
2. Miller MD, Krangel MS (1992) Biology and biochemistry of the cytokines: a family of chemotactic and inflammatory cytokines. Crit Rev Immunol 12:17–46
3. Matsushima K, Baldwin ET, Mukaida N (1992) Interleukin and MCAF: novel leukocyte recruitment and activating cytokines. Chem Immunol 51:236–265
4. Taub DD, Oppenheim JJ (1993) Review of the chemokine meeting the third international symposium of chemotactic cytokines. Cytokine 5:175–179

5. Hack CE, Arden LA, Thijs LG (1997) Role of cytokines in sepsis. Adv Immunol 66:101–195
6. Waugh DJ, Wilson C (2008) The interleukin-8 pathway in cancer. Clin Cancer Res 14: 6735–6741
7. Xu L, Fidler IJ (2000) Interleukin 8: an autocrine growth factor for human ovarian cancer. Oncol Res 12:97–106
8. Veltri RW, Miller MC, Zhao G et al (1999) Interleukin-8 serum levels in patients with benign prostatic hyperplasia and prostate cancer. Urology 53:139–147
9. Smith DR, Polverini PJ, Kunkel SL et al (1994) Inhibition of interleukin 8 attenuates angiogenesis in bronchogenic carcinoma. J Exp Med 179:1409–1415
10. Di Celle PF, Carbone A, Marchis D et al (1994) Cytokine gene expression in B-cell chronic lymphocytic leukemia: evidence of constitutive interleukin-8 (IL-8) mRNA expression and secretion of biologically active IL-8 protein. Blood 84:220–228
11. Green AR, Green VL, White MC et al (1997) Expression of cytokine messenger RNA in normal and neoplastic human breast tissue: identification of IL-8 as a potential regulatory factor in breast tumors. Int J Cancer 72:937–941
12. Brew R, Erikson JS, West DC et al (1997) Interleukin-8 as a growth factor for human colorectal carcinoma cells in vitro. Biochem Soc Trans 25:264S–268S
13. Tjiong MY, Van Der Vange N, Ten Kate FJ et al (1999) Increased IL-6 and IL-8 levels in cervicovaginal secretions of patients with cervical cancer. Gynecol Oncol 73:285–291
14. Yasumoto K, Okamoto S, Mukaida N et al (1992) Tumor necrosis factor alpha and interferon gamma synergistically induce interleukin 8 production in a human gastric cancer cell line through acting concurrently on AP-1 and NF-κB-like binding sites of the interleukin 8 gene. J Biol Chem 267:22506–22511
15. Kitadai Y, Takahashi Y, Haruma K et al (1999) Transfection of interleukin-8 increases angiogenesis and tumorigenesis of human gastric carcinoma cells in nude mice. Br J Cancer 81: 647–653
16. Jones SA (2005) Directing transition from innate to acquired immunity: defining a role for IL-6. J Immunol 175:3463–3468
17. Sansone P, Storci G, Tavolari S et al (2007) IL-6 triggers malignant features in mammospheres from human ductal breast carcinoma and normal mammary gland. J Clin Invest 12: 3988–4002
18. Gao SP, Mark KG, Leslie K et al (2007) Mutations in the EGFR kinase domain mediate STAT3 activation via IL-6 production in human lung adenocarcinomas. J Clin Invest 12:3846–3856
19. Coward J, Kulbe H, Chakravarty P et al (2011) Interleukin-6 as a therapeutic target in human ovarian cancer. Clin Cancer Res 18: 6083–6096
20. Singh RK, Sudhakar A, Lokeshwar BL (2010) Role of chemokines and chemokine receptors in prostate cancer development and progression. J Cancer Sci Ther 2:89–94
21. Mukaida N, Mahe Y, Matsushima K (1990) Cooperative interaction of NFκB and cis-regulatory enhancer binding protein-like factor binding elements in activating the interleukin-8 gene by pro-inflammatory cytokines. J Biol Chem 265:21128–21133
22. Kunsch C, Rosen CA (1993) NFκB subunit-specific regulation of the IL-8 promoter. Mol Cell Biol 13:6137–6146
23. Shimizu H, Mitomo K, Watanabe T et al (1990) Involvement of a NFκB-like transcription factor in the activation of the interleukin-6 gene by inflammatory lymphokines. Mol Cell Biol 2:561–568
24. Libermann TA, Baltimore D (1990) Activation of interleukin-6 gene expression through the NFκB transcription factor. Mol Cell Biol 5:2327–2334
25. Baeuerle PA, Baltimore D (1996) NFκB: ten years after. Cell 87:13–20
26. Baldwin AS (2001) The transcription factor NFκB and human disease. J Clin Invest 107:3–6
27. Yamamoto Y, Gaynor RB (2001) Therapeutic potential of inhibition of the NFκB pathway in the treatment of inflammation and cancer. J Clin Invest 107:135–142
28. Ghosh S, Karin M (2002) Missing pieces in the NFκB puzzle. Cell 109:S81–S96
29. Wan F, Lenardo MJ (2010) The nuclear signaling of NFκB: current knowledge, new insights, and future perspectives. Cell Res 20:24–33
30. Annunziata CM, Stavnes HT, Kleinberg L et al (2010) NFκB transcription factors are coexpressed and convey a poor outcome in ovarian cancer. Cancer 116:3276–3284
31. Leizer AL, Alvero AB, Fu HH et al (2011) Regulation of inflammation by the NFκB pathway in ovarian cancer stem cells. Am J Reprod Immunol 65:438–447
32. Hideshima T, Chauhan D, Richardson P et al (2002) NFκB as a therapeutic target in multiple myeloma. J Biol Chem 277:16639–16647
33. Richardson PG, Mitsiades C, Hideshima T et al (2005) Proteasome inhibition in the treatment of cancer. Cell Cycle 4:290–296

34. Shah JJ, Orlowski RZ (2009) Proteasome inhibitors in the treatment of multiple myeloma. Leukemia 23:1964–1979
35. Kuhn DJ, Orlowski RZ (2012) The immunoproteasome as a target in hematologic malignancies. Semin Hematol 49:258–262
36. Manna S, Singha B, Phyo SA et al (2013) Proteasome inhibition by bortezomib increases IL-8 expression in androgen-independent prostate cancer cells: the role of IKKα. J Immunol 191:2837–2846
37. Singha B, Gatla H, Manna S et al (2014) Proteasome inhibition increases recruitment of IKKβ, S536P-p65 and transcription factor EGR1 to interleukin-8 (IL-8) promoter, resulting in increased IL-8 production in ovarian cancer cells. J Biol Chem 289: 2687–2700

Chapter 28

Analysis of LPS-Induced, NFκB-Dependent Interleukin-8 Transcription in Kidney Embryonic Cell Line Expressing TLR4 Using Luciferase Assay

Tamara Yunusova, Mumtaz Akhtar, and Vladimir Poltoratsky

Abstract

Gene expression is orchestrated by a complex network of signal transduction pathways that typically originate on cell surface receptors and culminate in DNA-binding transcription factors, which translocate to the nucleus and bind cis-regulatory elements in promoter regions of genes, thereby inducing de novo synthesis of the nascent RNA transcripts and their splicing. Gene expression arrays monitor abundance of the matured, spliced cDNA, which undergoes additional posttranscriptional modifications that greatly affect the half-life of the cDNA. Thus, the relative abundance of cDNA is not necessarily commensurable with the activity of promoters of the corresponding genes. In contrast, reporter gene assays provide valuable insight into the regulation of gene expression at the level of transcription and allow for discerning the contribution of individual transcription factors into changes in gene expression. Here, we describe a robust reporter gene assay method that is useful for exploration of transcription regulatory network, which regulates gene expression in response to inflammation. The method is exemplified by using the promoter region of the prototypic pro-inflammatory chemokine interleukin-8 (IL-8, CXCL8), which plays an important role in immune response as well as carcinogenesis. Using the luciferase reporter gene assay, we analyze the activation status of the IL-8 promoter in lipopolysaccharide (LPS)-stimulated human embryonic kidney cells.

Key words Interleukin-8, TLR4, Gene expression, Luciferase assay, Promoter activity, NFκB signaling, AP-1 signaling

1 Introduction

A hallmark of the immune response is the migration of leukocytes from the bloodstream to the site of injury. Various chemotactic agonists direct this migration, and interleukin-8 (IL-8, CXCL8) is one of them. IL-8, a neutrophil chemotactic polypeptide, was purified in 1987 [1] and current research has demonstrated its role in inflammation [2–5]. In addition to its role in immune system response, IL-8 plays a significant role in carcinogenesis both indirectly, by promoting inflammation and recruiting leukocytes to

Ivana Vancurova (ed.), *Cytokine Bioassays: Methods and Protocols*, Methods in Molecular Biology, vol. 1172,
DOI 10.1007/978-1-4939-0928-5_28, © Springer Science+Business Media New York 2014

tumors, and directly, by enhancing the viability of endothelial cells and regulating angiogenesis [6].

IL-8 is a member of the CXC chemokine subfamily, which is characterized by the presence of an amino acid between two cysteine residues. The protein consists of 72 amino acids and has a molecular weight of 8 kDa. It is resistant to plasma peptidases, heat, and extreme pH levels, but can be inactivated via reduction of the disulfide bonds. X-ray crystallography studies have revealed that IL-8 forms dimers. The monomer consists of an N terminal domain that is attached by two disulfide bonds to a core structure containing three anti-parallel β strands followed by terminal alpha helix.

The IL-8 gene was localized to 4q12-q21 region [7, 8]. It spans approximately 4 kb and consists of four exons; the first three of the four exons code the protein. Single nucleotide polymorphism of IL-8 has been associated with inflammatory, cardiovascular, and autoimmune diseases [9–15]. Polymorphism of IL-8 has also been connected with several types of cancer [16–20].

IL-8 is expressed in a number of cells including leukocytes, endothelial cells, fibroblasts, keratinocytes, hepatocytes, human gastric cells, adipose tissues, and various malignant cells [14, 21–24]. The IL-8 expression can be induced by inflammatory signaling, hormones, hypoxia, acidosis, hyperglycemia, and oxidative and genotoxic stress [25–28]. In addition, the IL-8 gene expression can be elevated by upregulation of RAS oncogene and downregulation of PTEN, a tumor suppressor gene [29–31].

Expression of IL-8 is regulated predominantly on transcriptional level [32]. Two additional levels of regulation include regulation of mRNA stability and translation [28, 32–35]. Chromatin immunoprecipitation (ChIP) identified NFκB as a major transcription factor responsible for the inducible IL-8 expression [36–38]. The proinflammatory stimuli (e.g., LPS through TLR4) activate NFκB pathway through IκB kinase (IKK), which in turn phosphorylates IκB, an inhibitor of NFκB. IKK phosphorylation leads to degradation of IκBα, and subsequent translocation of NFκB to the nucleus, where it drives transcription of the IL-8 gene [39, 40]. Several additional cis-regulatory elements have been identified within the −133 to +1 bp region upstream the TATA box of the IL-8 promoter, including binding sites for AP-1, HIF1α, C/EBP, and NRF. Thus many additional transcription factors play an important role in the regulation of IL-8 gene expression (Fig. 1) [36–38].

In this chapter, we describe a method to determine the activation status of the IL-8 promoter using the luciferase reporter gene assay. The luciferase assay is widely used for assessing the role of the cis-acting transcription factors. As a reporter gene, luciferase is 1,000 times more sensitive than GFP fluorescence and has become a standard method for measuring the promoter activity. To quantify the promoter strength, the luciferase reporter gene is placed

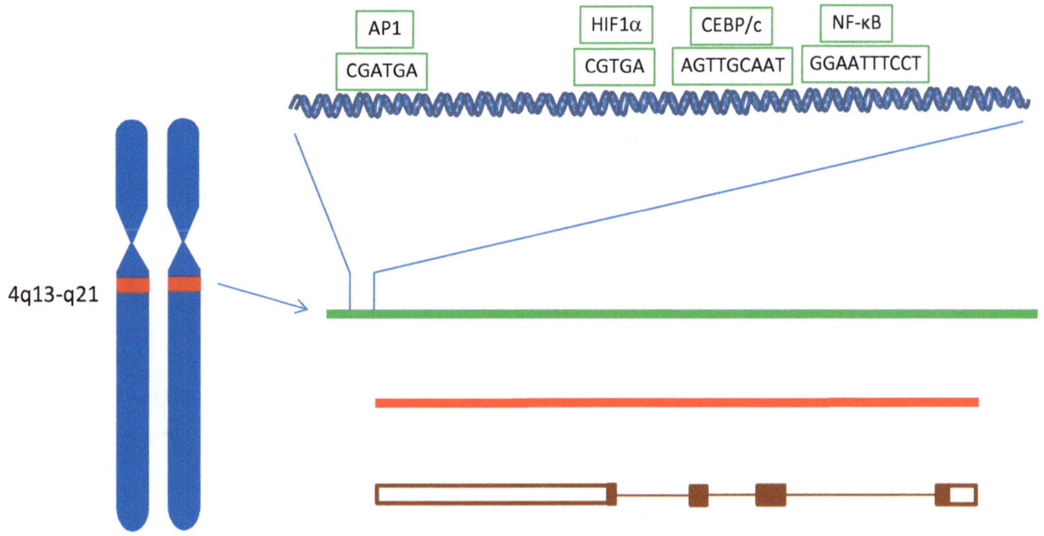

Fig. 1 Schematic illustration of human IL-8 promoter

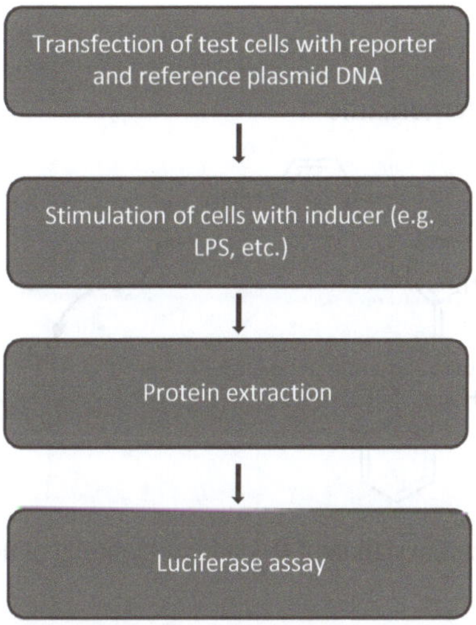

Fig. 2 Schematic illustration of the protocol

downstream of the promoter of interest. The luciferase genes were isolated from firefly (*Photinus pyralis*) or sea pansy (*Renilla reniformis*).

The major steps of the assay include: (a) transfection of test cells with the reporter and reference plasmid DNA; (b) stimulation of cells with inducer (e.g., LPS); (c) protein extraction; (d) luciferase assay (Fig. 2). Activity of the promoter is monitored by quantifying the reporter enzyme activity, subcloned downstream the promoter sequence. Luciferase reporters have higher sensitivity, low background, and wide dynamic range over traditional ones, including β-galactosidase, alkaline phosphatase, chloramphenicol acetyltransferase, and fluorescent proteins.

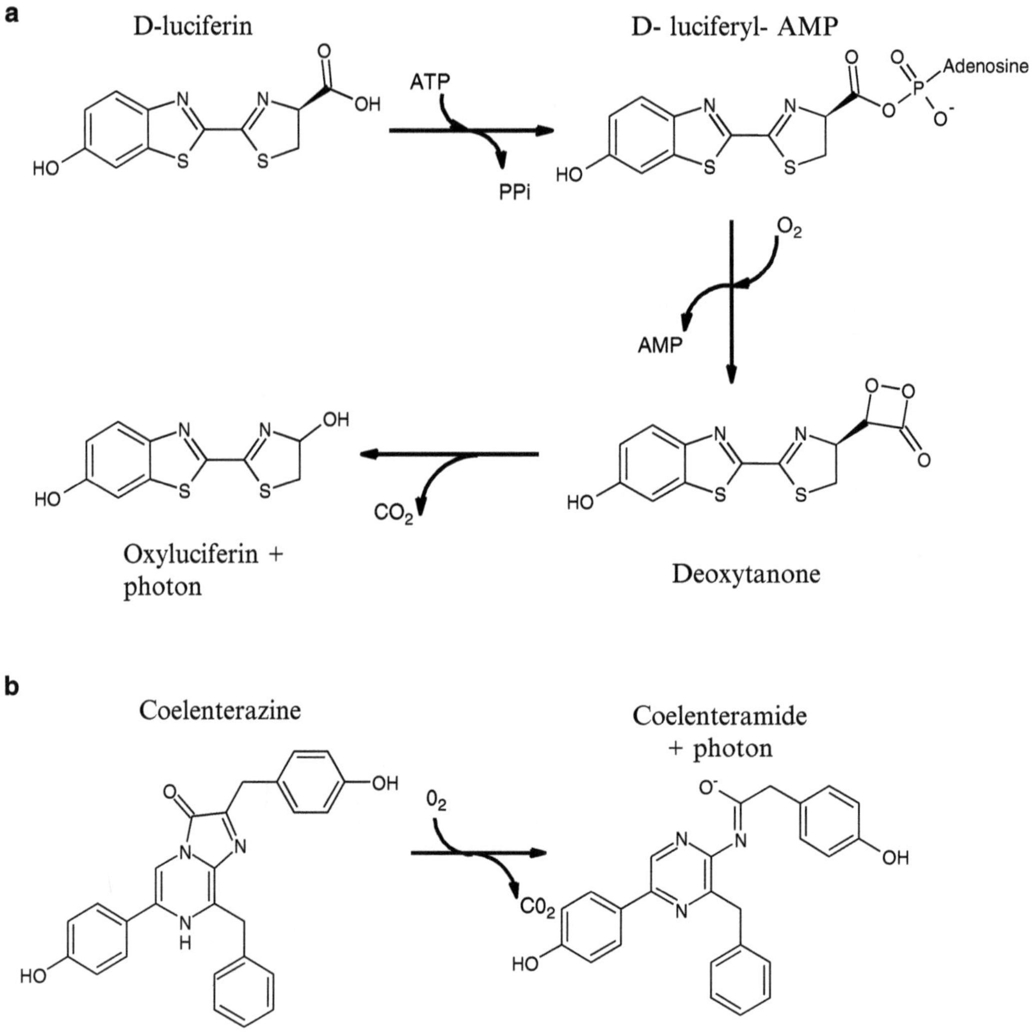

Fig. 3 Oxidative decarboxylation of D-luciferin by firefly pLuc luciferase (**a**) and of colentrazine by *Renilla* rLuc luciferase (**b**)

The firefly (pLuc) or *Renilla* (rLuc) luciferase enzymes catalyze the oxygenation of substrate in the presence of molecular oxygen, to yield product of the reaction and give off light. PLuc and rLuc luciferases have different chemical preferences and therefore their activities could be measured in the same tube. Pluc catalyzes the ATP-dependent carboxylative oxidation of D-luciferin in the presence of oxygen and Mg^{2+}, producing oxyluciferin and photon of a yellow light ($\lambda max = 475$ nm) (Fig. 3a). Rluc catalyzes ATP and Mg^{2+} independent oxidation of the coelenterazine to coelenteramide with a release of the photon of a blue light ($\lambda max = 560$ nm) (Fig. 3b).

The firefly luciferase activity is typically measured 16–72 h after transfection. The luminescence is measured in relative light units (RLU) that are proportional to the strength of the promoter. Activity of the test promoter is normalized to empty reporter vector (without the promoter) or to the reporter construct comprising *Renilla* luciferase placed under control of constitutive promoter.

2 Materials

2.1 DNA Constructs, Cells, Growth Media (See Note 1)

1. Vector pGL4.32_IL-8 encoding modified pLuc (luc2p) under control of IL-8 promoter (−272 to +46 bp) promoter (*see* **Note 2**).
2. Reference vector pGL4.74 encoding rLuc under control of constitutive HSV-TK promoter (*see* **Note 2**).
3. 293_TLR4 human embryonic kidney cells, expressing human TLR4, MD2, and CD14 (Invogene, Inc. Catalog # 293-htlr4md2cd14) (*see* **Note 3**).
4. Complete Dulbecco's Modified Eagle's Medium (DMEM): DMEM containing high glucose (4.5 g/L), and supplemented with 1 mM sodium pyruvate, 10 % of fetal bovine serum (FBS), 2 mM L-glutamine, 10 μg/mL blasticidin, 50 μg/mL hygromycin, 50 μg/mL penicillin, and 50 μg/mL streptomycin.
5. DMEM without antibiotics: DMEM as described above, but without antibiotics: blasticidin, hygromycin, penicillin, and streptomycin (*see* **Note 4**).
6. Tissue culture incubator.

2.2 Transfection and Induction Solutions (See Note 5)

1. 2 M $CaCl_2$: Dissolve 29.4 g $CaCl_2 \times 2H_2O$ in 100 mL of H_2O.
2. 2× HBS: Dissolve 8 g NaCl, 0.37 g KCl, 201.1 mg $Na_2HPO_4 \times 7H_2O$, 1.0 g dextrose, and 5.0 g Hepes in 400 mL of H_2O. Adjust pH to 7.05 with NaOH, add water to 500 mL, and sterilize solution by filtration through 0.22 μm filter. Store at room temperature.
3. 10 mM TE buffer: Mix 1 mL of 1 M Tris–HCl, pH 7.5, with 200 μL of 0.5 M EDTA and adjust water to 100 mL. Sterilize solution by filtration through 0.22 μm filter and store at room temperature.
4. Lipopolysaccharide (LPS) solution: Dissolve 10 mg of LPS in 1 mL of PBS. Sterilize solution by filtration through 0.45 μM filter and store at 4 °C.

2.3 Luciferase Solutions and Reagents

1. Luciferin solution 100× (100 mM): Dissolve 30.23 mg of D-luciferin in 10 mL of distilled water. Keep in frozen aliquots at −20 °C for short storage; for longer storage, keep at −80 °C in polypropylene tubes under nitrogen, protected from light.

2. ATP solution 100× (300 mM): Dissolve 165.4 mg of ATP disodium salt in 10 mL of distilled water. Aliquot and freeze for storage at −20 °C in polypropylene tubes.
3. 1.5 M $MgSO_4$ solution (100×): Dissolve 370 mg of $MgSO_4 \times 7H_2O$ in 10 mL of distilled water. Aliquot and store at +4 °C in polypropylene tubes.
4. 1 M Hepes buffer, pH 7.8: Dissolve 23.83 g Hepes in 80 mL of distilled water, adjust pH to 7.8 by sodium hydroxide solution and add distilled water to 100 mL.
5. PLuc luciferase assay buffer: 25 mM Hepes, pH 7.8, 15 mM $MgSO_4$, 4 mM EGTA, 100 μM coenzyme A, 1 mM DTT.
6. Colentrazine solution 100× (100 mM): Dissolve 4.23 mg of colentrazine in 10 mL of acidified ethanol (10 mL ethanol plus 0.2 mL of 3 M HCl). Keep in aliquots for short storage at −20 °C, and for longer storage at −80 °C in polypropylene tubes under nitrogen, protected from light.
7. RLuc luciferase substrate solution with quenching pLuc luciferase buffer: 220 mM sodium phosphate, pH 5.1, 1.1 M NaCl, 2.2 mM EDTA, 0.44 mg/mL BSA, 1.3 mM NaN_3, 1 μM colentrazine. Prepare fresh prior to use.
8. Cell lysis buffer: 100 mM Hepes, pH 7.8, 2 mM DTT, 4 mM EGTA, 0.2 % Triton X100, 10 % (v/v) glycerol plus protease inhibitors cocktail (Sigma).

3 Methods

3.1 Transfection Using $CaCl_2$ and Induction (See Note 5)

1. One day before transfection, plate 2–4 × 10^6 cells in 10 mL of DMEM without antibiotics in 10 cm tissue culture dishes. The cells should be at 60–80 % confluence at the time of transfection.
2. Next day, change medium to 8 mL with fresh DMEM without antibiotics.
3. To prepare DNA complexes in tissue culture hood, add 10 μg of DNA of pGL4.32_IL8 and 1 μg pGL4.7 to 64 μL of $CaCl_2$ and adjust volume to 500 μL with sterile, endotoxin- and nuclease-free H_2O. Mix gently. In separate 15 mL tube, add 500 μL of 2× HBS (*see* **Note 6**).
4. Place the HBS-containing tube on a Vortex on low speed and add drop-wise DNA solution with constant mixing.
5. Incubate without shaking for 20 min at room temperature (solution may appear cloudy due to the formation of the calcium phosphate-DNA complex) (*see* **Note** 7).

6. Gently mix solution and add it to plated cells drop wise. Mix by gently rocking the plate (*see* **Note 8**).
7. Incubate cells at 37 °C in a 5 % CO_2 humidified incubator for 5–16 h.
8. After incubation is complete, wash cells two times with DMEM medium or sterile PBS.
9. Trypsinize cells and plate in a 6-well dish in complete DMEM medium at density 2×10^5 cells per well.
10. Incubate at 37 °C in a CO_2 incubator.
11. After overnight incubation, add LPS to a final concentration of 100 ng/mL in complete DMEM medium. Culture cells for 8 h in CO_2 incubator. For one well containing 2 mL of media, add 20 μL of stock LPS solution (*see* **Note 9**).

3.2 Luciferase Reporter Assay

1. Wash cells with PBS and add 500 μL of the cell lysis buffer per well.
2. Incubate at room temperature for 15 min on a shaker platform at 200 rpm.
3. Transfer lysates to Eppendorf tubes, centrifuge at 10,000 × *g* for 5 min. Discard the cell debris and transfer supernatants to new tubes.
4. Prepare pLuc luciferase assay mix. Mix pLuc luciferase buffer with ATP and luciferin solution. Add 100 μL of luciferin and 100 μL of ATP solutions per 10 mL of pLuc luciferase assay buffer (*see* **Note 10**).
5. Mix 20 μL of the sample with 100 μL of the pLuc luciferase buffer in triplicates in opaque or white 96 well plate (*see* **Note 11**).
6. Incubate in dark at room temperature for 10 min.
7. Quantify the light emission of the firefly luciferase using luminometer.
8. Prepare rLuc luciferase substrate solution with quenching pLuc luciferase buffer (*see* **Note 10**).
9. Add 100 μL of prepared rLuc solution to tested wells.
10. Incubate in dark at room temperature for 10 min.
11. Quantify the light emission of the *Renilla* luciferase using luminometer.
12. Calculate the firefly luciferase activity, using activity of *Renilla* luciferase as an internal control for determining the transfection efficiency, and present results, as a normalized relative light units (RLU).

4 Notes

1. At the start, the experiments should include both negative and positive controls. As a negative control, we use the empty pGL4.10 vector that encodes luc2p under minimal promoter, and as a positive control, we use pGL4.10_SV40 vector encoding luc2p under SV40 promoter.
2. It is important to purify plasmids from the endotoxin. Investigators could use Endo-Free plasmid purification kits that are available from Qiagen and BIOMIGA companies. Another efficient alternative method of removing LPS from the purified plasmid is the extraction with Triton X-114 [41]. For one extraction, mix plasmid with TritonX-114 to a final concentration of 1.0 % (w/w) on ice and incubate on ice for 10 min. Then, incubate the samples at 45 °C for 10 min, and follow with centrifugation at 37 °C for 10 min. Then, transfer the upper phase to a fresh Eppendorf tube. After the third extraction, precipitate the DNA after addition of the equal volume of isopropanol and centrifuge for 10 min at 10,000×*g*. Discard the supernatant and wash the pellet twice with 70 % ethanol. After air dry is complete, dissolve the pellet in TE buffer.
3. 293_TLR cells are rapidly growing cells. They have to be replated regularly. Avoid keeping cells in confluent monolayer. This will decrease viability and transfection efficiency of cells.
4. It is important to keep cells during transfection in a medium without antibiotics. Increased accumulation of the antibiotics decreases viability of the transfected cells. Alternatively to calcium phosphate method of transfection, you can use commercially available transfection reagents, including Lipofectamine (Life Technologies), Fugene (Roche), and *Trans*It LT (Mirus). Other possibility is electroporation that requires expensive equipment. Different transfection methods have different efficiency in different cell types.
5. Solutions contaminated with lipopolysaccharide (LPS, endotoxin) will increase expression of IL-8. Therefore it is important to use double-distilled, endonuclease- and endotoxin-free water to prepare all solutions.
6. The ratios between pGL4.32_IL8 and reference pGL 4.74 vectors are determined empirically for different tested systems, depending on the level of IL-8 expression in tested cells.
7. The complexes stay stable for 2 h.
8. It is a good practice to transfect tested cells in parallel with plasmid encoding green fluorescent protein (GFP) and monitor the transformation efficiency by calculating ratio of GFP-positive to GFP-negative cells observed under a fluorescent microscope.

9. Individual times for expression depend on the cells used in the experiment.
10. The pLuc, rLuc substrates, and ATP solutions are unstable and have to be mixed right before assay.
11. Luciferase assay requires specific opaque or white plates. Do not use the transparent 96 well plates, because luminescence signals from neighboring wells would interfere.

References

1. Yoshimura T, Matsushima K, Tanaka S et al (1987) Purification of a human monocyte-derived neutrophil chemotactic factor that has peptide sequence similarity to other host defense cytokines. Proc Natl Acad Sci U S A 84:9233–9237
2. Takata H, Tomiyama H, Fujiwara M et al (2004) Cutting edge: expression of chemokine receptor CXCR1 on human effector CD8+ T cells. J Immunol 173:2231–2235
3. Gasser O, Missiou A, Eken C et al (2005) Human CD8+ T cells store CXCR1 in a distinct intracellular compartment and up-regulate it rapidly to the cell surface upon activation. Blood 106:3718–3724
4. Berahovich RD, Lai NL, Wei Z et al (2006) Evidence for NK cell subsets based on chemokine receptor expression. J Immunol 177: 7833–7840
5. Burke SM, Issekutz TB, Mohan K et al (2008) Human mast cell activation with virus-associated stimuli leads to the selective chemotaxis of natural killer cells by a CXCL8-dependent mechanism. Blood 111:5467–5476
6. Koch AE, Polverini PJ, Kunkel SL et al (1992) Interleukin-8 as a macrophage-derived mediator of angiogenesis. Science 258:1798–1801
7. Modi WS, Dean M, Seuanez HN et al (1990) Monocyte-derived neutrophil chemotactic factor (MDNCF/IL-8) resides in a gene cluster along with several other members of the platelet factor 4 gene superfamily. Hum Genet 84:185–187
8. O'Donovan N, Galvin M, Morgan JG (1999) Physical mapping of the CXC chemokine locus on human chromosome 4. Cytogenet Cell Genet 84:39–42
9. Andia DC, Letra A, Casarin RC et al (2012) Genetic analysis of the IL8 gene polymorphism (rs4073) in generalized aggressive periodontitis. Arch Oral Biol 58:211–217
10. Apostolakis S, Vogiatzi K, Amanatidou V et al (2009) Interleukin 8 and cardiovascular disease. Cardiovasc Res 84:353–360
11. Brennan FM, Zachariae CO, Chantry D et al (1990) Detection of interleukin 8 biological activity in synovial fluids from patients with rheumatoid arthritis and production of interleukin 8 mRNA by isolated synovial cells. Eur J Immunol 20:2141–2144
12. Garey KW, Jiang ZD, Ghantoji S et al (2010) A common polymorphism in the interleukin-8 gene promoter is associated with an increased risk for recurrent Clostridium difficile infection. Clin Infect Dis 51:1406–1410
13. Javor J, Bucova M, Cervenova O et al (2012) Genetic variations of interleukin-8, CXCR1 and CXCR2 genes and risk of acute pyelonephritis in children. Int J Immunogenet 39: 338–345
14. Lee EB, Zhao J, Kim JY et al (2007) Evidence of potential interaction of chemokine genes in susceptibility to systemic sclerosis. Arthritis Rheum 56:2443–2448
15. Mahanta A, Kakati S, Baruah S (2014) The association of IL-8-251T/A polymorphism with complicated malaria in Karbi Anglong district of Assam. Cytokine 65:210–216
16. Bondurant KL, Lundgreen A, Herrick JS et al (2013) Interleukin genes and associations with colon and rectal cancer risk and overall survival. Int J Cancer 132:905–915
17. Snoussi K, Mahfoudh W, Bouaouina N et al (2010) Combined effects of IL-8 and CXCR2 gene polymorphisms on breast cancer susceptibility and aggressiveness. BMC Cancer 10:283
18. Zabaleta J, Su LJ, Lin HY et al (2009) Cytokine genetic polymorphisms and prostate cancer aggressiveness. Carcinogenesis 30:1358–1362
19. Ahn MH, Park BL, Lee SH et al (2011) A promoter SNP rs4073T>A in the common allele of the interleukin 8 gene is associated with the development of idiopathic pulmonary fibrosis via the IL-8 protein enhancing mode. Respir Res 12:73
20. Kamali-Sarvestani E, Aliparasti MR, Atefi S (2007) Association of interleukin-8 (IL-8 or CXCL8) -251T/A and CXCR2 +1208C/T

gene polymorphisms with breast cancer. Neoplasma 54:484–489

21. Alvehus M, Simonyte K, Andersson T et al (2012) Adipose tissue IL-8 is increased in normal weight women after menopause and reduced after gastric bypass surgery in obese women. Clin Endocrinol (Oxf) 77:684–690
22. Bruun JM, Pedersen SB, Richelsen B (2001) Regulation of interleukin 8 production and gene expression in human adipose tissue in vitro. J Clin Endocrinol Metab 86:1267–1273
23. Murphy C, McGurk M, Pettigrew J et al (2005) Nonapical and cytoplasmic expression of interleukin-8, CXCR1, and CXCR2 correlates with cell proliferation and microvessel density in prostate cancer. Clin Cancer Res 11:4117–4127
24. Yuan A, Yang PC, Yu CJ et al (2000) Interleukin-8 messenger ribonucleic acid expression correlates with tumor progression, tumor angiogenesis, patient survival, and timing of relapse in non-small-cell lung cancer. Am J Respir Crit Care Med 162:1957–1963
25. Imamura R, Konaka K, Matsumoto N et al (2004) Fas ligand induces cell-autonomous NF-kappaB activation and interleukin-8 production by a mechanism distinct from that of tumor necrosis factor-alpha. J Biol Chem 279: 46415–46423
26. Waugh DJ, Wilson C (2008) The interleukin-8 pathway in cancer. Clin Cancer Res 14: 6735–6741
27. Kasahara T, Mukaida N, Yamashita K et al (1991) IL-1 and TNF-alpha induction of IL-8 and monocyte chemotactic and activating factor (MCAF) mRNA expression in a human astrocytoma cell line. Immunology 74:60–67
28. Shi Q, Xiong Q, Le X et al (2001) Regulation of interleukin-8 expression by tumor-associated stress factors. J Interferon Cytokine Res 21: 553–566
29. Maxwell PJ, Coulter J, Walker SM et al (2013) Potentiation of inflammatory CXCL8 signalling sustains cell survival in PTEN-deficient prostate carcinoma. Eur Urol 64:177–188
30. Sparmann A, Bar-Sagi D (2004) Ras-induced interleukin-8 expression plays a critical role in tumor growth and angiogenesis. Cancer Cell 6:447–458
31. Wislez M, Fujimoto N, Izzo JG et al (2006) High expression of ligands for chemokine receptor CXCR2 in alveolar epithelial neoplasia induced by oncogenic kras. Cancer Res 66: 4198–4207
32. Hoffmann E, Dittrich-Breiholz O, Holtmann H et al (2002) Multiple control of interleukin-8 gene expression. J Leukoc Biol 72:847–855
33. Kuhns DB, Gallin JI (1995) Increased cell-associated IL-8 in human exudative and A23187-treated peripheral blood neutrophils. J Immunol 154:6556–6562
34. Mukaida N, Okamoto S, Ishikawa Y et al (1994) Molecular mechanism of interleukin-8 gene expression. J Leukoc Biol 56:554–558
35. Roebuck KA (1999) Regulation of interleukin-8 gene expression. J Interferon Cytokine Res 19:429–438
36. Jung YD, Fan F, McConkey DJ et al (2002) Role of P38 MAPK, AP-1, and NF-kappaB in interleukin-1beta-induced IL-8 expression in human vascular smooth muscle cells. Cytokine 18:206–213
37. Kunsch C, Lang RK, Rosen CA et al (1994) Synergistic transcriptional activation of the IL-8 gene by NF-kappa B p65 (RelA) and NF-IL-6. J Immunol 153:153–164
38. Yasumoto K, Okamoto S, Mukaida N et al (1992) Tumor necrosis factor alpha and interferon gamma synergistically induce interleukin 8 production in a human gastric cancer cell line through acting concurrently on AP-1 and NF-kB-like binding sites of the interleukin 8 gene. J Biol Chem 267:22506–22511
39. Chang TP, Vancurova I (2013) NFκB function and regulation in cutaneous T-cell lymphoma. Am J Cancer Res 3:433–445
40. DiDonato JA, Mercurio F, Karin M (2012) NF-kappaB and the link between inflammation and cancer. Immunol Rev 246:379–400
41. Cotten M, Baker A, Saltik M et al (1994) Lipopolysaccharide is a frequent contaminant of plasmid DNA preparations and can be toxic to primary human cells in the presence of adenovirus. Gene Ther 1:239–246

Chapter 29

Chromatin Immunoprecipitation Analysis of Bortezomib-Mediated Inhibition of NFκB Recruitment to IL-1β and TNFα Gene Promoters in Human Macrophages

Shannon Sanacora, Tzu-Pei Chang, and Ivana Vancurova

Abstract

Interleukin-1β (IL-1) and tumor necrosis factor-α (TNF) are important pro-inflammatory cytokines involved in the mediation of the immune response, inflammation, tissue repair, and tumor progression. Regulation of IL-1 and TNF expression is mediated at the level of transcription by the transcription factor NFκB. Inhibition of NFκB activity by the proteasome inhibitor bortezomib (BZ) has been used as a front-line therapy in multiple myeloma and other hematological malignancies. In this chapter, we describe a protocol that uses chromatin immunoprecipitation (ChIP) to analyze the NFκB recruitment to endogenous IL-1 and TNF promoters in BZ-treated human macrophages. Corresponding to the BZ-suppressed mRNA levels of IL-1 and TNF, we show that BZ inhibits p65 NFκB recruitment to IL-1 and TNF promoters. This study specifically uses U937 macrophages, but the protocol could be easily modified to analyze the regulation of NFκB recruitment in other cell types.

Key words NFκB, IL-1β, TNFα, Macrophages, Bortezomib

1 Introduction

Macrophages play a pivotal role in the immune response by initiating, sustaining, and resolving inflammation. Induced by microbial invasion and/or tissue damage, macrophages initiate the production and release of both pro-inflammatory and anti-inflammatory cytokines. IL-1β and TNFα are two key pro-inflammatory cytokines involved in immunity and inflammation as well as cancer progression. Derived from the cell wall of gram-negative bacteria, lipopolysaccharide (LPS) may be used to activate macrophages to induce the production of these cytokines. Upon LPS activation, cytokine expression is induced through a cascade of events and regulated at the level of transcription [1–8].

The transcription factor NFκB family plays a crucial role in the transcriptional regulation of a variety of genes involved in the immune and inflammatory responses, cell survival, proliferation,

Ivana Vancurova (ed.), *Cytokine Bioassays: Methods and Protocols*, Methods in Molecular Biology, vol. 1172, DOI 10.1007/978-1-4939-0928-5_29, © Springer Science+Business Media New York 2014

and differentiation [9–13]. Constitutive activation of NFκB is implicated in numerous types of cancer, chronic inflammation, and angiogenesis [14–19]. Synthesis of pro-inflammatory cytokines, such as IL-1 and TNF, is mediated by NFκB; therefore, transcriptional regulation of such genes provides a potential therapeutic target.

The NFκB family is comprised of five subunits including p65 (Rel A), Rel B, c-Rel, p50, and p52 [9–13]. These subunits form specific homo- or heterodimers, imperative for target DNA binding. In unstimulated cells, these inactive NFκB dimers are often found in the cytoplasm bound to the inhibitory protein IκBα [9–11]. Induced by extracellular stimuli, such as inflammatory cytokines, IκBα is phosphorylated by the enzymes of IκB kinase (IKK) complex, ubiquitinated, and selectively degraded by the 26S proteasome, thus resulting in the release of NFκB [9–13]. Upon release, NFκB translocates to the nucleus and binds to the promoter sequences of NFκB-dependent pro-inflammatory and pro-survival genes, thus inducing their transcription.

Inhibition of NFκB activity and NFκB-dependent transcription by the 26S proteasome inhibitor bortezomib (BZ) has been used as a frontline therapy in multiple myeloma and other hematological malignancies [20–23]. However, our recent studies have demonstrated that BZ has a gene-specific effect on the regulation of NFκB-controlled genes, depending on the composition of the recruited NFκB complexes, and their posttranslational modifications [24–28].

In this chapter, we describe a protocol that uses chromatin immunoprecipitation (ChIP) to analyze the BZ regulation of NFκB p65 recruitment to IL-1β and TNFα promoters in human macrophages. We show that in LPS-stimulated human U937 macrophages, recruitment of NFκB p65 to IL-1β and TNFα promoters is suppressed by BZ, and corresponds with the reduced mRNA levels of IL-1β and TNFα. The main points of this protocol are: (1) Cell culture and incubation of differentiated U937 macrophages with BZ; (2) ChIP analysis of NFκB p65 recruitment to NFκB-dependent promoters using p65 antibody; and (3) Quantitative real-time PCR analysis using primers for IL-1β and TNFα gene promoters.

2 Materials

2.1 Cell Culture

1. U937 cells obtained from the American Type Culture Collection.
2. RPMI 1640 medium supplemented with 10 % fetal bovine serum (FBS), 2 mM glutamine, 1 mL sodium pyruvate, and penicillin–streptomycin solution.
3. Phorbol 12-myristate 23-acetate (PMA) dissolved at 1 mg/mL in dimethyl sulfoxide (DMSO) and stored at −80 °C.

4. Lipopolysaccharide (LPS) is dissolved at 1 mg/mL in sterile water and stored at −80 °C.
5. Bortezomib (BZ) dissolved at 10 mM in DMSO and stored at −20 °C.
6. Phosphate buffered saline (PBS), pH 7.4.
7. 75 cm^2 culture flasks.
8. Standard 6-well plates with clear flat bottom.
9. 2 mL microcentrifuge tubes.

2.2 Chromatin Immunoprecipitation

2.2.1 Protein–DNA Cross-Linking, Cell Lysis, and DNA Shearing

1. 37 % formaldehyde solution: Aliquot in sterile 15 mL tubes in the laminar flow cabinet and store at room temperature.
2. 2.5 M glycine solution: Dissolve 3.75 g of glycine in 20 mL of deionized water and store at room temperature.
3. 100 mM phenylmethylsulfonyl fluoride (PMSF): Dissolve 17 mg of PMSF in 1 mL of absolute ethanol. Store at −20 °C (*see* **Note 1**).
4. Protease inhibitor cocktail for mammalian cell extracts containing pepstatin A, bestatin, leupetin, aprotinin, 4-[2-aminoethyl]-benzenesulfonyl fluoride [AEBSF], and transepoxysuccinyl-L-leucylamido [4-guanidino] butane (E-64) (Sigma) (*see* **Note 2**).
5. SDS lysis buffer: 1 % (w/v) SDS, 10 mM EDTA and 50 mM Tris–HCl, pH 8.1.
6. ChIP dilution buffer: 0.01 % (w/v) SDS, 1.1 % (v/v) Triton X-100, 1.2 mM EDTA, 16.7 mM Tris–HCl, pH 8.1, and 167 mM NaCl.
7. 15 mL centrifuge tubes.
8. 1.5 mL microcentrifuge tubes.
9. Vibra cell sonicator.

2.2.2 Preclearing, Immunoprecipitation, and Washing

1. Protein A/G PLUS-agarose.
2. Immunoprecipitating antibody NFκB p65.
3. Low-salt immune complex wash buffer: 0.1 % (w/v) SDS, 1 % (v/v) Triton X-100, 2 mM EDTA, 20 mM Tris–HCl, pH 8.1, and 150 mM NaCl. Adjust pH to 8.1. Store at 4 °C. Add protease inhibitors (2 mM PMSF and 2 % protease inhibitor cocktail) just before use.
4. High-salt immune complete wash buffer: 0.1 % (w/v) SDS, 1 % (v/v) Triton X-100, 2 mM EDTA, 20 mM Tris–HCl, pH 8.1, and 500 mM NaCl. Adjust pH to 8.1. Store at 4 °C. Add protease inhibitors (2 mM PMSF and 2 % protease inhibitor cocktail) just before use.
5. LiCl immune complex was buffer: 0.25 M LiCl, 1 % (v/v) IGEPAL CA-630, 1 % (w/v) deoxycholic acid (Na salt), 0.1 %

(w/v) SDS, 1 mM EDTA and 10 mM Tris–HCl, pH 8.1. Store at 4 °C. Add protease inhibitors (2 mM PMSF and 2 % protease inhibitor cocktail) just before use.

6. 1× Tris–EDTA (TE) buffer: 1 mM EDTA and 10 mM Tris–HCl, pH 8.1. Adjust pH to 8.1. Store at 4 °C.

2.2.3 Elution and Reverse Cross-Linking

1. Elution buffer: 1 % SDS and 0.1 M NaHCO3. Add 100 μL of 10 % SDS and 100 μL of 1 M NaHCO3 to 0.8 mL of deionized water to make 1 mL of elution buffer. Each sample requires 500 μL of elution buffer. Prepare fresh before reaction; each sample requires 500 μL of elution buffer (*see* **Note 3**).
2. 5 M NaCl: Dissolve 29.2 g of NaCl in 100 mL deionized water. Store at 4 °C.
3. 0.5 M EDTA: Dissolve 18.6 g of ethylenediaminetetraacetic acid (EDTA) disodium salt dihydrate in 100 mL of deionized water. Store at 4 °C.
4. 1 M Tris–HCl, pH 6.5: Dissolve 12.1 g of Tris base in 80 mL of deionized water. Adjust pH to 6.5 with 6N HCl. Bring total volume to 100 mL. Filter and store at 4 °C.
5. Proteinase K solution, 20 mg/mL (Invitrogen).
6. 1.5 mL microcentrifuge tubes.

2.2.4 DNA Extraction

1. Phenol: chloroform: isoamyl alcohol, 25:24:1, pH 8.0.
2. Chloroform: isoamyl alcohol, 24:1.
3. Ethanol, absolute.
4. 70 % ethanol: Add 70 mL of absolute ethanol to deionized water to make up a volume of 100 mL. Store at 4 °C.
5. Gen-Elute-LPA.
6. 3 M sodium acetate (NaOAc): Dissolve 6.15 g of anhydrous NaOAc in 25 mL of deionized water. Adjust pH to 5.2 using glacial acetic acid. This stock solution can be stored at room temperature.
7. Nuclease-free water.
8. 1.5 mL microcentrifuge tubes.

2.3 Real-Time PCR

1. iQ SYBR Green Supermix (Bio-Rad).
2. iCycler iQ PCR Plates, 96-well (Bio-Rad).
3. iCycler iQ Optical Tape (Bio-Rad).
4. Nuclease-free water.
5. Specific primers (10 μM working stock solutions) for different promoter regions.
6. 1.5 mL microcentrifuge tubes.

3 Methods

In this section, we describe the protocol for analysis of NFκB recruitment to IL-1β and TNFα promoters using ChIP in U937 macrophages. However, this protocol can be easily modified for other cells as well. First, the promoter region of specific gene is obtained from NCBI Web site (http://www.ncbi.nlm.nih.gov/pubmed/) or (http://www.ncbi.nlm.nih.gov/gene/). By searching the name of the gene, the information about accession number, genomic location, DNA sequence, and transcript can be obtained (*see* **Note 4**). Then, we use the TFSEARCH Web site (http://www.cbrc.jp/research/db/TFSEARCH.html) to search the putative transcription factor binding sites. After entering the DNA sequence and choosing the classification, many putative binding sites of different transcriptional factors appear; sites with the score >85 should be considered (*see* **Note 5**). For designing the ChIP primers, there are two primer-designing Web sites that can be used: http://www.idtdna.com/Scitools/Applications/Primerquest/ or http://www.yeastgenome.org/cgi-bin/web-primer (*see* **Notes 6** and 7). The recruitment is then analyzed by quantitative real-time PCR. On average, this protocol can be accomplished within 4–6 days. Figure 1 illustrates the principle of the ChIP assay. Figure 2 illustrates IL-1β and TNFα gene expression through mRNA analysis. Figure 3 illustrates the recruitment of the NFκB subunit p65 to the promoter sites of IL-1β and TNFα.

3.1 Cell Culture

1. Grow U937 monocytic cells in complete RPMI medium to a concentration 1×10^6 cells/mL in tissue culture incubator at 37 °C in 5 % CO_2 humidified atmosphere.
2. Induce differentiation of U937 cells into macrophages by adding PMA at a final concentration 10 ng/mL. Mix well but gently. Transfer the cells to a 6-well plate; each well will receive 5 mL of the above cell suspension.
3. Incubate the plate 1 day in tissue culture incubator to allow cell differentiation (induced by 24 h treatment with PMA).
4. After 24 h, add BZ at a final concentration of 100 nM and/or add LPS at a final concentration of 1 μg/mL. Incubate the cells in tissue culture incubator for the desired times.

3.2 Chromatin Immunoprecipitation

3.2.1 Protein–DNA Cross-Linking, Cell Lysis, and DNA Shearing

1. At desired times, cross-link proteins to DNA by adding 135 μL of 37 % formaldehyde stock solution to 5 mL of cell culture so that the final concentration of formaldehyde is 1 %. Perform this step in the laminar flow cabinet.
2. Incubate the cells with formaldehyde for 15 min in tissue culture incubator (37 °C, 5 % CO_2 humidified atmosphere). Be consistent with the fixation conditions for all time points (*see* **Note 8**).

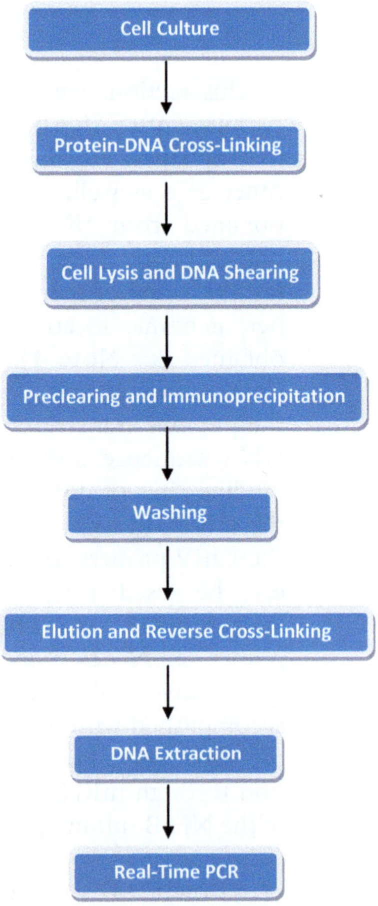

Fig. 1 Schematic illustration of the ChIP assay. First, cells are fixed with formaldehyde and lysed. Subsequently, chromatin is fragmented to around 500 bp fragments by sonication. The transcription factor–chromatin complex is then immunoprecipitated by using specific antibody. After DNA purification, the enrichment of particular DNA sequence is quantified by real-time PCR

3. Stop the formaldehyde induced cross-linking by adding 250 μL of 2.5 M glycine solution into each well so that the final concentration of glycine is 0.125 M.
4. Scrape the cells from the wells and collect them in 15 mL centrifuge tubes. Centrifuge at 1,700 × *g* for 4 min in a refrigerated centrifuge (*see* **Note 9**).
5. Carefully aspirate the supernatant and remove as much medium as possible.
6. Add 2 mL of ice-cold PBS containing protease inhibitors into the tubes and centrifuge at 1,700 × *g* for 4 min in a refrigerated centrifuge. Repeat this wash step one more time.
7. Resuspend the pellets in 250 μL of SDS lysis buffer and transfer to a prechilled 1.5 mL microcentrifuge tube. Incubate on ice for 10 min to aid the cell lysis.

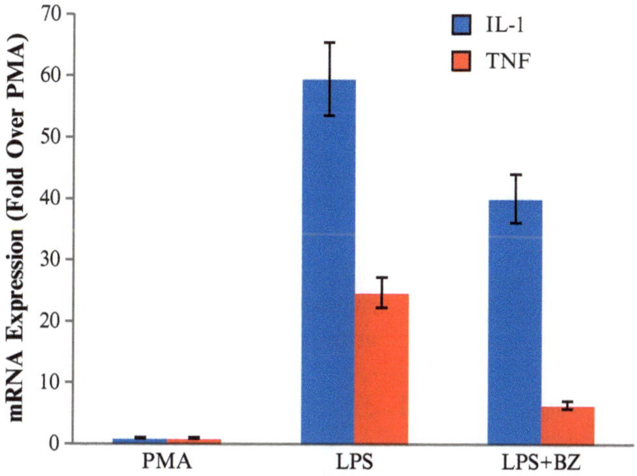

Fig. 2 mRNA analysis of IL-1β and TNFα in BZ-treated human macrophages. U937 cells were differentiated with 10 ng/mL PMA for 24 h. Cells were then treated 2 h with BZ (100 nM), followed by LPS stimulation (1 μg/mL, 6 h). mRNA levels were measured and quantified by real-time PCR. The data are presented as a fold over PMA mRNA levels without LPS stimulation or BZ treatment

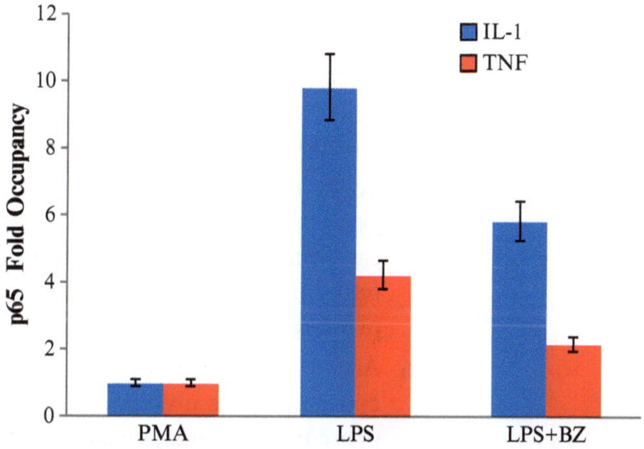

Fig. 3 ChIP analysis of NFκB p65 recruitment to IL-1β and TNFα promoters in U937 macrophages. U937 cells were differentiated with 10 ng/mL PMA for 24 h. Cells were then treated 2 h with BZ (100 nM), followed by LPS stimulation (1 μg/mL, 6 h). The NFκB p65 binding to endogenous IL-1β and TNFα promoters was measured by ChIP analysis and quantified by real-time PCR. The data are presented as the change in occupancy over the human IGX1A (SA Biosciences) sequence control

8. Sonicate the lysates on ice to shear DNA into 400–500 bp fragments. Sonication conditions: four 10-s pulses followed by 30-s rest periods at output 40 (*see* **Note 10**).
9. Centrifuge the sonicated samples for 10 min at 12,000×*g* at 4 °C (*see* **Note 9**).

10. Transfer the supernatants to new prechilled 2 mL microcentrifuge tubes and discard the pellets.
11. Dilute the sonicated cell supernatants to a final volume of 2 mL by adding 1.75 mL of ChIP dilution buffer containing the protease inhibitors.
12. Set aside a portion of the diluted cell supernatant (100 μL) to quantify the amount of DNA present in different samples. This sample is considered to be your input/starting material and needs to have the protein–DNA cross-links reversed by heating at 65 °C for 4 h (*see* **step 4** of Subheading 3.2.3).

3.2.2 Preclearing, Immunoprecipitation, and Washing

1. To reduce the nonspecific background, add 75 μL of protein A/G PLUS-agarose slurry to 1.9 mL of the diluted cell supernatant (from **step 11** of Subheading 3.2.1) and incubate for 1 h at 4 °C while rotating the tubes on a shaker (*see* **Note 11**).
2. Pellet the agarose beads by brief centrifugation at 150×*g* at 4 °C and carefully collect the supernatant in a prechilled 2 mL microcentrifuge tube.
3. Add 5 μg of immunoprecipitating p65 antibody per reaction and incubate overnight at 4 °C with constant rotation in a shaker (*see* **Note 12**).
4. The next day, add 75 μL of protein A/G PLUS-agarose slurry to each sample and incubate for 2 h at 4 °C with rotation to collect the specific antibody–protein complexes (*see* **Notes 11** and **13**).
5. Pellet the agarose beads with bound protein complexes by gentle centrifugation (150×*g* at 4 °C for 1 min). Carefully discard the supernatants containing unbound, nonspecific protein–DNA. The A/G agarose pellets now contain the specific protein–antibody complexes.
6. Add 1 mL of ice-cold low-salt immune complex wash buffer containing the protease inhibitors to the tubes containing the protein A/G agarose antibody–protein complexes. Incubate the complexes at 4 °C for 5 min with constant rotation.
7. Pellet the agarose beads by centrifugation (150×*g*, 1 min, 4 °C). Carefully discard the supernatants and keep the pellets on ice.
8. Add 1 mL of ice-cold high-salt immune complex wash buffer containing the protease inhibitors to protein A/G agarose antibody–protein complexes and incubate 5 min at 4 °C while rotating on a shaker.
9. Centrifuge the agarose beads at 150×*g* for 1 min at 4 °C. Carefully discard the supernatants and add 1 mL of ice-cold LiCl immune complex wash buffer containing the protease inhibitors to the pellets. Incubate the beads at 4 °C for 5 min with constant rotation.

10. Centrifuge the agarose beads at 150×*g* for 1 min at 4 °C. Carefully discard the supernatants and add 1 mL of ice-cold 1× TE buffer. Incubate the beads at 4 °C for 5 min with constant rotation. Centrifuge as described above and carefully remove the supernatant as much as possible.

3.2.3 Elution and Reverse Cross-Linking

1. Add 250 μL of freshly prepared elution buffer to the pelleted A/G agarose antibody–protein complexes. Vortex briefly (5 s) and incubate at room temperature for 10 min with constant rotation.
2. Pellet down the agarose beads by centrifuging at 150×*g* for 1 min at room temperature. Carefully transfer the supernatant (eluate) to a new, labeled 1.5 mL microcentrifuge tube.
3. Add 250 μL of freshly prepared elution buffer to the pellets. Repeat **steps 1** and **2** described above to elute the remaining proteins from the agarose beads. Centrifuge at 150×*g* for 1 min at room temperature and transfer the supernatant to 1.5 mL microcentrifuge tube already containing 250 μL of supernatant from **step 2**. You should now have 500 μL of eluate per reaction.
4. Add 20 μL of 5 M NaCl to the combined eluates (500 μL) and 4 μL of 5 M NaCl to the input/starting material (100 μL) from **step 12** of Subheading 3.2.1 to reverse the protein–DNA cross-links. Incubate the complexes 4 h in a 65 °C water bath.
5. After 4 h, add 10 μL of 0.5 M EDTA, 20 μL of 1 M Tris–HCl, pH 6.5, and 1 μL of 20 mg/mL proteinase K to the combined eluates. Incubate for 1 h at 45 °C.
6. For the input sample, add 2 μL of 0.5 M EDTA, 4 μL of 1 M Tris–HCl, pH 6.5, and 0.2 μL of 20 mg/mL proteinase K and incubate for 1 h at 45 °C.

3.2.4 DNA Extraction

1. Add equal volume of phenol: chloroform: isoamyl alcohol (25:24:1) to the microcentrifuge tube containing the eluate with protein–DNA cross-links reversed. Vortex vigorously for 5 s. Centrifuge at 12,000×*g* for 3 min at room temperature. Collect the top (aqueous) layer.
2. Add equal volume of chloroform: isoamyl alcohol (24:1) to the aqueous layer from **step 1** (above) in a microcentrifuge tube. Vortex vigorously for 5 s. Centrifuge at 12,000×*g* for 3 min at room temperature. Collect again the top (aqueous) layer.
3. Add twice the volume of 100 % (absolute) ethanol, 1/10th volume of 3 M NaOAc, pH 5.2, and 2 μL of GenElute-LPA to the aqueous layer from **step 2** in a microcentrifuge tube. Mix by inverting the tube. Keep the microcentrifuge tube at −20 °C overnight to allow DNA precipitation (*see* **Note 14**).
4. The next day, centrifuge the microcentrifuge tube at 12,000×*g* for 30 min at 4 °C. Collect the DNA containing pellet.

5. Wash the pellet with 100 μL of ice-cold 70 % ethanol. Centrifuge at 12,000 × *g* for 10 min at 4 °C. Collect the pellet. Try to remove as much ethanol as possible.
6. Allow the pellet to air dry.
7. Dissolve the DNA pellet in 50 μL of nuclease-free water. Perform real-time PCR as described below.

3.3 Real-Time PCR

3.3.1 Design and Dilution of the Primers

1. Design all PCR primers with the optimum annealing temperature at 55 °C.
2. Prepare a 100 μM stock of both forward and reverse primers. From the 100 μM primer stocks, prepare 10 μM working stock solutions for PCR reaction (*see* **Note 15**).
3. Each PCR reaction has a volume of 25 μL and will use 1 μL of each 10 μM primer so that the final concentration of each primer in PCR reaction is 0.4 μM.
4. The input DNA should be diluted five times in nuclease-free water before performing PCR reaction. Each PCR reaction (25 μL volume) for the input sample will use 5 μL of five times diluted input DNA.

3.3.2 Setting Up the Real-Time PCR

1. Set up the PCR reaction on ice.
2. Each reaction using immunoprecipitated DNA has a total volume of 25 μL and is set up as follows (*see* **Note 16**):
   (a) SYBR Green Supermix: 12.5 μL.
   (b) Nuclease-free water: 7.5 μL.
   (c) Forward primer (10 μM): 1.0 μL.
   (d) Reverse primer (10 μM): 1.0 μL.
   (e) DNA sample: 3.0 μL.
3. Each reaction using input DNA has a total volume of 25 μL and is set up as follows:
   (a) SYBR Green Supermix: 12.5 μL.
   (b) Nuclease-free water: 5.5 μL.
   (c) Forward primer (10 μM): 1.0 μL.
   (d) Reverse primer (10 μM): 1.0 μL.
   (e) Input DNA sample (five times diluted): 5.0 μL.
4. After loading the master-mix and the DNA into the wells of the PCR plate, seal the plate carefully using an optical tape and continue with the real-time PCR reaction in the thermal cycler.
5. Each reaction for both the immunoprecipitated and input samples should include a positive and negative control (*see* **Note 17**).

4 Notes

1. PMSF is unstable in aqueous environment. It is essential that it is dissolved in absolute alcohol (ethanol, methanol, or isopropanol); it will not freeze at −20 °C. Add PMSF to buffers in the final working concentration of 2 mM just before use.
2. This protease inhibitor cocktail contains protease inhibitors with a broad specificity for the inhibition of serine, cysteine and aspartic proteases, and aminopeptidases. It should be stored at −20 °C, and added to buffers just before use.
3. Prepare stock solutions of 10 % SDS and 1 M $NaHCO_3$ by dissolving 5 g of SDS in 50 mL of deionized water, and 4.2 g of $NaHCO_3$ in 50 mL deionized water, respectively. These stock solutions can be stored at room temperature.
4. It is important to get all information about the genomic sequence of promoter region, mRNA, and protein. Copy and save the genomic sequence that is near the transcription start site (TSS), especially the 2,000 bp region upstream and 500 bp region downstream of the TSS.
5. A score higher than 85 indicates binding of the transcription factor to the sequence.
6. For designing the ChIP primers, select 150 nucleotides upstream and 150 nucleotides downstream from the transcription factor (TF) binding site. Below are some general rules for designing the ChIP primers:
   (a) Primer length (bp): Optimum, 20; Minimum, 18; Maximum, 22.
   (b) Tm (°C): Optimum, 55; Minimum, 53; Maximum, 57.
   (c) GC content (%): Optimum, 50; Minimum, 40; Maximum, 60.
   (d) The size of the PCR fragment should be between 100 bp and 200 bp.

   Design at least two sets of primers for each TF binding site, since the primers not always work properly.
7. The reverse copy of the primer or a sequence can be reversed manually or using the Web site http://www.thelabnotebook.com/sequence.php?seq=1.
8. Formaldehyde is a reversible protein–DNA cross-linking agent that preserves protein–DNA interactions in cells.
9. It is helpful to always centrifuge the tubes in one position (for example, when using 1.5 mL microcentrifuge tubes, position them with cap snaps facing towards the center of the rotor). This way you can always expect the pellets to be at the same place.

10. Make sure to keep the samples on ice at all times in between the shearing. In addition, it is also helpful to place the tip of the sonicator a beaker filled with ice for 30 s in between successive 10-s pulses to ensure that the tip is not overheated. When using a new sonicator, it is important to calibrate it so that the sheared DNA fragments are between 400 bp and 500 bp in size.
11. Make sure Protein A/G PLUS-agarose slurry is completely resuspended before adding.
12. In this protocol, we use 2.5 μL of p65 NFκB antibody (200 μg/0.1 mL) to obtain 5 μg of antibody per each immunoprecipitation reaction.
13. At this point, the agarose beads bind the antibody–TF–DNA complexes.
14. The presence of LPA during ethanol precipitation results in complete recovery of fragments larger than 20 base pairs. The nucleic acid-LPA coprecipitate is visible upon addition of ethanol.
15. The primers for amplifying the NFκB binding sites in IL-1β and TNFα promoters were designed to anneal optimally at 55 °C. Prepare a 10 μM working stock solution of each primer and store at −20 °C.
16. It is convenient to prepare a master-mix by mixing all components (for desired amount of reactions) except for the DNA. Prepare the master-mix in a 1.5 mL microcentrifuge tube. Aliquot 22 μL of the master-mix into each well of the PCR plate. Add 3 μL of immunoprecipitated DNA sample into each well.
17. The positive control primers provide a control for successful chromatin immunoprecipitation and gene transcription. In this protocol, we use the ChIP-qPCR Human GAPDH Proximal Promoter Positive Control primer (GPH10001C(+)01A, Qiagen). The negative primer provides a reference of the amount of nonspecific genomic DNA that co-immunoprecipitates during the procedure. In this protocol, we use ChIP-qPCR Human 1GX1A Negative Control primer (GPH 00001C(−)01A, Qiagen).

References

1. Dinarello CA (2011) Interleukin-1 in the pathogenesis and treatment of inflammatory diseases. Blood 117:3720–3732
2. Lopez-Castejon G, Brough D (2011) Understanding the mechanism of IL-1β secretion. Cytokine Growth Factor Rev 22:189–195
3. Kasza A (2013) IL-1 and EGF regulate expression of genes important in inflammation and cancer. Cytokine 62:22–23
4. Marusic J, Podlipnik C, Jevsevar S et al (2012) Recognition of human tumor necrosis factor α (TNF-α) by therapeutic antibody fragment: energetics and structural features. J Biol Chem 287:8613–8620
5. Barbara JA, Van Ostade X, Lopez A (1996) Tumor necrosis factor α (TNF-α): the good, the bad, and potentially very effective. Immunol Cell Biol 74:434–443

6. Vilcek J, Lee TH (1991) Tumor necrosis factor: new insights into the molecular mechanisms of its multiple actions. J Biol Chem 266:7313–7316
7. Guoqing C, Goeddel DV (2002) TNF-R1 signaling: a beautiful pathway. Science 296: 1634–1635
8. Mancino A, Lawrence T (2010) NFκB and tumor-associated macrophages. Clin Cancer Res 16:784–789
9. Baeuerle PA, Baltimore D (1996) NFκB: ten years after. Cell 87:13–20
10. Baldwin AS (2001) The transcription factor NFκB and human disease. J Clin Invest 107:3–6
11. Yamamoto Y, Gaynor RB (2001) Therapeutic potential of inhibition of the NFκB pathway in the treatment of inflammation and cancer. J Clin Invest 107:135–142
12. Ghosh S, Karin M (2002) Missing pieces in the NFκB puzzle. Cell 109:S81–S96
13. Wan F, Lenardo MJ (2010) The nuclear signaling of NFκB: current knowledge, new insights, and future perspectives. Cell Res 20:24–33
14. Prasad S, Ravindran J, Aggarwal BB (2010) NFκB and cancer: how intimate is the relationship. Mol Cell Biochem 336:25–37
15. Li Q, Withoff S, Verma IM (2005) Inflammation-associated cancer: NFκB is the lynchpin. Trends Immunol 26:318–325
16. Grivennikov SI, Greten FR, Karin M (2010) Immunity, inflammation, and cancer. Cell 140: 883–899
17. Ghosh S, Hayden MS (2008) New regulators of NFκB in inflammation. Nat Rev Immunol 8:837–848
18. Hayden MS, Ghosh S (2012) NFκB, the first quarter-century: remarkable progress and outstanding questions. Genes Dev 26:203–234
19. Fan Y, Dutta J, Gupta N et al (2008) Regulation of programmed cell death by NFκB and its role in tumorigenesis and therapy. Adv Exp Med Biol 615:223–250
20. Hideshima T, Chauhan D, Richardson P et al (2002) NFκB as a therapeutic target in multiple myeloma. J Biol Chem 277:16639–16647
21. Richardson PG, Mitsiades C, Hideshima T et al (2005) Proteasome inhibition in the treatment of cancer. Cell Cycle 4:290–296
22. McConkey DJ, Zhu K (2008) Mechanisms of proteasome inhibitor action and resistance in cancer. Drug Resist Updat 11:164–169
23. Shah JJ, Orlowski RZ (2009) Proteasome inhibitors in the treatment of multiple myeloma. Leukemia 23:1964–1979
24. Vu HY, Juvekar A, Ghosh C et al (2008) Proteasome inhibitors induce apoptosis of prostate cancer cells by inducing nuclear translocation of IκBα. Arch Biochem Biophys 475:156–163
25. Ghosh CC, Ramaswami S, Juvekar A et al (2010) Gene specific repression of proinflammatory cytokines in stimulated human macrophages by nuclear IκBα. J Immunol 185: 3685–3693
26. Juvekar A, Manna S, Ramaswami S et al (2011) Bortezomib induces nuclear translocation of IκBα resulting in gene-specific suppression of NFκB-dependent transcription and induction of apoptosis in CTCL. Mol Cancer Res 9:183–194
27. Manna S, Singha B, Phyo SA et al (2013) Proteasome inhibition by bortezomib increases IL-8 expression in androgen-independent prostate cancer cells: the role of IKKα. J Immunol 191:2837–2846
28. Singha B, Gatla H, Manna S et al (2014) Proteasome inhibition increases recruitment of IKKβ, S536P-p65 and transcription factor EGR1 to interleukin-8 (IL-8) promoter, resulting in increased IL-8 production in ovarian cancer cells. J Biol Chem 289:2687–2700

Chapter 30

Analysis of TGFβ1 and IL-10 Transcriptional Regulation in CTCL Cells by Chromatin Immunoprecipitation

Tzu-Pei Chang, Myra Kim, and Ivana Vancurova

Abstract

The immunosuppressive cytokines transforming growth factor β1 (TGFβ1) and interleukin-10 (IL-10) regulate a variety of biological processes including differentiation, proliferation, tissue repair, tumorigenesis, inflammation, and host defense. Aberrant expression of TGFβ1 and IL-10 has been associated with many types of autoimmune and inflammatory disorders, as well as with many types of cancer and leukemia. Patients with cutaneous T cell lymphoma (CTCL) have high levels of malignant CD4+ T cells expressing IL-10 and TGFβ1 that suppress the immune system and diminish the antitumor responses. The transcriptional regulation of TGFβ1 and IL-10 expression is orchestrated by several transcription factors, including NFκB. However, while the transcriptional regulation of pro-inflammatory and anti-apoptotic genes by NFκB has been studied extensively, much less is known about the NFκB regulation of immunosuppressive genes. In this chapter, we describe a protocol that uses chromatin immunoprecipitation (ChIP) to analyze the transcriptional regulation of TGFβ1 and IL-10 by measuring recruitment of NFκB p65, p50, c-Rel, Rel-B, and p52 subunits to TGFβ1 and IL-10 promoters in human CTCL Hut-78 cells.

Key words Chromatin immunoprecipitation, Cutaneous T cell lymphoma, IL-10, NFκB, TGFβ1, Transcriptional regulation

1 Introduction

TGFβ1 and IL-10 are immunosuppressive cytokines that have been linked to immune defects associated with malignancy and autoimmune disorders, increased susceptibility to infections, as well as chronic inflammatory conditions. Cutaneous T cell lymphoma (CTCL) represents a heterogeneous group of non-Hodgkin lymphomas characterized by skin invasive T cells that exhibit Th2-like phenotype by secreting the immunosuppressive cytokines IL-10 and TGFβ1 [1–6]. In patients with Sézary syndrome, an erythrodermic leukemic variant of CTCL, the skin CTCL cells contain a high level of constitutive NFκB activity, which is responsible for the resistance to apoptosis of these cells [7–12]. In addition to the increased expression of anti-apoptotic and pro-inflammatory genes

Ivana Vancurova (ed.), *Cytokine Bioassays: Methods and Protocols*, Methods in Molecular Biology, vol. 1172, DOI 10.1007/978-1-4939-0928-5_30,

in CTCL cells, resulting in their increased proliferation and survival, NFκB also induces expression of IL-10 and TGFβ1, thus contributing to the immunosuppressive nature of CTCL [13–16]. However, while the NFκB regulation of anti-apoptotic and pro-inflammatory genes has been studied extensively, relatively very little is known about the NFκB regulation of immunosuppressive genes.

The mammalian NFκB family consists of p65 (RelA), c-Rel, Rel-B, p50 (NFκB1), and p52 (NFκB2) subunits. These proteins form various combinations of homodimers and heterodimers that regulate their target genes by binding to the specific κB enhancer sites [17–19]. The signaling pathways that mediate NFκB activation can be broadly divided into the canonical and noncanonical pathways. The canonical pathway is engaged by ligands for antigen and cytokine receptors, and leads to the cytoplasmic degradation of the NFκB inhibitor, IκBα, resulting in the nuclear translocation of p50/65 and p50/c-Rel heterodimers. The noncanonical NFκB pathway is initiated by stimulation of different signaling molecules and leads to the activation of p52/RelB dimers [20–23]. While the cytoplasmic pathways leading to nuclear translocation and activation of NFκB dimers have been studied extensively, much less is known about the nuclear events regulating NFκB-dependent transcription [24–26].

Regulation of IL-10 and TGFβ1 expression is complex, and is controlled by several transcription factors and regulators, including NFκB. In vitro studies using the electrophoretic mobility shift assay (EMSA) in CTCL Hut-78 cells have indicated that the NFκB-binding site in human IL-10 promoter is regulated predominantly by p50/50 homodimers [15, 16]. However, more recent chromatin immunoprecipitation (ChIP) experiments have shown that the ability of specific genes to selectively recruit certain NFκB dimers cannot be predicted based on the in vitro EMSA data or over-expression experiments. For example, previous studies from our laboratory have demonstrated that while the p65/p50 heterodimers are the most predominant NFκB dimers in CTCL Hut-78 cells, they regulate only a subset of NFκB dependent genes, such as cIAP1 and cIAP2, while Bcl2 is regulated mainly by p50/50 homodimers [27–29].

In this chapter, we describe a protocol that uses ChIP to analyze the regulation of IL-10 and TGFβ1 expression by NFκB subunits in CTCL Hut-78 cells. The main points of this protocol are: (1) identification of the putative NFκB-binding sites within the IL-10 and TGFβ1 promoter regions and designing specific primers for the ChIP assay; (2) ChIP analysis of NFκB recruitment to IL-10 and TGFβ1 promoter regions using p65, p50, c-Rel, Rel-B, and p52 NFκB antibodies; and (3) quantitative real-time PCR analysis using primers for human TGFβ1 and IL-10 gene promoters.

2 Materials

2.1 Cell Culture

1. CTCL Hut-78 cells (American Type Culture Collection).
2. RPMI complete medium: RPMI supplemented with 10 % fetal bovine serum (FBS), 2 mM glutamine, 1 mM sodium pyruvate and 1 % penicillin-streptomycin solution.
3. 75 cm^2 culture flasks.
4. 6-well plates with clear flat bottom.

2.2 Chromatin Immunoprecipitation

1. 37 % formaldehyde solution: Aliquot in sterile 15 mL tubes in the laminar flow cabinet, and store at a room temperature.
2. 2.5 M glycine solution: Dissolve 3.75 g of glycine in 20 mL of deionized water, and store at room temperature.
3. Phosphate buffer saline (PBS, pH 7.4).
4. 100 mM phenylmethylsulfonyl fluoride (PMSF): Dissolve 17 mg of PMSF in 1 mL of absolute ethanol. Store at −20 °C (*see* **Note 1**).
5. Protease inhibitor cocktail for mammalian cell extracts containing pepstatin A, bestatin, leupeptin, aprotinin, 4-[2-aminoethyl]-benzenesulfonyl fluoride [AEBSF], and trans-epoxysuccinyl-L-leucylamido [4-guanidino] butane (E-64) (*see* **Note 2**).
6. SDS lysis buffer: 1 % (w/v) SDS, 10 mM EDTA, 50 mM Tris–HCl, pH 8.1.
7. ChIP dilution buffer: 0.01 % (w/v) SDS, 1.1 % (v/v) Triton X-100, 1.2 mM EDTA, 16.7 mM Tris–HCl, pH 8.1, 167 mM NaCl.
8. Protein A/G PLUS-Agarose.
9. Immunoprecipitating antibodies: Antibodies recognizing human NFκB p65, NFκB p50, c-Rel, Rel-B, and p52 subunits.
10. Low-salt immune complex wash buffer: 0.1 % (w/v) SDS, 1 % (v/v) Triton X-100, 2 mM EDTA, 20 mM Tris–HCl, pH 8.1, 150 mM NaCl. Adjust pH to 8.1. Store at 4 °C. Add protease inhibitors (2 mM PMSF and 2 % (v/v) protease inhibitor cocktail) just before use.
11. High-salt immune complex wash buffer: 0.1 % (w/v) SDS, 1 % (v/v) Triton X-100, 2 mM EDTA, 20 mM Tris–HCl, pH 8.1, 0.5 M NaCl. Adjust pH to 8.1. Store at 4 °C. Add protease inhibitors (2 mM PMSF and 2 % protease inhibitor cocktail) just before use.
12. LiCl immune complex wash buffer: 0.25 M LiCl, 1 % (v/v) IGEPAL CA-630, 1 % (w/v) deoxycholic acid (Na salt), 0.1 % (w/v) SDS, 1 mM EDTA, 10 mM Tris–HCl, pH 8.1.

Adjust pH to 8.1. Store at 4 °C. Add protease inhibitors (2 mM PMSF and 2 % protease inhibitor cocktail) just before use.

13. 1× Tris-EDTA (TE) buffer: 1 mM EDTA, 10 mM Tris–HCl, pH 8.1. Adjust pH to 8.1. Store at 4 °C.
14. Elution buffer: 1 % SDS, 0.1 M $NaHCO_3$. Add 100 μL of 10 % SDS and 100 μL of 1 M $NaHCO_3$ to 800 μL of deionized water to make 1 mL of elution buffer. Prepare fresh before each reaction; each sample requires 500 μL of elution buffer (*see* **Note 3**).
15. 5 M NaCl: Dissolve 29.2 g of NaCl in 100 mL of deionized water. Store at 4 °C.
16. 0.5 M EDTA: Dissolve 18.6 g of ethylenediaminetetraacetic-acid (EDTA) disodium salt dihydrate in 100 mL of deionized water. Store at 4 °C.
17. 1 M Tris–HCl, pH 6.5: Dissolve 12.1 g of Tris base in 80 mL of deionized water. Adjust pH to 6.5 with 6 N HCl. Bring the total volume to 100 mL, filter, and store at 4 °C.
18. Proteinase K solution, 20 mg/mL.
19. Phenol: chloroform: isoamyl alcohol mixture, 25:24:1, pH 8.0.
20. Chloroform: isoamyl alcohol mixture, 24:1.
21. Ethanol, absolute.
22. 70 % Ethanol: Add 70 mL of absolute ethanol to deionized water to make up a volume of 100 mL. Store at 4 °C.
23. Linear polyacrylamide (LPA), 25 mg/mL.
24. 3 M sodium acetate (NaOAc): Dissolve 6.15 g of anhydrous NaOAc in 25 mL of deionized water. Adjust pH to 5.2 using glacial acetic acid. This stock solution can be stored at room temperature.
25. Nuclease-free water.

2.3 Real-Time Polymerase Chain Reaction

1. SYBR Green Supermix.
2. Real-Time PCR Plates, 96 well.
3. Optical tape.
4. Specific primers (10 μM working stock solutions) for different promoter regions.
5. Nuclease-free water.

3 Methods

In this section, we describe the protocol for analysis of NFκB recruitment to IL-10 and TGFβ1 promoters using chromatin immunoprecipitation (ChIP) in CTCL Hut-78 cells.

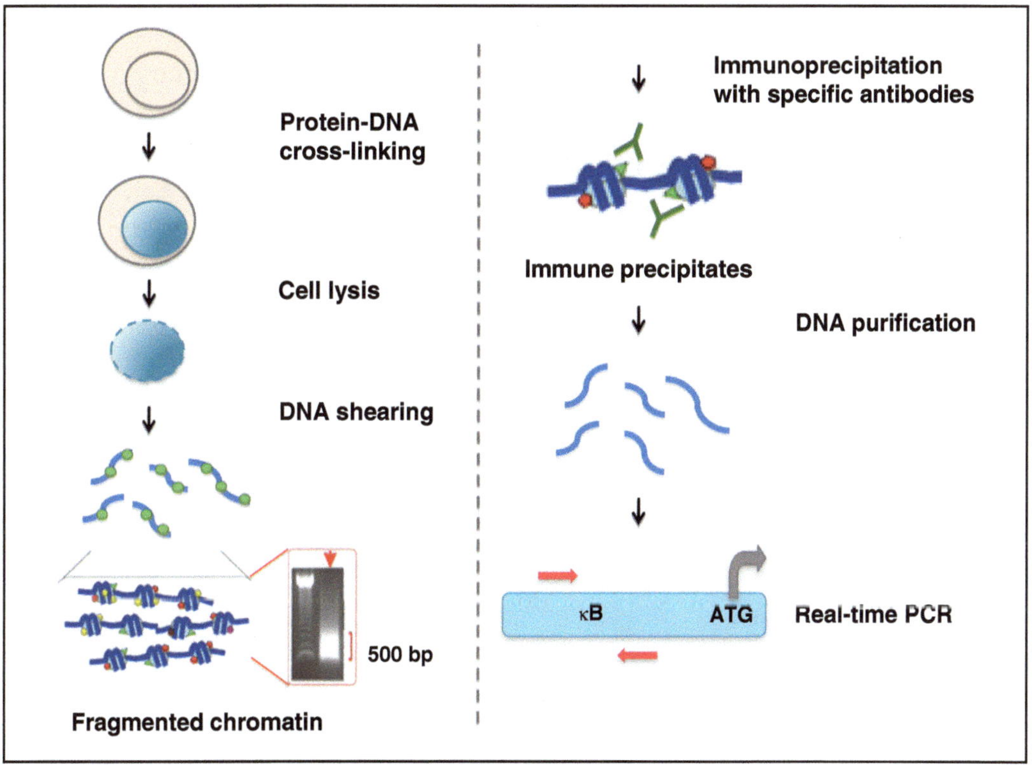

Fig. 1 Schematic illustration of the ChIP assay. First, cells are fixed with formaldehyde and lysed. Subsequently, chromatin is fragmented to around 500 bp fragments by sonication. The transcription factor–chromatin complexes are immunoprecipitated by using specific antibodies. After DNA purification, the enrichment of particular DNA sequence is quantified by real-time PCR

However, this protocol can be easily modified for other cells as well. First, the promoter region of specific genes (IL-10 and TGFβ) is obtained from NCBI website (http://www.ncbi.nlm.nih.gov/pubmed/) or (http://www.ncbi.nlm.nih.gov/gene/). By searching the name of the gene, the information about accession number, genomic location, DNA sequence, and transcript can be obtained (*see* **Note 4**). Then, we use the TFSEARCH website (http://www.cbrc.jp/research/db/TFSEARCH.html) to search for the putative transcription factor-binding sites. After entering the DNA sequence and choosing the classification, many putative binding sites of different transcriptional factors appear; sites with the score >85 should be considered (*see* **Note 5**). For designing the ChIP primers, there are two primer-designing websites that can be used: http://www.idtdna.com/Scitools/Applications/Primerquest/ or http://www.yeastgenome.org/cgi-bin/web-primer (*see* **Notes 6** and **7**). The recruitment is then analyzed by quantitative real-time PCR. On average, this protocol can be accomplished within 4–6 days. Figure 1 illustrates the principle of the ChIP assay. Figure 2 illustrates the NFκB-binding sites in human IL-10 and TGFβ promoters. Figure 3 illustrates the recruitment of individual NFκB subunits to these sites.

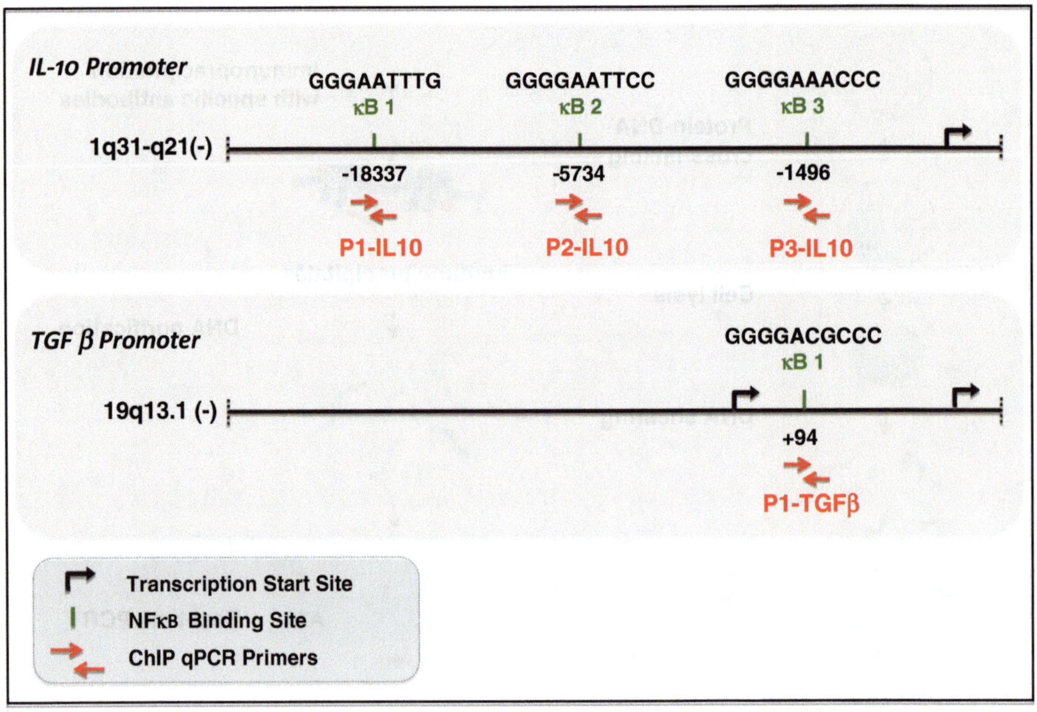

Fig. 2 The putative NFκB binding sites in IL-10 and TGFβ promoters. IL-10 promoter locates in the genomic region 1q31-q21(–) and contains three putative NFκB-binding sites: κB 1 (–188,337), κB 2 (–5,734), and κB 3 (–1,496). Three specific ChIP primers, P1-IL10, P2-IL10, and P3-IL10 were designed to analyze the NFκB recruitment to IL-10 promoter. TGFβ1 locates in the genomic region 19q13.1(–) and contains only one putative NFκB-binding site, κB 1 (+94). The specific ChIP primer P1-TGFβ1 was designed to analyze the NFκB recruitment to TGFβ1 promoter

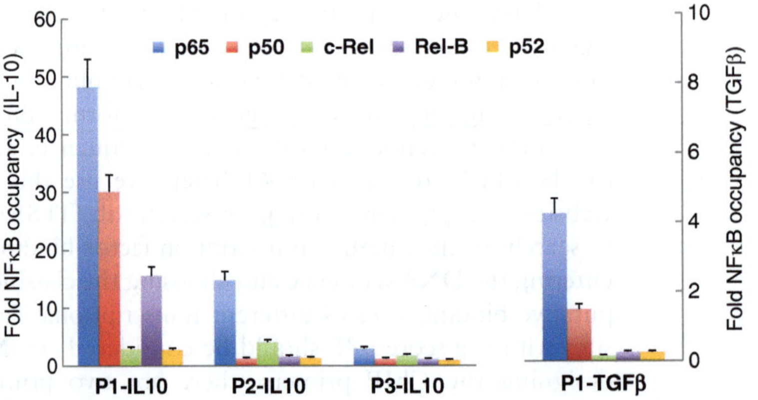

Fig. 3 ChIP analysis of p65, p50, c-Rel, Rel-B, and p52 recruitment to IL-10 and TGFβ1 promoters in Hut-78 cells. Recruitment of NFκB subunits p65, p50, c-Rel, Rel-B, and p52 to IL-10 and TGFβ1 promoters was analyzed by ChIP and quantified by real-time PCR. The data are presented as the change in occupancy over the human IGX1A sequence control

3.1 Cell Culture and Chromatin Immunoprecipitation (ChIP)

1. Grow Hut-78 cells in RPMI complete medium to 1×10^6 cells/mL concentration. Incubate in a humidified 5 % CO_2 atmosphere at 37 °C overnight. At desired times, cross-link proteins to DNA by adding 135 μL of 37 % formaldehyde stock solution to 5 mL of cell culture, so that the final concentration of formaldehyde is 1 %. Perform this step in the laminar flow cabinet.
2. Incubate cells with formaldehyde for 15 min in culture incubator (37 °C, 5 % CO_2). Be consistent with the fixation condition for all time points (*see* **Note 8**).
3. Neutralize the formaldehyde induced cross-linking by adding 250 μL of 2.5 M glycine solution into each well so that the final concentration of glycine is 0.125 M.
4. Collect cells into 15 mL centrifuge tubes and then centrifuge at 1,700 × *g* for 5 min in a refrigerated centrifuge (*see* **Note 9**).
5. Carefully remove the supernatant as much as possible. Resuspend the cell pellets in 2 mL of ice-cold PBS containing the protease inhibitors (2 mM PMSF and 2 % protease inhibitor cocktail) and centrifuge at 1,700 × *g* for 5 min in a refrigerated centrifuge. Repeat this wash step one more time.
6. Carefully remove the supernatant as much as possible. Resuspend the cell pellets in 250 μL of SDS lysis buffer and transfer to a prechilled 1.5 mL microcentrifuge tube. Incubate on ice for 10 min to aid the cell lysis.
7. Sonicate the cell lysates on ice to shear DNA into 400–500 bp fragments. Sonication conditions: four 10-s pulses followed by 30-s rest periods at output 40 (*see* **Note 10**).
8. Centrifuge the sonicated samples for 10 min at 12,000 × *g* at 4 °C (*see* **Note 9**). Transfer the supernatants to new prechilled 2 mL microcentrifuge tubes and discard the pellets.
9. Dilute the sonicated cell supernatants to a final volume of 2 mL by adding 1.75 mL of ChIP dilution buffer containing the protease inhibitors (2 mM PMSF and 2 % protease inhibitor cocktail).
10. Set aside a portion of the diluted cell supernatant (100 μL) to quantify the amount of DNA present in the sample. This sample is considered to be your input/starting material and needs to have the protein-DNA cross-links reversed by heating at 65 °C for 4 h (*see* **step 26** of Subheading 3.1).
11. To reduce the nonspecific background, add 80 μL of Protein A/G PLUS-Agarose slurry to 1.9 mL of the diluted cell supernatant (from **step 9**, above) and incubate for 2 h at 4 °C while rotating the tubes in a rotator (*see* **Note 11**).

12. Pellet the Agarose beads by a brief centrifugation at 150×*g* at 4 °C, and carefully collect the supernatant in a prechilled 2 mL microcentrifuge tube.
13. Add 5 μg of specific immunoprecipitating antibody per reaction and incubate overnight at 4 °C with constant rotation in a rotator (*see* **Note 12**).
14. The next day, add 50 μL of Protein A/G PLUS-Agarose slurry to each sample and incubate for 2 h at 4 °C with rotation to collect the specific antibody–protein complexes (*see* **Notes 11** and **13**).
15. Pellet the Agarose beads with bound protein complexes by gentle centrifugation (150×*g* at 4 °C for 1 min). Carefully discard the supernatants containing unbound, nonspecific protein–DNA complexes. The A/G Agarose pellets should now contain only the specific antibody–protein–DNA complexes.
16. Add 1 mL of ice-cold low-salt immune complex wash buffer containing protease inhibitors (2 mM PMSF and 2 % protease inhibitor cocktail) to the tubes containing the Protein A/G Agarose–antibody–protein complexes. Incubate the complexes at 4 °C for 5 min with constant rotation.
17. Pellet Agarose beads by centrifugation (150×*g* at 4 °C for 1 min). Carefully discard supernatants and keep the pellets on ice.
18. Add 1 mL of ice-cold high-salt immune complex wash buffer containing the protease inhibitors to protein A/G Agarose–antibody–protein complexes and incubate at 4 °C for 5 min with constant rotation.
19. Pellet the Agarose beads by centrifugation (150×*g* at 4 °C for 1 min). Carefully discard the supernatants and keep the pellets on ice.
20. Add 1 mL of ice-cold LiCl wash buffer containing the protease inhibitors to the bead pellets. Incubate the beads at 4 °C for 5 min with constant rotation.
21. Centrifuge the beads at 150×*g* at 4 °C for 1 min. Carefully discard the supernatants.
22. Add 1 mL of ice-cold 1× TE buffer and incubate the beads at 4 °C for 5 min with constant rotation. Centrifuge as described above and carefully remove the supernatant as much as possible.
23. Add 250 μL of freshly prepared Elution Buffer to the pelleted A/G Agarose–antibody–protein complexes. Vortex briefly (5–10 s) and incubate at room temperature for 15 min with constant rotation.
24. Pellet down the Agarose beads by centrifuging at 150×*g* for 1 min at room temperature. Carefully transfer the supernatant (eluate) to a new, labeled 1.5 mL microcentrifuge tube.

25. Add 250 μL of freshly prepared Elution Buffer to the pellets. Repeat **step 23** described above to elute the remaining proteins from the Agarose beads. Centrifuge at 150 × *g* for 1 min at room temperature and transfer the supernatant (eluate) to 1.5 mL microcentrifuge tube already containing 250 μL of supernatant from the **step 24**. You should now have 500 μL of the eluate per reaction.
26. Add 20 μL of 5 M NaCl to the combined eluates (500 μL) and 4 μL of 5 M NaCl to the input/starting material (100 μL, from **step 10** of Subheading 3.1) to reverse the protein-DNA cross-links. Incubate the complexes for 4 h in a 65 °C water bath.
27. After 4 h, add 10 μL of 0.5 M EDTA, 20 μL of 1 M Tris–HCl, pH 6.5, and 1 μL of 20 mg/mL Proteinase K to the combined eluates and incubate for 1 h at 45 °C.
28. For the input sample, add 2 μL of 0.5 M EDTA, 4 μL of 1 M Tris–HCl, pH 6.5, and 0.2 μL of 20 mg/mL Proteinase K and incubate for 1 h at 45 °C.
29. Add equal volume of phenol:chloroform:isoamyl alcohol (25:24:1, pH 8.0) to the microcentrifuge tubes containing the eluate and the input sample. Vortex vigorously for 5 s. Centrifuge at 12,000 × *g* for 3 min at room temperature. Collect the top (aqueous) layers into new microcentrifuge tubes.
30. Add equal volume of chloroform:isoamyl alcohol (24:1) to the aqueous layers from **step 29** (above). Vortex vigorously for 5 s. Centrifuge at 12,000 × *g* for 3 min at room temperature. Collect again the top (aqueous) layers into new microcentrifuge tubes.
31. Add twice the volume of 100 % (absolute) ethanol, 1/10th volume of 3 M NaOAc, pH 5.2, and 2 μL of LPA to the aqueous layers from **step 30** (above). Mix by inverting the tubes. Keep the microcentrifuge tubes at −20 °C for overnight to allow DNA precipitation (*see* **Note 14**).
32. The next day, centrifuge the microcentrifuge tubes at 12,000 × *g* for 30 min at 4 °C. Collect the DNA containing pellets.
33. Wash the pellets with 100 μL of ice-cold 70 % ethanol. Centrifuge at 12,000 × *g* for 10 min at 4 °C. Collect the pellets. Try to remove as much ethanol as possible. Allow to air dry until the white pellets become invisible.
34. Dissolve the DNA pellets in 50 μL of nuclease-free water and store the samples at −20 °C. Perform real-time PCR as described below.

3.2 Real-Time Polymerase Chain Reaction

1. Prepare a 100 μM stock of each primer using nuclease-free water, for both the forward and reverse primers. From the 100 μM primer stocks, prepare 10 μM working stock solutions for PCR reaction using the nuclease-free water (*see* **Note 15**).
2. Each PCR reaction has a volume of 25 μL and will use 1 μL of each 10 μM primer, so that the final concentration of each primer in PCR reaction is 0.4 μM.
3. The input DNA should be diluted 3 times in nuclease-free water before performing PCR reaction. Each PCR reaction (25 μL volume) for the input sample will use 3 μL of 3 times diluted input DNA.
4. Set up the PCR reactions on ice.
5. Each reaction using immunoprecipitated DNA has a total volume of 25 μL, and is set up as follows (*see* **Note 16**):
   (a) SYBR® Green Supermix: 12.5 μL
   (b) Nuclease-free water: 7.5 μL
   (c) Forward primer (10 μM): 1.0 μL
   (d) Reverse primer (10 μM): 1.0 μL
   (e) Immunoprecipitated DNA sample: 3.0 μL.
6. Each reaction using input DNA has a total volume of 25 μL, and is set up as follows (*see* **Note 17**):
   (a) SYBR® Green Supermix: 12.5 μL
   (b) Nuclease-free water: 7.5 μL
   (c) Forward primer (10 μM): 1.0 μL
   (d) Reverse primer (10 μM): 1.0 μL
   (e) Input DNA sample (3 times diluted): 3.0 μL.
7. After loading the master-mix and the DNA into the wells of the PCR plate, seal the plate carefully using an optical tape and continue with the real-time PCR reaction in the thermal cycler.
8. Each reaction for both the immunoprecipitated and input samples should include a positive and negative control (*see* **Note 18**).

4 Notes

1. PMSF is unstable in aqueous environment. It is essential that it is dissolved in absolute alcohol (ethanol, methanol, or isopropanol); it will not freeze at −20 °C. Add PMSF to buffers in the final working concentration of 2 mM just before use.
2. This protease inhibitor cocktail contains protease inhibitors with a broad specificity for the inhibition of serine, cysteine, and aspartic proteases, and aminopeptidases. It should be stored at −20 °C, and added to buffers just before use.

3. Prepare stock solutions of 10 % SDS and 1 M $NaHCO_3$ by dissolving 5 g of SDS in 50 mL of deionized water, and 4.2 g of $NaHCO_3$ in 50 mL deionized water, respectively. These stock solutions can be stored at room temperature.
4. It is important to get all information about the genomic sequence of promoter region, mRNA, and protein. Copy and save the genomic sequence that is near the transcription start site (TSS), especially the 2,000 bp region upstream and 500 bp region downstream of the TSS.
5. A score higher than 85 indicates binding of the transcription factor to the sequence.
6. For designing the ChIP primers, select 150 nucleotides upstream and 150 nucleotides downstream from the TF-binding site. Below are some general rules for designing the ChIP primers:
   (a) Primer length (bp): Optimum, 20; Minimum, 18; Maximum, 22.
   (b) Tm (°C): Optimum, 55; Minimum, 53; Maximum, 57.
   (c) GC content (%): Optimum, 50; Minimum, 40; Maximum, 60.
   (d) The size of the PCR fragment should be between 100 bp and 200 bp.

   Design at least two sets of primers for each TF-binding site, since the primers not always work properly.
7. The reverse copy of the primer or a sequence can be reversed manually or using the website http://www.thelabnotebook.com/sequence.php?seq=1.
8. Formaldehyde is a reversible protein-DNA cross-linking agent that preserves protein–DNA interactions in cells.
9. It is helpful always to centrifuge the tubes in one position (for example, when using 1.5 mL microcentrifuge tubes, position them with cap snaps facing towards the center of the rotor). This way you can always expect the pellets to be at the same place.
10. Make sure to keep the samples on ice at all times in between the shearing. In addition, it is also helpful to place the tip of the sonicator in a beaker filled with ice for 30 s in between successive 10-s pulses to ensure that the tip is not overheated. When using a new sonicator, it is important to calibrate it so that the sheared DNA fragments are between 400 bp and 500 bp in size.
11. Make sure Protein A/G PLUS-Agarose slurry is completely resuspended before adding.
12. In this protocol, we used 2.5 μL each of p65 NFκB antibody (200 μg/0.1 mL), p50 antibody (200 μg/0.1 mL), c-Rel

antibody (200 μg/0.1 mL), Rel-B antibody (200 μg/0.1 mL) and p52 antibody (200 μg/0.1 mL) to obtain 5 μg of antibody per each immunoprecipitation reaction.

13. At this point, the Agarose beads bind the antibody–TF–DNA complexes.
14. The presence of LPA during ethanol precipitation results in complete recovery of fragments larger than 20 base pairs. The nucleic acid-LPA coprecipitate is visible upon addition of ethanol.
15. The primers for amplifying the NFκB-binding sites in IL-10 and TGFβ1 promoters were designed to anneal optimally at 55 °C. Prepare a 10 μM working stock solution of each primer and store at −20 °C.
16. It is convenient to prepare a master-mix by mixing all components (for desired amount of reactions) except for the DNA. Prepare the master-mix in a 1.5 mL microcentrifuge tube. Aliquot 22 μL of the master-mix into each well of the PCR plate. Add 3 μL of immunoprecipitated DNA sample into each well.
17. Prepare a master-mix containing all of the above except the input DNA in a 1.5 mL microcentrifuge tube. Aliquot 22 μL of the master-mix into each well of the PCR plate. Add 3 μL of input DNA sample (3 times diluted) into each well.
18. The positive control primers provide a control for successful chromatin immunoprecipitation and gene transcription. In this protocol, we used the ChIP-qPCR Human GAPDH Proximal Promoter Positive Control primer (GPH10001C(+)01A). The negative primer provides a reference of the amount of nonspecific genomic DNA that co-immunoprecipitates during the procedure. In this protocol, we used ChIP-qPCR Human 1GX1A Negative Control primer (GPH 00001C(−)01A).

Acknowledgements

This work was supported by NIH grant CA173452 to I. Vancurova.

References

1. Berger CL, Tigelaar R, Cohen J et al (2005) Cutaneous T-cell lymphoma: malignant proliferation of T-regulatory cells. Blood 105: 1640–1647
2. Querfeld C, Rosen ST, Guitart J et al (2005) The spectrum of cutaneous T-cell lymphomas: new insights into biology and therapy. Curr Opin Hematol 12:273–278
3. Saed G, Fivenson DP, Naidu Y et al (1994) Mycosis fungoides exhibits a Th1-type cell-mediated cytokine profile whereas Sezary syndrome expresses a Th2-type profile. J Invest Dermatol 103:29–33
4. Chong BF, Wilson AJ, Gibson HM et al (2008) Immune function abnormalities in peripheral blood mononuclear cell cytokine expression

differentiates stages of cutaneous T-cell lymphoma/mycosis fungoides. Clin Cancer Res 14:646–653

5. Abraham RM, Zhang Q, Odum N et al (2011) The role of cytokine signaling in the pathogenesis of cutaneous T-cell lymphoma. Cancer Biol Ther 12:1019–1022
6. Krejsgaard T, Odum N, Geisler C et al (2012) Regulatory T cells and immunodeficiency in mycosis fungoides and Sezary syndrome. Leukemia 26:424–432
7. Qin JZ, Nestle FO, Häffner A et al (1997) Cutaneous T cell lymphoma cells contain constitutive NFκB complexes. J Invest Derm 108:225
8. Döbbeling U, Qin JZ, Dummer R et al (1998) Suppressors of constitutive NFκB activities in CTCL cells induce apoptosis. Arch Dermatol Res 290:58
9. Izban KF, Ergin M, Qin JZ et al (2000) Constitutive expression of NFκB is a characteristic feature of mycosis fungoides: implications for apoptosis resistance and pathogenesis. Hum Pathol 31:1482–1490
10. Sors A, Jean-Louis F, Pellet C et al (2006) Down-regulating constitutive activation of the NFκB canonical pathway overcomes the resistance of cutaneous T-cell lymphoma to apoptosis. Blood 107:2354–2363
11. Sors A, Jean-Louis F, Bégué E et al (2008) Inhibition of IκB kinase subunit 2 in cutaneous T-cell lymphoma down-regulates NFκB constitutive activation, induces cell death, and potentiates the apoptotic response to antineoplastic chemotherapeutic agents. Clin Cancer Res 14: 901–911
12. Kiessling MK, Klemke CD, Kaminski MM et al (2009) Inhibition of constitutively activated NFκB induces reactive oxygen species- and iron-dependent cell death in cutaneous T-cell lymphoma. Cancer Res 69:2365–2374
13. Giri DK, Aggarwal BB (1998) Constitutive activation of NFκB causes resistance to apoptosis in human cutaneous T cell lymphoma HuT-78 cells. Autocrine role of tumor necrosis factor and reactive oxygen intermediates. J Biol Chem 273:14008–14014
14. O'Connell MA, Cleere R, Long A et al (1995) Cellular proliferation and activation of NFκB are induced by autocrine production of TNFα in the human T lymphoma line HuT-78. J Biol Chem 270:7399–7404
15. Mori N, Prager D (1997) Activation of the interleukin-10 gene in the human T lymphoma line HuT-78: identification and characterization of NFκB binding sites in the regulatory region of the interleukin-10 gene. Eur J Haematol 59:162–170
16. Mori N, Prager D (1998) Interleukin-10 gene expression and adult T-cell leukemia. Leuk Lymphoma 29:239–248
17. Hayden MS, Ghosh S (2012) NFκB, the first quarter-century: remarkable progress and outstanding questions. Genes Dev 26:203–234
18. Wan F, Lenardo MJ (2009) Specification of DNA binding activity of NFκB proteins. Cold Spring Harb Perspect Biol 1:a000067
19. Hayden MS, Ghosh S (2008) Shared principles in NFκB signaling. Cell 132:3344–3362
20. Yamamoto Y, Gaynor RB (2001) Therapeutic potential of inhibition of the NFκB pathway in the treatment of inflammation and cancer. J Clin Invest 107:135–142
21. Aggarwal BB (2004) NFκB: the enemy within. Cancer Cell 6:203–208
22. Naugler WE, Karin M (2008) NFκB and cancer-identifying targets and mechanisms. Curr Opin Genet Dev 18:19–26
23. Sun SC (2012) The noncanonical NFκB pathway. Immunol Rev 246:125–140
24. Hoffmann A, Natoli G, Ghosh G (2006) Transcriptional regulation via the NFκB signaling module. Oncogene 25:6706–6716
25. Wan F, Lenardo MJ (2010) The nuclear signaling of NFκB: current knowledge, new insights, and future perspectives. Cell Res 20:24–33
26. Natoli G (2012) NFκB and chromatin: ten years on the path from basic mechanisms to candidate drugs. Immunol Rev 246:183–192
27. Juvekar A, Ramaswami S, Manna S et al (2012) Electrophoretic mobility shift assay analysis of NFκB transcriptional regulation by nuclear IκBα. Methods Mol Biol 809:49–62
28. Juvekar A, Manna S, Ramaswami S et al (2011) Bortezomib induces nuclear translocation of IκBα resulting in gene-specific suppression of NFκB-dependent transcription and induction of apoptosis in CTCL. Mol Cancer Res 9: 183–194
29. Chang TP, Vancurova I (2013) NFκB function and regulation in cutaneous T-cell lymphoma. Am J Cancer Res 3:433–445

Chapter 31

Radiolabeled Semi-quantitative RT-PCR Assay for the Analysis of Alternative Splicing of Interleukin Genes

Felitsiya Shakola, Stephen Byrne, Kainaat Javed, and Matteo Ruggiu

Abstract

Alternative splicing evolved as a very efficient way to generate proteome diversity from a limited number of genes, while at the same time modulating posttranscriptional events of gene expression—such as stability, turnover, subcellular localization, binding properties, and general activity of both mRNAs and proteins. Since the vast majority of human genes undergo alternative splicing, it comes to no surprise that interleukin genes also show extensive alternative splicing. In fact, there is a growing body of evidence indicating that alternative splicing plays a central role in modulating the pleiotropic functions of cytokines, and aberrant expression of alternatively spliced interleukin mRNAs has been linked to disease. However, while several interleukin splice variants have been described, their function is still poorly understood. This is particularly relevant, since alternatively spliced cytokine isoforms can act both as disease biomarkers and as candidate entry points for therapeutic intervention. In this chapter we describe a protocol that uses radiolabeled semi-quantitative RT-PCR to efficiently detect, analyze, and quantify alternative splicing patterns of cytokine genes.

Key words RNA, Gene expression, Alternative splicing, Reverse transcription, PCR, RT-PCR

1 Introduction

A healthy, functional immune system is able to adopt a rapid and effective response to changing environmental conditions in order to identify and interact with foreign molecules and to ultimately protect the organism from disease-causing agents. Activated macrophages and neutrophils employ an intricate network of signaling molecules, chemical messages, and secreted proteins that are released upon injury or infection—cytokines and interleukins. The diverse members of this group of proteins can either elicit and stimulate an inflammatory response or inhibit it by regulating the proliferation, differentiation, or function of different immune cell types.

A convenient, rapid, efficient, and tightly regulated mode of fine-tuning gene expression and generate proteomic diversity is alternative splicing of pre-mRNA [1]. Alternative splicing occurs

Ivana Vancurova (ed.), *Cytokine Bioassays: Methods and Protocols*, Methods in Molecular Biology, vol. 1172, DOI 10.1007/978-1-4939-0928-5_31, © Springer Science+Business Media New York 2014

when distinct mRNA molecules are generated from the same primary transcript by the combinatorial inclusion and/or skipping of specific exons [2, 3]. This form of gene expression regulation requires the interaction of trans-acting factors (small nuclear ribonucleoprotein particles or snRNPs; splicing factors) with cis-acting sequences (splice sites; exonic and intronic enhancers and inhibitors)—much like RNA polymerase employs different transcription factors binding to different promoter sequences. Splicing patterns include the selective inclusion or skipping of cassette exons, mutually exclusive exons, competing splice sites, and polyadenylation signals. Unlike transcription, however, alongside transcript abundance alternative splicing can also alter the structure, stability, and turnover of transcripts, and the proteins they encode, thus controlling the binding properties, enzymatic activity, protein stability, posttranslational modifications, and intracellular localization of a large number of protein isoforms originating from a single gene [4, 5]. About 94 % of human genes undergo alternative splicing [6], and about 50 % of disease-causing genetic mutations affect pre-mRNA splicing [7, 8]. Alternative splicing networks play an important role as a source of protein complexity in the immune system [9–11]. For example, alternative splicing can downregulate gene expression by creating protein isoforms that are unstable or not functional, as well as creating isoforms of cytokine and their receptors with antagonistic functions [12].

The functional role of alternative splicing in cytokine biology both in normal conditions and in disease is only recently becoming to be elucidated. An example of intense research on cytokine splicing is interleukin 7 (*IL-7*; Fig. 1). IL-7 is known to play an important role in the development of B and T cells [13–16]. While in healthy individuals it stimulates the proliferation of antitumor reactive cells, IL-7 overexpression can lead to a variety of malignancies [17–20]. Other studies performed with the canonical, full-length form of IL-7 point toward a role in arthritis pathology [21], lymphopenia [22, 23], and host versus graft disease [24].

Recent research focusing on the role of alternatively spliced cytokine isoforms in disease revealed that the IL-7 splice variant IL-7δ5 stimulates human breast cancer cell proliferation [25] probably by altering binding to the IL-7 receptor (IL-7R) resulting in activating of a different signaling pathway, phosphorylation of STAT5, and tumor cell survival [26]. Furthermore, IL-7δ5-mediated phosphorylation of STAT5 in CD4+ and CD8+ T cells stimulates thymocyte maturation and T-cell survival [27], suggesting a role of IL-7δ5 in leukemia. Indeed seven more splice variants of IL-7 have been identified in leukemic cells from children with acute lymphoblastic leukemia, although their biological function remains to be elucidated [19]. IL-7 transcripts lacking exon 4 are the predominant transcript in peripheral blood mononuclear cells from patients with multiple sclerosis [28]. Interestingly, a soluble,

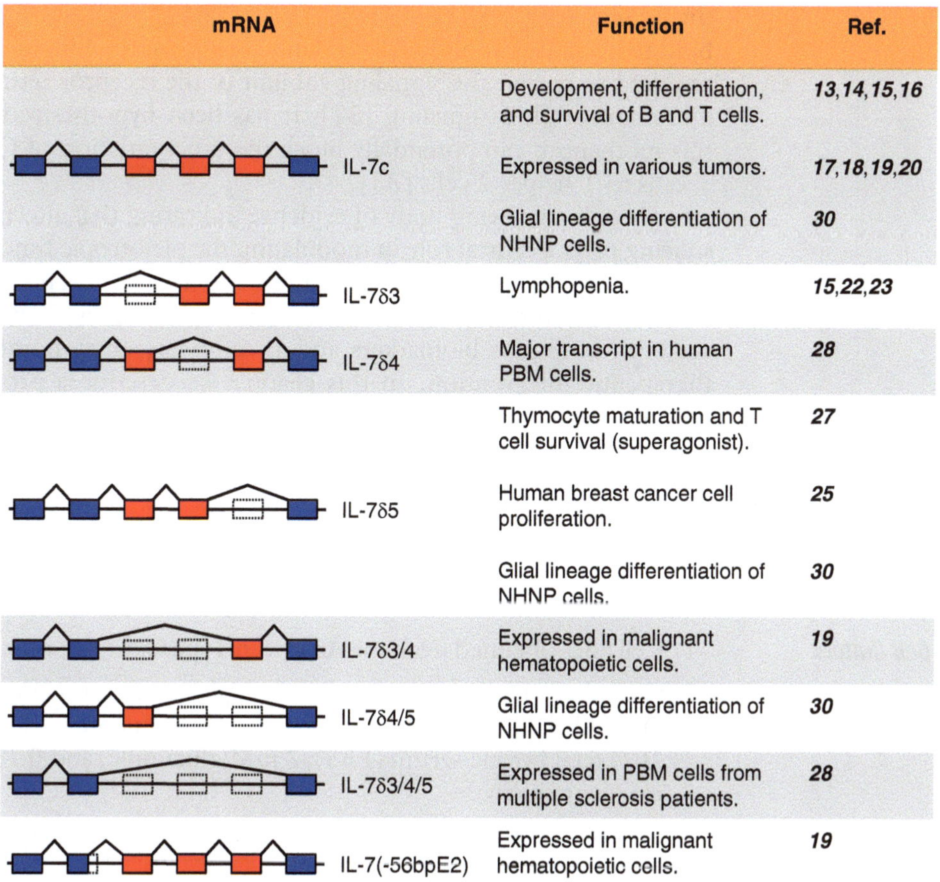

Fig. 1 The various splice isoforms of IL-7, their function, and occurrence. Schematic representation of the human *IL-7* gene. Located on chromosome 8 q21.12, *IL-7* is comprised of six exons and five introns spanning 73 Kb. Constitutive exons 1, 2, and 6 are indicated as *blue boxes*, while alternatively spliced exons 3, 4, and 5 are indicated as *red boxes*. Introns and exons are not drawn to scale. IL-7c: canonical isoform including all six exons. The letter δ indicates alternatively skipped exons. *NHNP cell* normal human neuronal progenitor cells, *PBM cell* peripheral blood mononuclear cell, *−56bpE2* splice isoform that uses an alternative 5′ splice site in exon 2; only one isoform is shown, although the alternative 5′ splice site is also present in isoforms skipping exon 4, and exons 3/4/5. Relevant references are indicated

alternatively spliced isoform of IL-7R is associated with an increased risk of developing multiple sclerosis [29]. The importance of this splice event in disease pathology, however, is still unclear. IL-7 splice variants have also been demonstrated to play a key role in brain development, mediating differentiation of glial cells from human neural progenitor cells [30], evidence suggesting that alternative splicing can regulate pleiotropic cytokine functions outside of the immune system (Fig. 1).

Alternative splicing can also lead to production of cytokines with antagonistic functions. An example of that is IL-6 and its splice isoform IL-6δ4. The two subunits of IL-6R only associate in the presence of IL-6 [31]. The smaller IL-6δ4 isoform is unable to bind to one subunit of the IL-6R, and absence of the binding sites required to recruit the signaling subunit of the receptor results in inhibition of IL-6 signaling [32]. It has been hypothesized that this mechanism can potentially block the differentiation of CD4+ T cells to T helper 2 cells [33].

There is a growing body of evidence indicating that alternative splicing plays a central role in modulating the pleiotropic functions of cytokines, both in normal conditions and in disease. This is particularly relevant, since alternatively spliced cytokine isoforms can act both as disease biomarkers and as candidate entry points for therapeutic intervention. In this chapter we describe a protocol that uses radiolabeled reverse transcriptase polymerase chain reaction (RT-PCR) to efficiently detect, analyze, and quantify alternative spicing isoforms of cytokine mRNAs.

2 Materials

2.1 Cell Culture

1. Cell lines obtained from the American Type Culture Collection.
2. Complete Dulbecco's modified Eagle medium (DMEM): DMEM medium supplemented with high glucose containing 10 % fetal bovine serum (FBS), 2 mM glutamine, and 0.1 mg/mL gentamicin.
3. Phosphate buffered saline (PBS), pH 7.4 without calcium and magnesium.
4. 10 cm culture dishes.
5. Fluid aspiration system or vacuum pump.
6. Cell scrapers.
7. 1.5 mL microcentrifuge tubes.

2.2 RNA Extraction

1. TRIzol Reagent (Life Technologies) for RNA extraction.
2. 1 mL syringes.
3. 25G disposable hypodermic needles.
4. 1.5 mL microcentrifugation tubes.
5. Chloroform.
6. Isopropyl alcohol.
7. 75 % ethanol: Add 75 mL of absolute ethanol to deionized water to make up a volume of 100 mL. Store at 4 °C.
8. RNase-free water.
9. DNase I.
10. RNase inhibitor 20 U/μL.
11. UltraPure phenol:chloroform:isoamyl alcohol (25:24:1, v/v, pH 8.0).

12. Chloroform:isoamyl alcohol (24:1).
13. 3 M sodium acetate (NaOAc): Dissolve 6.15 g of anhydrous NaOAc in 25 mL of deionized water. Adjust pH to 5.2 using glacial acetic acid. This stock solution can be stored at room temperature.
14. Glycogen solution: Dissolve 5 mg of glycogen in 1 mL of RNase-free water.
15. Molecular grade absolute ethanol.

2.3 Reverse Transcription

1. RevertAid First strand cDNA synthesis kit (Thermo Scientific; *see* **Note 1**), containing M-MuLV Reverse Transcriptase 200 U/μL and oligo $d(T)_{18}$ 0.5 mg/mL.
2. Thermocycler PCR machine.
3. PCR tubes (0.2–0.5 mL, depending on PCR machine).
4. RNase-free water.

2.4 ^{32}P 5′-End Labeling and Purification of Primer and DNA Molecular Weight Marker

1. Appropriate personal protective equipment and plexiglass screens and containers for radioactivity work with beta emitters.
2. ATP, [γ-^{32}P]-6000 Ci/mmol, 10 mCi/mL.
3. T4 polynucleotide kinase.
4. Specific DNA primers for each alternative splicing event; 5 μM in RNase-free water (*see* **Note 2**).
5. ΦX174 DNA/*HinfI* dephosphorylated markers (Promega; *see* **Note 3**).
6. RNase-free water.
7. 1.5 mL microcentrifuge tubes.
8. Illustra MicroSpin G-25 column (GE Healthcare).

2.5 Reverse Transcriptase Polymerase Chain Reaction to Determine the Linear Range of Amplification

2.5.1 Reverse Transcriptase Polymerase Chain Reaction

1. Forward DNA primer 5 μM.
2. Reverse DNA primer 5 μM.
3. Radiolabeled forward DNA primer.
4. dNTP mix, 10 mM each.
5. AmpliTaq Gold® DNA polymerase (Life technologies).
6. 10× PCR Buffer: 100 mM Tris–HCl, pH 8.0, 500 mM KCl, 15 mM $MgCl_2$.
7. Thermocycler PCR machine.
8. PCR tubes (0.2 mL).
9. RNase-free water.
10. 2× RNA loading dye: 95 % formamide; 0.025 % sodium dodecyl sulfate (SDS); 0.025 % bromophenol blue; 0.025 % xylene cyanol FF; 0.025 % ethidium bromide; 0.5 mM EDTA.

2.5.2 Electrophoresis of PCR Products on Denaturing Polyacrylamide Gel

1. Hoefer SE660 vertical gel electrophoresis apparatus or equivalent, including glasses, 1 mm thick combs and spacers (*see* **Note 4**).
2. High voltage power supply.
3. 40 % acrylamide stock solution (acrylamide:bisacrylamide 19:1).
4. 10 % ammonium persulfate (APS; w/v).
5. *N*,*N*,*N*′,*N*′-Tetramethylethylenediamine (TEMED), electrophoresis grade.
6. Electrophoresis grade urea, ultrapure.
7. 0.5 M EDTA, pH 8.0: In a 500 mL beaker dissolve 93.06 g of EDTA in 300 mL of ultrapure deionized water. Adjust pH to 8.0 with glacial acetic acid. Bring volume to 500 mL with ultrapure deionized water and filter sterilize.
8. 10× TBE buffer: In a large beaker dissolve 108 g of Tris base and 55 g of boric acid in 800 mL of ultrapure deionized water; add 40 mL of 0.5 M EDTA, pH 8.0. Bring volume to 1 L with ultrapure deionized water. The solution can be stored at room temperature (*see* **Note 5**).
9. 10 % acrylamide, 8 M urea solution in 1× TBE: In a large beaker combine 240.24 g of electrophoresis grade urea, 125 mL of 40 % acrylamide stock solution (19:1), and 50 mL of 10× TBE stock solution. Stir to dissolve with a baked stir bar. Bring volume to 500 mL with ultrapure deionized water. Store solution at room temperature shielded from light.
10. 6 % acrylamide, 8 M urea gel in 1× TBE (final volume: 30 mL): In a small beaker combine 18 mL of 10 % acrylamide, 8 M urea solution in 1× TBE; 12 mL of 8 M urea in 1× TBE; 200 μL of 10 % APS; 10 μL of TEMED (*see* **Note 6**).
11. 2× RNA loading dye.
12. Heat block preset to 95 °C.
13. Flat-tipped micropipette gel loading tips.
14. Dishwashing liquid detergent.
15. 70 % ethanol (v/v).
16. 50 mL Falcon tube.

2.5.3 Gel Drying

1. Gel fixing solution: 50 % methanol (v/v); 10 % acetic acid (v/v) in water.
2. Gel opening device or thin spatula.
3. Clear plastic lab wrap.
4. Whatman 3MM paper.
5. Glass or plastic tray.
6. Funnel.

7. Orbital shaker.
8. Gel dryer with vacuum pump.

2.5.4 Quantification of PCR Products

1. Geiger counter.
2. Baby powder.
3. Quantification imaging screen.
4. Quantification imaging device and software.
5. Software for statistical analysis such as Excel (Microsoft) or Prism (GraphPad).

3 Methods

In this section, we describe the protocol for the detection, analysis, and quantification of alternatively spliced mRNA isoforms by semi-quantitative RT-PCR from total RNA extracted from tissue culture cell lines. However, this protocol can be easily modified and used for other applications, such as RNA analysis from tissue samples.

Sequence information for a specific transcript can be obtained from the NCBI website (http://www.ncbi.nlm.nih.gov/gene/). A search by gene name will provide gene ID number, official full name, chromosomal location, and genomic sequence. By clicking on "Nucleotide" on the menu on the right handside of the page, it is then possible to access the sequence of single mRNA molecules, including specific splice variants of that gene. After selecting a specific mRNA to analyze, by selecting the "Pick Primers" option on the right handside of the page it is possible to access Primer-BLAST: a primer-designing software that allows to find primers specific for our PCR template. Given that we are looking at alternatively spliced molecules, it is important that primers are selected to span exon-exon junctions (*see* **Note 7**). Alternatively, a specific sequence can be analyzed using a primer design software such as AmplifX. In this case, the location of exon-exon junction onto the mRNA sequence has to be known in advance (*see* **Note 8**).

Most RT-PCR assays used for the detection of alternatively spliced cytokine mRNAs rely on densitometric analysis of band intensity for the quantification of PCR products resolved by gel electrophoresis [19, 27, 28, 30, 34, 35], a technique of poor accuracy and reproducibility due to the low dynamic range of ethidium bromide gels. In this chapter, we describe a PCR-based technique that relies on radioactively labeled PCR products for quantification. The major advantages of semi-quantitative RT-PCR over real-time RT-PCR include:

1. The ability to detect multiple alternatively spliced isoforms in a single reaction using a single pair of amplifying primers.
2. An internal control (a housekeeping gene such as tubulin, actin, GAPDH, or ribosomal RNA) to compensate for variation in

the amounts of input template RNA is not needed, since each alternatively spliced isoform represents a percentage of the total.

3. The ability to detect alternatively spliced exons that may be too small in size to allow the design of a primer or primer pair within that exon. In fact, semi-quantitative RT-PCR has been used to efficiently detect and quantify exons as small as 24 nt [36].

Conversely, real-time RT-PCR combines amplification and detection in a single step [37], while semi-quantitative RT-PCR requires separation of amplified products by DNA gel electrophoresis, and data analysis through quantification imaging devices such as PhosphorImager (Molecular Dynamics) or Typhoon (GE Healthcare). Moreover, semi-quantitative RT-PCR offers a limited dynamic range and requires more optimization steps, such as the determination of exponential phase of the PCR reaction—the linear range, for each PCR product of that target gene [37–39]. This step is particularly critical, since shorter PCR products will be preferentially amplified within the same PCR reaction. However, the sensitivity—defined as the lower detection limit of the assay, of the two methodologies is similar [40], and splice isoforms representing as little as 1 % of the total can be efficiently detected and quantified [36]. Finally, while real-time RT-PCR is more accurate, it is also more expensive.

Figure 2 describes the major steps involved in this protocol. On average, the entire protocol can be accomplished in 3 days: RNA extraction, reverse transcription, and RT-PCR for determination of linear range of amplification can all be accomplished in Day 1; gel electrophoresis, and gel drying can be accomplished in Day 2; analytical RT-PCR and gel electrophoresis can be accomplished in Day 3 (Fig. 2). If the number of samples to be processed and the number of alternatively spliced events to be analyzed (that is, the number of different PCR reactions for each RT sample) is not very large, steps in Day 1 and Day 2 can be performed in 1 day.

Figure 3 illustrates a typical radiolabeled semi-quantitative RT-PCR experiment. A first radiolabeled semi-quantitative RT-PCR experiment is performed in order to identify the linear range of amplification for a specific alternative splicing event (Fig. 3a, b). The linear range of amplification takes place in the PCR reaction when reaction components are still in excess. During this linear phase of amplification PCR products are accumulating at a constant rate, making quantification in this phase reliable, since the amount of PCR product in the reaction is expected to accumulate exponentially and, in ideal conditions, double at every cycle. On the other hand, the amount of PCR product measured beyond the exponential phase (i.e. during the plateau phase) is extremely sensitive to very little variations in initial amplification of the target, making quantification of products in this phase of the reaction curve unreliable [37].

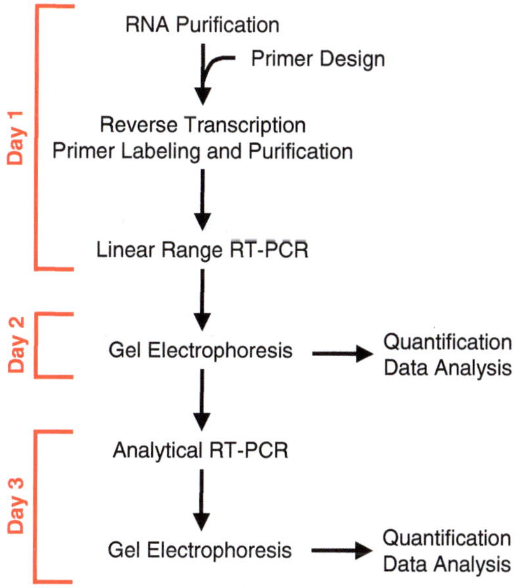

Fig. 2 Schematic flowchart and timeline of RT-PCR assay. The entire assay can be usually completed in 3 days. If needed, steps in Day 1 and Day 2 can be performed in 1 day

To determine the linear range of amplification, a PCR reaction is performed at increasing cycle numbers—or, more conveniently, a fixed amount of PCR reaction from a single tube is collected at different PCR cycles, and the radiolabeled PCR products are resolved by electrophoresis on a denaturing polyacrylamide gel (Fig. 3a; *see* ref. 41). After exposing the processed gel to the screen of a quantification imaging device, PCR products are quantified with computer software. Cycle number can be plotted against the signal obtained in order to identify the exponential (linear) range and the plateau phase. If cycle number is plotted against the log of the signal, samples in linear range amplification should lay on a straight line (Fig. 3b). Subsequently, a second, analytical radiolabeled semi-quantitative RT-PCR experiment is performed at a PCR cycle corresponding to the previously determined linear range of amplification (Fig. 3c). Finally, PCR products corresponding to different alternatively spliced isoforms are quantified and plotted either as percentage of total (Fig. 3d), or as fold change normalized to a control sample (Fig. 3e).

3.1 Cell Culture

1. Grow cells in complete DMEM medium. Incubate in a humidified 5 % CO_2 atmosphere at 37 °C until subconfluent (*see* **Note 9**).
2. Wash cells once with 5 mL of ice-cold PBS. Discard PBS using a fluid aspiration system.

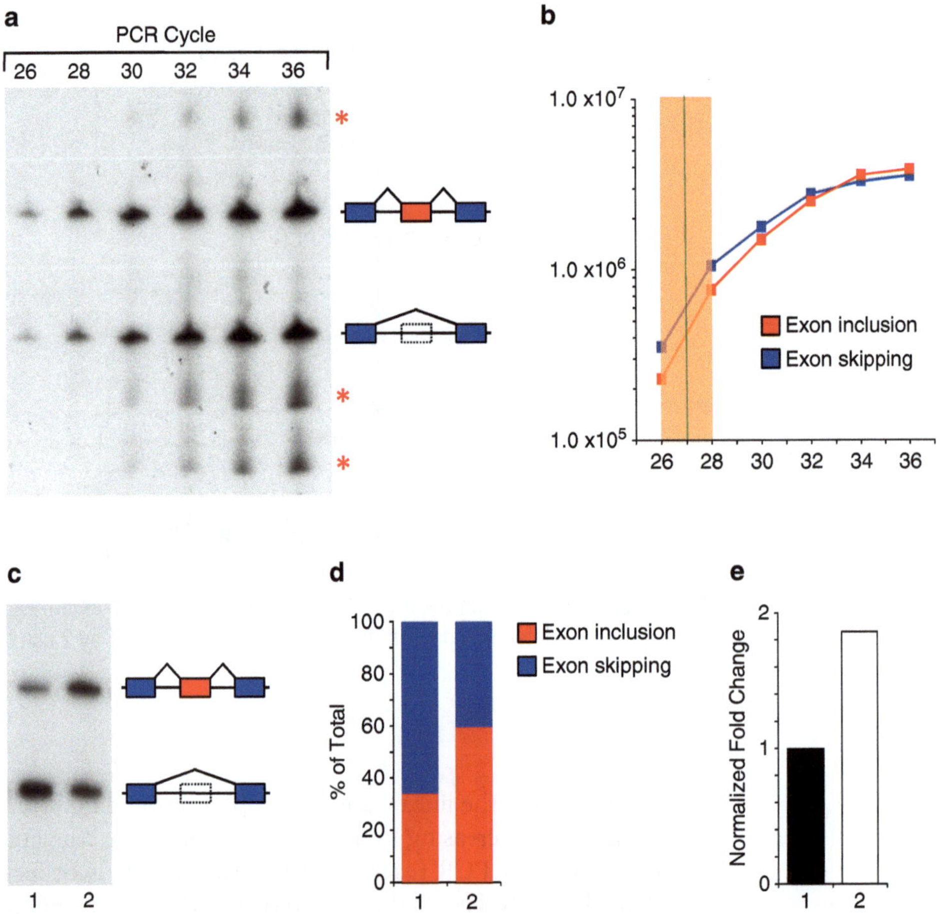

Fig. 3 Example of radiolabeled semi-quantitative RT-PCR assay for detection and quantification of alternatively spliced mRNA isoforms. (**a**) Autoradiogram of semi-quantitative RT-PCR for the identification of the linear range of amplification. A single RT sample was amplified, and a fixed volume of PCR reaction was collected at the indicated PCR cycles; PCR products were resolved on a denaturing polyacrylamide gel. Cartoons of the two alternatively spliced isoforms are shown on the *right*; constitutive exons are shown as *blue boxes*, while the alternatively spliced exon is shown as a *red box*. *Red asterisks* indicate "shadow" or "ghosts" bands with aberrant migration due to PCR product overload. (**b**) Phosphorimager quantification of the gel shown in (**a**). The intensity of the bands corresponding to the two alternatively spliced isoforms is plotted in Log scale versus the PCR cycle number. The linear range of amplification, which in this case happens between cycle 26 and 28, is indicated as an *orange rectangle*. For analytical RT-PCR, 27 PCR cycles were chosen (*dotted green line*). (**c**) Autoradiogram of radiolabeled analytical semi-quantitative RT-PCR for the same alternatively spliced event as in (**a**) from two different tissue culture samples. PCR conditions were the same as in (**a**), but 27 PCR cycles were used. Note that equal amounts of radiolabeled PCR products were loaded in each lane. (**d**) Quantification of the experiment in (**c**), where each alternatively spliced isoform for sample 1 and sample 2 is shown as percentage of the total for that sample. (**e**) Quantification of the experiment in (**c**), where exon inclusion is shown as fold change relative to sample 1

3. Collect cells by adding 0.5 mL of ice-cold PBS to the plate; scrape cells, and transfer cells into labeled 1.5 mL microcentrifuge tubes by using a P1000 pipette.
4. Pellet cells by centrifugation for 5 min at 250×*g*. Discard supernatant using a fluid aspiration system.

3.2 RNA Extraction

3.2.1 Homogenization and Cell Lysis

1. Homogenize the cell pellet using 0.5 mL of TRIzol Reagent; pass the cell lysate several times through a P1000 pipette. Add another 0.5 mL of TRIzol reagent to the tube (total volume: 1 mL).
2. Using a 1 mL disposable syringe, pass the cell lysate five times through a 25G disposable needle; this step will help the lysis and will shear genomic DNA (*see* **Note 10**).
3. Incubate the homogenized samples 5 min at room temperature to allow complete dissociation of nucleoprotein complexes.
4. Add 200 μL of chloroform per 1 mL of TRIzol Reagent. Shake tubes vigorously with a vortex for a few seconds, and incubate at room temperature on the bench for 2–3 min.
5. Centrifuge samples at <12,000 × *g* for 15 min at 4 °C. Following centrifugation, the mixture separates into a lower red, phenol-chloroform phase, an interphase, and a colorless aqueous upper phase (*see* **Note 11**).

3.2.2 RNA Precipitation

1. Without touching the interphase, transfer the aqueous phase into a labeled fresh 1.5 mL microcentrifuge tube.
2. Precipitate the RNA by adding 0.5 mL of isopropyl alcohol per 1 mL of TRIzol Reagent used for the initial homogenization.
3. Incubate samples 10 min at room temperature and centrifuge at <12,000 × *g* for 10 min at 4 °C.
4. Remove supernatant with fluid aspiration system, and wash RNA pellet by adding 1 mL of 75 % ethanol per 1 mL of TRIzol reagent used for the initial homogenization. Invert the tubes a few times by hand, and centrifuge them at <8,000 × *g* for 5 min at 4 °C.
5. Carefully remove supernatant with fluid aspiration system (*see* **Note 12**), and air dry the pellet for 10 min at room temperature on the bench.
6. Resuspend pellet in 40 μL of RNase-free water.

3.2.3 DNase I Digestion

1. Set up the DNase I digestion on ice.
2. Each reaction has a final volume of 50 μL and is set up as follows (*see* **Note 13**):
   (a) 10× TURBO DNase buffer: 5 μL
   (b) RNase Inhibitor 20 U/μL: 0.5 μL
   (c) TURBO DNase 2 U/μL: 3 μL
   (d) RNA: 40 μL
   (e) RNase-free water: 1.5 μL
3. Incubate samples at 37 °C for 30 min to 1 h.

3.2.4 Phenol/Chloroform Extraction

1. Bring sample volume to 180 μL by adding 130 μL of RNase-free water.
2. Add equal volume of phenol:chloroform:isoamyl alcohol (25:24:1) to each microcentrifuge tube containing the RNA samples. Vortex vigorously for 5 s. Centrifuge at 12,000 × *g* for 5 min at room temperature. Collect the top (aqueous) layer into a new microcentrifuge tube.
3. Add equal volume of chloroform:isoamyl alcohol (24:1) to the aqueous layer from step 1. Vortex vigorously for 5 s. Centrifuge at 12,000 × *g* for 5 min at room temperature. Collect again the top (aqueous) layer into a new microcentrifuge tube.
4. To each sample add 20 μL of 3 M sodium acetate, pH 5.2 (final concentration: 0.3 M).

3.2.5 RNA Precipitation

1. To each sample add 2 μL of glycogen solution.
2. Precipitate the RNA by adding 500 μL (equals to 2.5 volumes) of absolute ethanol.
3. Vortex vigorously for 5 s. Centrifuge at 12,000 × *g* for 15 min at 4 °C.
4. Remove supernatant with fluid aspiration system, and wash RNA pellet by adding 500 μL of 75 % ethanol. Invert the tubes a few times by hand, and centrifuge them at 12,000 × *g* for 5 min at 4 °C.
5. Carefully remove supernatant with fluid aspiration system (*see* **Note 14**), and air dry the pellet for 10 min at room temperature on the bench.
6. Resuspend pellet in 50 μL of RNase-free water.
7. Determine the RNA concentration by reading the optical density at 260 nm with a spectrophotometer (*see* **Note 15**).

3.3 Reverse Transcription

1. Set up the reverse transcription reaction on ice, using 500 ng of total RNA per sample.
2. Aliquot the equivalent of 500 ng of total RNA in PCR tubes. Add RNase-free water to 11 μL. To each tube add 1 μL of oligo $d(T)_{18}$. Final volume: 12 μL.
3. Denature RNA by incubating samples 5 min at 65 °C in PCR machine. Chill at 4 °C.
4. Each reaction has a final volume of 20 μL and is set up as follows:
   (a) 5× reaction buffer: 4 μL
   (b) RiboLock™ RNase Inhibitor: 1 μL
   (c) 10 mM dNTP mix: 2 μL
   (d) M-MuLV Reverse Transcriptase: 1 μL

5. Incubate samples at 42 °C for 1 h in PCR machine.
6. Each reverse transcriptase reaction should include a "no enzyme" negative control for genomic DNA contamination.
7. Bring final volume to 50 μL with water (1 μL of reverse transcriptase reaction is the equivalent of 10 ng of starting total RNA). If not to be used immediately, store samples at −20 °C.

3.4 ^{32}P Labeling and Purification of Primer and DNA Molecular Weight Marker

3.4.1 ^{32}P Labeling of Primer and DNA Molecular Weight Marker

1. Resuspend DNA primers to 5 μM in RNase-free water.
2. Set up the labeling reaction on ice, using 10 pmoles of DNA primer for each reaction.
3. Each reaction has a final volume of 10 μL and is set up as follows:
   (a) 10× T4 polynucleotide kinase buffer: 1 μL
   (b) Forward primer 5 μM: 2 μL
   (c) ATP, [γ-^{32}P]-6000 Ci/mmol, 10 mCi/mL: 6 μL
   (d) T4 polynucleotide kinase (10 U/μL): 1 μL
4. Incubate samples at 37 °C for 45 min.
5. Stop the labeling reaction by incubating the sample at 95 °C for 10 min and chill on ice.
6. Add 20 μL of RNase-free water (final volume: 30 μL).
7. Also label the same way 100 ng of ΦX174/*Hinf I* dephosphorylated DNA molecular weight marker.

3.4.2 Column Purification of 5′-End Labeled Primer and DNA Molecular Weight Marker

1. Resuspend the resin in the G-25 column by vortexing.
2. Loosen the cap one-quarter turn and twist off the bottom closure.
3. Place the column in the supplied collection tube.
4. Spin for 1 min at 735 × *g*.
5. Place the column into a fresh 1.5 mL microcentrifuge tube.
6. Slowly apply appropriate volume of sample to the top-center of the resin, being careful not to disturb the resin bed.
7. Spin column for 2 min at 735 × *g*. The purified sample is collected at the bottom of the 1.5 mL microcentrifuge tube.
8. Dispose appropriately of the radioactive column.
9. Dilute the radioactively labeled DNA molecular weight marker sample 1:50 in nuclease-free water.
10. Store the purified, labeled samples at −20 °C in plexiglass container for radioactivity work with beta emitters until needed (*see* **Note 16**).

3.5 Reverse Transcriptase Polymerase Chain Reaction to Determine the Linear Range of Amplification

3.5.1 Reverse Transcriptase Polymerase Chain Reaction

1. Set up the reverse transcription polymerase chain reaction (RT-PCR) on ice.
2. The final volume of each RT-PCR reaction will vary depending on the number of samples that will be collected at different PCR cycles for each RT sample (5 μL of PCR reaction per sample every 2 cycles) as a total volume of 50 μL and is set up as follows:
   (a) 10× Gold Taq Buffer: 5 μL
   (b) 10 mM dNTP mix: 1 μL
   (c) Forward primer (5 μM): 2 μL
   (d) Reverse primer (5 μM): 2 μL
   (e) ^{32}P, 5′-end labeled and purified forward primer: 2 μL
   (f) RT reaction: 7.5 μL
   (g) AmpliTaq Gold® DNA polymerase 5 U/μL: 0.5 μL
   (h) Nuclease-free water: 30 μL.
3. The PCR conditions should be as follows: a single 5 min denaturation cycle at 95 °C; 30–35 cycles where each cycle is: 20 s denaturation at 95 °C; 20 s annealing at 50–55 °C; 30 s extension at 72 °C.
4. Collect 5 μL of PCR reaction from each sample every 2 cycles, starting at cycle number 20, and transfer into a fresh PCR tube.
5. Add 15 μL of 2× RNA loading dye to each sample, for a total volume of 20 μL.
6. Store samples at −20 °C until ready to be loaded.

3.5.2 Electrophoresis of PCR Products on Denaturing Polyacrylamide Gel

1. Wash glass plates, spacers, and combs in a diluted solution of dishwashing liquid detergent; rinse thoroughly in deionized water. To prevent water spots, rinse the plates with 70 % ethanol and allow them to air dry (*see* **Note 17**).
2. Assemble gel glasses and spacers, and pour a 10 % polyacrylamide, 8 M urea denaturing gel.
3. With the aid of an automatic pipettor (*see* **Note 18**) and a 10 mL pipette, slowly but steadily pour the gel solution in the assembled gel plates. Once the solution reaches the top of the glasses, insert the comb. Leave the remainder of the gel solution in the 50 mL Falcon tube to check for polymerization. Once the solution has completely polymerized (~15–20 min), leave the gel for another 5–10 min (*see* **Note 19**). It is important to ensure that the partitions between the teeth of the comb are completely formed (*see* **Note 20**).
4. Prepare sufficient amounts of 1× TBE running buffer.

5. Once the gel has polymerized, carefully remove the comb and with the aid of a 10–20 mL syringe with needle thoroughly rinse the wells with 1× TBE to remove urea residues (*see* **Note 21**).
6. Add 1× TBE running buffer to the top and bottom running chambers. Prerun the gel at 25–45 mA for 40 min (*see* **Note 22**).
7. Prepare samples to be loaded: load 5 μL of diluted DNA marker, adjust volume with 2× RNA loading dye so that the volume of DNA marker loaded on the gel is the same as the PCR samples (20 μL). Heat denature samples 5 min at 95 °C, and chill on ice.
8. Turn the power supply off. Thoroughly rinse wells with 1× TBE as at **step** 7. With the aid of a flat-tipped micropipette loading tip, load samples on gel (*see* **Note 23**).
9. Run the gel at 25–45 mA the appropriate gel distance to optimize the resolution of the bands of interest. For a 6 % denaturing acrylamide gel, the apparent migration of loading dyes is 26 bp for bromophenol blue and 106 bp for xylene cyanol.

3.5.3 Gel Drying

1. Turn the power supply off and drain the buffer. The bottom buffer will be particularly radioactive and should be handled accordingly.
2. Put the gel face down, slide the spacers out of the gel, and with the help of a gel opening device slowly pry the top plate up, starting from a corner. Make sure that the gel remains attached to the bottom plate.
3. Put the bottom plate with the gel on a glass or plastic tray; add gel fixing solution to cover the gel, cover tray with clear plastic lab wrap, and gently rock on orbital shaker for 30 min. Part of the gel may detach from the glass plate.
4. Carefully take the glass plate with the gel on it out of the tray and put on flat surface over clean bench paper. Make sure that the gel doesn't slide off the plate.
5. Trim two pieces of Whatman 3MM paper to the size of the gel; soak one piece of paper in gel fixing solution, and carefully lay it over the gel. Carefully pick up the paper starting from a corner, making sure that the gel is securely attached to the paper. Slowly lift the paper with the gel attached to it, turn it over, and lay them "gel side up" on clean bench paper. Now the loading side of the gel is face up.
6. Cover the gel with plastic wrap; put the second piece of dry, Whatman 3MM paper at the bottom. With the help of scissors, trim the paper and plastic wrap in excess, and dry at 80 °C using a vacuum-driven gel drier (*see* **Note 24**).
7. With the help of a funnel recover the used fixing solution. Fixing solution can be reused multiple times.

3.5.4 Quantification of PCR Products

1. Take gel out of drying apparatus. Switch off drying apparatus. With the help of a Geiger counter, check the radioactivity of the gel. The plastic wrap will help prevent the gel from sticking to the imaging device screen (*see* **Note 25**).
2. Expose gel to quantification imaging screen for 3 h.
3. Quantify gel bands using imaging device and software (*see* **Note 26**).
4. Determine the linear range of amplification of each PCR product.
5. Pick a PCR cycle number that is within the linear range of amplification for all splice products.

3.6 Analytical Reverse Transcription Polymerase Chain Reaction

3.6.1 Reverse Transcription Polymerase Chain Reaction

1. Set up the RT-PCR reaction on ice, using 3 μL of reverse transcriptase product (cDNA) per sample (equivalent to 30 ng of starting RNA) generated in step in Subheading 3.3 and stored at −20 °C.
2. Each reaction has a total volume of 20 μL and is set up as follows:
   (a) 10× Gold Taq Buffer: 2 μL
   (b) 10 mM dNTP mix: 0.4 μL
   (c) Forward primer (5 μM): 0.8 μL
   (d) Reverse primer (5 μM): 0.8 μL
   (e) ^{32}P, 5′-end labeled and purified forward primer: 0.8 μL
   (f) RT reaction: 3 μL
   (g) AmpliTaq Gold® DNA polymerase 5 U/μL: 0.2 μL
   (h) Nuclease-free water: 12 μL.
3. Each PCR reaction should include a "no RT" negative control and a "water-only" negative control as PCR templates instead of RT reaction.
4. The PCR conditions should be as follows: a single 5 min denaturation cycle at 95 °C; *x* cycles (where *x* is the cycle number determined to be within the linear range of amplification) where each cycle is: 20 s denaturation at 95 °C; 20 s annealing at 50–55 °C; 30 s extension at 72 °C; a single 7 min extension cycle at 72 °C.
5. Add 2 volumes (equals to 40 μL) of 2× RNA loading dye to each sample, for a total volume of 60 μL.

3.6.2 Electrophoresis of PCR Products on Denaturing Polyacrylamide Gel

1. Same as in Subheading 3.5.2. Load 7.5 μL (equivalent to 1/8th of total volume) of each sample per lane (*see* **Note 27**).
2. Store samples at −20 °C in plexiglass container for radioactivity work with beta emitters until needed.
3. Follow same procedure as in Subheading 3.5.2.

4. Dry gel as in Subheading 3.5.3.
5. Expose gel to quantification imaging screen.
6. Quantify PCR products as in Subheading 3.5.4.
7. After gel quantification with imaging device, it may be useful to run a new, normalized gel so that the amount of radioactivity is the same for each lane (Fig. 3c).

4 Notes

1. Other reverse transcription enzymes, such as SuperScript III by Life Technologies work equally well. However, since SuperScript III works at 50 °C instead of 42, this reverse transcriptase is preferable when strong RNA secondary structures are expected.
2. The concentration of the DNA primers should be determined by reading the optical density of primers diluted in water with a spectrophotometer. For a 1-cm path length, the optical density at 260 nm (OD_{260}) equals 1.0 for a 30–37 μg/mL solution of DNA primers.
3. The 20 DNA fragments range in size from 24 to 726 bp.
4. The best results are obtained with a cooled system, which allows PCR products to run in a straight front, limiting the "smiling" migration of products and dyes on the gel.
5. Undissolved clumps may be made to dissolve by placing the bottle of buffer in a hot water bath. Although the buffer can be sterilized by autoclaving, this is not necessary.
6. TEMED has to be added last, immediately before pouring the gel. The gel can be used immediately, or store up to 24 h at room temperature, or 48 h at 4 °C. To prevent the gel from drying during storage, cover the top of the gel and the comb with a paper towel dampened with 1× TBE.
7. Size of the PCR products will vary depending on the size of the alternatively spliced exon(s). However, when possible product size should be kept small, i.e. below 500 bp. Moreover, the difference in size between PCR products and, subsequently, amplification efficiency should be taken into account as well. Since smaller PCR products will be preferentially amplified, having products that are very different in size will make identification of a PCR cycle number that falls within the linear range of the reaction for all PCR products more difficult.
8. Publicly available online resources about genome annotation of alternative splicing include the USCS Genome Browser (http://genome.ucsc.edu/), Fast DB (http://www.fast-db.com/fastdb2/frame.html), EASANA (https://www.easana.com/),

ECgene (http://genome.ewha.ac.kr/ECgene/), SpliceInfo (http:// spliceinfo.mbc.nctu.edu.tw/), ProSplicer (http:// prosplicer.mbc.nctu.edu.tw/), the Burge Laboratory Web Server (http://hollywood.mit.edu/), and other databases listed on EURASNET (http://www.eurasnet.info/tools/asdatabases).

9. Cells should not be allowed to reach confluency, since that may affect gene expression and alternative splicing patterns.
10. After homogenization and before addition of chloroform, samples can be stored at –80 °C for several weeks.
11. Although usually not necessary, glycogen can be added as a carrier if low yields of RNA are expected.
12. Be very gentle in aspirating the supernatant, since at this stage the pellet will detach itself from the minicentrifuge tube very easily.
13. These conditions will remove up to 30 μg of genomic DNA.
14. Be very gentle in aspirating the supernatant, since at this stage the pellet will detach itself from the minicentrifuge tube very easily.
15. For a 1-cm pathlength, the optical density at 260 nm (OD_{260}) equals 1.0 for a 40 μg/mL solution of RNA. Calculate the OD_{260}/OD_{280} ratio for an indication of nucleic acid purity. Pure RNA has an OD_{260}/OD_{280} ratio of ~2.0. Low ratios could be caused by protein or phenol contamination. The expected yield of RNA will vary from cell line to cell line, but a subconfluent 10 cm petri dish should give 10–50 μg of total RNA.
16. Steps in Subheadings 3.4.1 and 3.4.2 can be accomplished while the reverse transcription reaction from step in Subheading 3.3 is running.
17. To prevent glass contamination with skin oils, wear talc-free gloves and handle plates by their edges.
18. A 60 mL syringe will work just as well.
19. If you are planning to run the gel the same day you are also doing the RT-PCR reaction, you can prepare the gel the day before.
20. As polymerization proceeds, the change in refractive index between the polymerized and nonpolymerized acrylamide in the gel will become visible as a "Schlieren line" around the edges of the comb.
21. If the wells are not properly washed, this will result in bands appearing elongated, in the shape of a drop. Figure 3a represents an example of a gel in which the wells were not properly washed, resulting in PCR products migrating as elongated bands. Note that the elongated appearance of the bands will not interfere with quantification, as shown in Fig. 3b. In Fig. 3c the wells were properly washed, resulting in PCR products migrating correctly.

22. The amount of current will vary depending on the size of the gel. For an 18×16 cm, 1 mm thick gel, use 25 mA.
23. Load samples quickly to avoid formation of urea precipitates.
24. Make sure that the gel is completely dry before opening the dryer. If the drier is opened too soon, the gel may crack, or shrink, or break.
25. If the amount of radioactivity is particularly weak, gently peel off the plastic wrap. Put some baby powder on the gel, eliminate excess powder, and expose gel to imaging screen for quantification.
26. Experiment shown in Fig. 3a shows the appearance of "shadow" or "ghost" bands, indicated with red asterisks. When cloned and sequenced, these shadow bands appear identical in size and nucleotide composition to the expected PCR products. Shadow bands, therefore, are not the result of amplification of new alternatively spliced isoforms, but rather aberrantly migrating PCR products that appear when an excessive amount of PCR product is present in the gel. Note that the shadow bands become apparent at PCR cycles that are no longer within the linear range of the reaction.
27. It may be convenient to adjust the final volume of sample to be loaded on the gel to 20 μL by adding a further 12.5 μL of 2× RNA loading dye to each sample, as in **step** 7 in Subheading 3.5.2.

References

1. Wang ET, Sandberg R, Luo S et al (2008) Alternative isoform regulation in human tissue transcriptomes. Nature 456:470–476
2. Maniatis T, Tasic B (2002) Alternative pre-mRNA splicing and proteome expansion in metazoans. Nature 418:236–243
3. Black DL (2003) Mechanisms of alternative pre-messenger RNA splicing. Annu Rev Biochem 72:291–336
4. Stamm S, Ben-Ari S, Rafalska I et al (2005) Function of alternative splicing. Gene 344:1–20
5. Braunschweig U, Gueroussov S, Plocik AM et al (2013) Dynamic integration of splicing within gene regulatory pathways. Cell 152:1252–1269
6. Pan Q, Shai O, Lee LJ et al (2008) Deep surveying of alternative splicing complexity in the human transcriptome by high-throughput sequencing. Nat Genet 40:1413–1415
7. Singh RK, Cooper TA (2012) Pre-mRNA splicing in disease and therapeutics. Trends Mol Med 18:472–482
8. Lopez-Bigas N, Audit B, Ouzounis C et al (2005) Are splicing mutations the most frequent cause of hereditary disease? FEBS Lett 579: 1900–1903
9. Lynch KW (2004) Consequences of regulated pre-mRNA splicing in the immune system. Nat Rev Immunol 4:931–940
10. Martinez NM, Pan Q, Cole BS et al (2012) Alternative splicing networks regulated by signaling in human T cells. RNA 18:1029–1040
11. Martinez NM, Lynch KW (2013) Control of alternative splicing in immune responses: many regulators, many predictions, much still to learn. Immunol Rev 253:216–236
12. Sahoo A, Im SH (2010) Interleukin and interleukin receptor diversity: role of alternative splicing. Int Rev Immunol 29:77–109
13. Goodwin RG, Lupton S, Schmierer A et al (1989) Human interleukin 7: molecular cloning and growth factor activity on human and murine B-lineage cells. Proc Natl Acad Sci U S A 86:302–306
14. Welch PA, Namen AE, Goodwin RG et al (1989) Human IL-7: a novel T cell growth factor. J Immunol 143:3562–3567

15. Fry TJ, Mackall CL (2002) Interleukin-7: from bench to clinic. Blood 99:3892–3904
16. Krawczenko A, Kieda C, Dus D (2005) The biological role and potential therapeutic application of interleukin 7. Arch Immunol Ther Exp (Warsz) 53:518–525
17. Maeurer MJ, Walter W, Martin D et al (1997) Interleukin-7 (IL-7) in colorectal cancer: IL-7 is produced by tissues from colorectal cancer and promotes preferential expansion of tumour infiltrating lymphocytes. Scand J Immunol 45:182–192
18. Trinder P, Seitzer U, Gerdes J et al (1999) Constitutive and IFN-gamma regulated expression of IL-7 and IL-15 in human renal cell cancer. Int J Oncol 14:23–31
19. Korte A, Moricke A, Beyermann B et al (1999) Extensive alternative splicing of interleukin-7 in malignant hematopoietic cells: implication of distinct isoforms in modulating IL-7 activity. J Interferon Cytokine Res 19:495–503
20. Cosenza L, Gorgun G, Urbano A et al (2002) Interleukin-7 receptor expression and activation in nonhaematopoietic neoplastic cell lines. Cell Signal 14:317–325
21. Hartgring SA, Bijlsma JW, Lafeber FP et al (2006) Interleukin-7 induced immunopathology in arthritis. Ann Rheum Dis 65 Suppl 3:iii69–iii74
22. Fry TJ, Connick E, Falloon J et al (2001) A potential role for interleukin-7 in T-cell homeostasis. Blood 97:2983–2990
23. Huang HW, Chiang YJ, Hung SI et al (2007) An IL-7 splicing-defect lymphopenia mouse model revealed by genome-wide mutagenesis. J Biomed Sci 14:169–181
24. Chung B, Dudl E, Toyama A et al (2008) Importance of interleukin-7 in the development of experimental graft-versus-host disease. Biol Blood Marrow Transplant 14:16–27
25. Pan D, Liu B, Jin X et al (2012) IL-7 splicing variant IL-7delta5 induces human breast cancer cell proliferation via activation of PI3K/Akt pathway. Biochem Biophys Res Commun 422:727–731
26. Kroemer RT, Kroncke R, Gerdes J et al (1998) Comparison of the 3D models of four different human IL-7 isoforms with human and murine IL-7. Protein Eng 11:31–40
27. Vudattu NK, Magalhaes I, Hoehn H et al (2009) Expression analysis and functional activity of interleukin-7 splice variants. Genes Immun 10:132–140
28. Rane L, Vudattu N, Bourcier K et al (2010) Alternative splicing of interleukin-7 (IL-7) and interleukin-7 receptor alpha (IL-7Ralpha) in peripheral blood from patients with multiple sclerosis (MS). J Neuroimmunol 222:82–86
29. Lundmark F, Duvefelt K, Iacobaeus E et al (2007) Variation in interleukin 7 receptor alpha chain (IL7R) influences risk of multiple sclerosis. Nat Genet 39:1108–1113
30. Moors M, Vudattu NK, Abel J et al (2010) Interleukin-7 (IL-7) and IL-7 splice variants affect differentiation of human neural progenitor cells. Genes Immun 11:11–20
31. Paonessa G, Graziani R, De Serio A et al (1995) Two distinct and independent sites on IL-6 trigger gp 130 dimer formation and signalling. EMBO J 14:1942–1951
32. Bihl MP, Heinimann K, Rudiger JJ et al (2002) Identification of a novel IL-6 isoform binding to the endogenous IL-6 receptor. Am J Respir Cell Mol Biol 27:48–56
33. Rincon M, Anguita J, Nakamura T et al (1997) Interleukin (IL)-6 directs the differentiation of IL-4-producing CD4+ T cells. J Exp Med 185:461–469
34. Zhang XY, Zhang HJ, Zhang Y et al (2006) Identification and expression analysis of alternatively spliced isoforms of human interleukin-23 receptor gene in normal lymphoid cells and selected tumor cells. Immunogenetics 57:934–943
35. Guo R, Li Y, Ning J et al (2013) HnRNP A1/A2 and SF2/ASF regulate alternative splicing of interferon regulatory factor-3 and affect immunomodulatory functions in human non-small cell lung cancer cells. PLoS One 8:e62729
36. Ruggiu M, Herbst R, Kim N et al (2009) Rescuing Z+ agrin splicing in Nova null mice restores synapse formation and unmasks a physiologic defect in motor neuron firing. Proc Natl Acad Sci U S A 106:3513–3518
37. Wong ML, Medrano JF (2005) Real-time PCR for mRNA quantitation. Biotechniques 39:75–85
38. Gibson UE, Heid CA, Williams PM (1996) A novel method for real time quantitative RT-PCR. Genome Res 6:995–1001
39. Horner RM (2006) Relative RT-PCR: determining the linear range of amplification and optimizing the primers:competimers ratio. CSH Protoc 2006
40. Bastien P, Procop GW, Reischl U (2008) Quantitative real-time PCR is not more sensitive than "conventional" PCR. J Clin Microbiol 46:1897–1900
41. Rio DC, Ares MJ, Hannon GJ et al (2010) Polyacrylamide gel electrophoresis of RNA. Cold Spring Harb Protoc 2010:pdb.prot5444

INDEX

Ivana Vancurova (ed.), *Cytokine Bioassays: Methods and Protocols*, Methods in Molecular Biology, vol. 1172, DOI 10.1007/978-1-4939-0928-5, © Springer Science+Business Media New York 2014

H

I

K

L

M

N

O

P

R

S

MIX
Papier aus verantwortungsvollen Quellen
Paper from responsible sources
FSC® C105338

If you have any concerns about our products,
you can contact us on
ProductSafety@springernature.com

In case Publisher is established outside the EU,
the EU authorized representative is:
Springer Nature Customer Service Center GmbH
Europaplatz 3, 69115 Heidelberg, Germany

Printed by Libri Plureos GmbH
in Hamburg, Germany